Railway imperialism in China

Manchester University Press

STUDIES IN IMPERIALISM

When the 'Studies in Imperialism' series was founded by Professor John M. MacKenzie more than thirty years ago, emphasis was laid upon the conviction that 'imperialism as a cultural phenomenon had as significant an effect on the dominant as on the subordinate societies'. With well over a hundred titles now published, this remains the prime concern of the series. Cross-disciplinary work has indeed appeared covering the full spectrum of cultural phenomena, as well as examining aspects of gender and sex, frontiers and law, science and the environment, language and literature, migration and patriotic societies, and much else. Moreover, the series has always wished to present comparative work on European and American imperialism, and particularly welcomes the submission of books in these areas. The fascination with imperialism, in all its aspects, shows no sign of abating, and this series will continue to lead the way in encouraging the widest possible range of studies in the field. 'Studies in Imperialism' is fully organic in its development, always seeking to be at the cutting edge, responding to the latest interests of scholars and the needs of this ever-expanding area of scholarship.

To buy or to find out more about the books currently available in this series, please go to: https://manchesteruniversitypress.co.uk/series/studies-in-imperialism/

Railway imperialism in China

A political biography

Zheng Yangwen

鄭揚文

Manchester University Press

Published by Manchester University Press
Oxford Road, Manchester, M13 9PL

www.manchesteruniversitypress.co.uk

British Library Cataloguing-in-Publication Data
A catalogue record for this book is available from the British Library

ISBN 978 1 5261 9525 8 hardback

First published 2026

EU authorised representative for GPSR:
Easy Access System Europe, Mustamäe tee 50, 10621 Tallinn, Estonia
gpsr.requests@easproject.com

Typeset by Newgen Publishing UK

This book is dedicated to my father
Zheng Guicun (1936–2025)
who instilled curiosity in me

Contents

Figures

Acknowledgements

I want to thank Professor Xu Yue [徐跃] for hosting my visit to Sichuan University, which made my research at the Sichuan Provincial Archive much easier. I am very grateful to John Carroll at the University of Hong Kong and James St Andre at the Chinese University of Hong Kong for facilitating my visit to their libraries and the newly opened Sheng Xuanhuai Archive. Melanie Aspey, Natalie Attwood, Justin Cavernelis-Frost and Carl Copper at the Rothschild archive in London went out of their way to assist me; so did Robert Bickers at Historical Photographs of China, Michelle Barnes at Jardine Matheson Archive and Danielle Sellers at Royal Engineers Museum. I am very grateful to Jennifer Ning Chang and Huimin Lai who hosted my visit to the Institute of Modern History in Taipei and the staff at the Institute's Archive, Guo Tingyi and Fu Sinian Libraries. During COVID-19, Cai Jixiang, a researcher at Academic Sinica, conducted research for me remotely. COVID-19 forced me to frequent online archives and rely on digitised newspapers – a blessing in disguise as they are primary sources neglected by earlier generation of scholars. I particularly want to thank Peter Crush, a railway enthusiast and collector who shared his articles and images with me. My colleagues have been most helpful. Peter Gatrell and Christian Goeschel shared material. Philipp Roessner, Jean-Marc Dreyfuss and Ana Akulich checked translations of German, French and Russian texts. Mark Baker read the manuscript, Steven Pierce and Natalie Zacek helped me decipher Reader reports. Angela Roberts, George Odysseos, Olivia Mitchell and Sarah Butler copyedited the manuscript, and Richard Atherton, our phenomenal graphics designer, polished up the illustrations. My final thanks go to Emma Brennan and Meredith Carroll, the 'paragon of efficiency' who has laboured hard for the speedy publication of this monograph.

Zheng Yangwen 鄭揚文
Manchester, April 2025

Introduction

On 21 July 2021, in the midst of a global pandemic, CNN reported that a 'maglev bullet train that can reach speeds of 600 kilometres per hour (373 miles per hour) made its debut in Qingdao'.[1] This bullet train was the latest edition of China's high-speed rail franchise, which was initiated with the *Harmony* in 2007. Achieving speeds of 250 km/h, the *Harmony* was succeeded a decade later by the *Renaissance*, which clocked in an impressive 350 km/h. These impressive achievements have not been limited to the Chinese borders. The post-Mao regime exported its brand of high-speed rail to Turkey, Southeast Asia, Africa, Mexico and Latin America, building what Will Doig has called a 'high speed empire'.[2] Through the Belt and Road Initiative (BRI), Chinese trains now zigzag across Central Asia and Eastern Europe to reach Germany, Holland, Belgium, France and Spain. China, it seems, is conquering the world through rail in much the same way as Britain, Russia, France, Belgium, Germany, Japan and the United States had done to late Qing China (1842–1911). Railways served as 'engines for empire' as these colonial powers used it to carve out their spheres of influence and control.[3] In a dramatic reversal, China has gone from destroying foreign railways built on Chinese soil, as in 1877, to now using the very same engines to out-perform the West in their own game.

Why did the late Qing resist the introduction of railways to China during its period of reform when it embraced other modernisation projects? How exactly did Britain, Russia, France, Belgium, Germany, Japan and the United States use railways to divide and rule the late Qing? To what extent did the actions of these foreign powers instruct and inspire successive Chinese leaders and their endeavours? These are questions that remain unanswered or underexplored in existing scholarship. *Railway Imperialism in China: A Political Biography* charts the political genealogy of railways from their introduction in the heyday of railway imperialism to their transformation into a tool for both Chinese nation-making and furthering China's global ambitions.[4] It systematically scrutinises the intense rivalry among foreign powers that accompanied its introduction and development. Though dismissed by generations of historians as Confucian conservatism, the late Qing's resistance to railways directs us to what are now considered legitimate reasons. This book advances a model in the study of railway imperialism, synthesising how and why railways had become the preferred instrument for empire-making. More importantly, it considers both the immediate and long-term legacies of the political battles that surrounded railway imperialism, in which the Chinese Communist Party (CCP) has been the ultimate beneficiary. By allowing railways to take centre stage, this book sheds new light on the scramble for China during the Age of Empire. The railways offer a unique and insightful window through which we can measure the dynamics and complexity of modern and contemporary China.

Railways in the history of modern China

Despite the centrality of railways to late Qing politics and the emergence of modern China, the railway has been surprisingly marginal in the historiography. There are a number of dated general surveys, written in the early twentieth century. Percy Kent's excellent account of the origin and development of railways from the 1870s to the end of the Russo-Japanese War (1904–05) provides an exemplary groundwork, albeit without footnotes.[5] Mongton Chih Hsu details the divided nature of railway inauguration, delineating the different lines built by various foreign powers and the Chinese themselves by 1914.[6] Hsu's work highlights the plight of the Chinese and thus the problem with railway imperialism. More surveys have been published since the 1930s, providing updates on railway development during the Republic of China or ROC era (1912–49).[7] Scholars from China and Taiwan have compiled sources; works by Zeng Kunhua and Mi Rucheng are very informative and must-reads.[8] China's railways have also inspired popular interests.[9]

Despite the availability of these surveys and general accounts, Chinese railways have inspired few analytical studies. As such, there is only one historiographical debate, with the discourse centred on the question of railway imperialism versus economic benefits. E-tu Zen Sun best articulated the contours of the debate with her 1954 book, which postulated that the development of railways in China can be studied from three perspectives. First, it can be 'treated as a problem of finance', which would focus on matters surrounding foreign loans; second, it can be 'analysed as a phase of advance in technological civilisation'; and third, as the 'steel prop' of state structure, which points to efforts at modernisation and state making.[10] All three issues are intimately associated with the presence and activities of foreign powers, as late Qing China had neither the capital nor the expertise needed to embark on such projects. With such issues, scholars have taken opposing views. Wayne Altree, Chi-ming Hou and John Schrecker, for example, believe that railroads financed and built by foreigners forced China to modernise. These interpretations were challenged by E-tu Zen Sun, Joseph Esherick and Conrad Thomason, who highlight the detrimental effects of railway imperialism on Chinese society and sovereignty.[11] The tide was turning by the 1980s. Ernest Liang, for instance, discusses the impact of railway on agricultural development.[12] Ralph Huenemann presents a systematic study of pre-WWII railway economics and quantified its benefit to China in 1984.[13] This line of argument was taken further by Elisabeth Köll, who approaches the study of three railways from the perspective of a business institution, focusing on freight and passenger traffic, the impact on the transportation of goods and agriculture, technical education, the resultant process of urbanisation and regional benefit.[14]

Aside from this debate, existing scholarship on the subject is compartmentalised and scattered. Most are devoted to studies of individual lines. Some have received more attention, while others have not generated academic interest at all. Histories of Chinese railways often begin with the controversial 9.5-mile Wusong, the first line built on Chinese soil in 1876. Constructed by foreign merchants led by Jardine Matheson, it was dismantled in 1877. Although the Wusong is mentioned in works on late Qing reform and foreign enterprises, scholarship devoted to the Wusong is limited to a few articles.[15] David Pong argued in 1973 that its dismantling was a result of Confucian patriotism. Sandra Sturdevant disagreed in 1976, proposing that the Chinese action was aimed at preventing the foreigners from 'gaining more "rights" than they already enjoyed'.[16] Hsien-chun Wang, however, suggests that it was the Qing's unsettled railway policy and 'unreformed fiscal system' that led to its demise.[17] The Wusong has also received popular attention.[18]

Equally controversial is the barely 8-mile Tangshan to Xugezhuang or Tangxu railway, built to transport coal from the Kaiping coal mine in the early 1880s. The first line sanctioned by the Qing, it has generated very little academic interest – David Pletcher is one of the very few who mentioned it in his book on American expansion.[19] Yet, this line is extremely significant as it marked the beginning of the Qing's efforts at railway development. The Tangxu would grow to become the mother of all railroads in north China, as it would stretch to Tianjin, Beijing and Manchuria. Arthur Rosenbaum remains the only scholar who has authored two superlative articles about Tangxu's expansion, whereas Peter Crush, a railway enthusiast and collector, has published a pictorial survey.[20]

If the Tangxu is little known and barely studied, even less recognised and researched is the Xiyuan, a French narrow-gauge railway, hardly 2 miles long, that was built within the imperial palace for Empress Dowager Cixi in 1888. Many are not aware of its existence, nor of the several other lines built for her: the Western Tombs railway among them. Once again, it is Peter Crush who has provided an article on the Xiyuan railway.[21] The Wusong, Tangxu and Xiyuan deserve scholarly attention for the new light they shed, not just on the politics of railway development and late Qing reform but also on how the Qing's cautious approach in the 1880s was interrupted by the intense rivalry among foreign powers, which turned into the scramble for railway concessions in the 1890s and 1900s.

Britain, Russia, Germany, France, Belgium, Japan and the United States fought to finance, construct and control Qing China's railways. The Russian-built Chinese Eastern Railway or CER in Manchuria has received more attention than others, generating many surveys since the 1920s.[22] They are informative but are also dated narrative accounts, lacking academic rigour. Tan Guiluan has recently published an excellent volume on the construction and management of the CER.[23] The CER is often discussed in works on Russian expansion to East Asia and Sino-Soviet relations.[24] Nikita Vul and Michael Walker have written about the armed conflicts fought over the CER, whereas Bruce Elleman has published much on the line's role in Sino-Soviet diplomacy.[25] The Russian communities that sprouted up along the CER and around the stations, a microcosm of the empire, have become a subject of fascination.[26] David Wolff has authored an excellent book on the making of Russian Harbin, with the railway at its heart.[27]

In contrast with the CER, the German-constructed Jiaoji railway in Shandong province has received little recognition, despite often being mentioned in works on German imperialism and its legacies in China, which tend to focus on the colony of Qingdao.[28] The beating heart of the German colonial machine in East Asia, the Jiaoji has been side-lined, as scholarship remains limited to a handful of chapters. In his seminal work, *The Devil's Handwriting*, George Steinmetz examines the making of the Jiaoji railway in a chapter on Qingdao, whereas Klaus Mühlhahn has analysed various dimensions of German colonialism in Shandong.[29] Elisabeth Köll has written a chapter on the Jinpu railway, and she discussed its development and success as a business institution in her book.[30]

Japanese railway imperialism has received more attention. The Russo-Japanese War cut the CER in half, and Japan, as the victor, received the southern segment and renamed it the South Manchurian Railway or SMR. As the engine of the Japanese empire in China, the SMR is often mentioned in scholarship on Japanese imperialism.[31] Ramon Meyers devoted a chapter to Japanese colonialism in Manchuria with a focus on the SMR, S. C. M. Paine highlighted the origin and significance of Japanese railway encroachment, whereas Yoshihisa Tak Matsusaka has published an accomplished volume on the making of Japanese Manchuria with the SMR at its centre.[32] C. Walter Young studied Japanese control and jurisdiction along the SMR.[33] The SMR was serviced by a Research Department. John Young examined its research activities, whereas Ito Takeo authored a memoir about his tenure there.[34]

The French-built Dianyue railway is often considered alongside French imperialism in China and Indochina.[35] Of the few general surveys, Robert Lee's work remains the most comprehensive on French railway imperialism.[36] Amaury Lorin has produced excellent studies on Paul Doumer, governor-general of Indochina, who was instrumental in the making of the Dianyue, whereas Vatthana Pholsena and Selda Altan have published articles on the fate and politics of the coolies who worked on the railway.[37] Jean-François Rousseau investigated the railway's failure as an imperial project, while Florence Bretelle-Establet has argued that French presence in Yunnan ultimately benefited the province.[38] Anglo-French railway rivalry is also a topic of research.[39]

Britain was the first imperial power to initiate railways in China. Despite laying fewer tracks, Britain retained a form of dominance by issuing railway loans, achieving a hegemony that was unchallenged by others. Nevertheless, E-tu Zen Sun and D. McLean remain two of the few scholars who have researched British railway imperialism specifically; otherwise, discussions sit within discourse on British foreign policy in China.[40] Belgium, too, financed and built important lines – the Luhan and the Bianluo railways. Sadly, these have not generated much research. The United States tried, several times, to assert its railway vision in China, starting with Ulysses S. Grant.[41] They secured the contract for the Canton to Hankou or Yuehan railroad, but not much has been written about their endeavours.[42] Clarence Davis and Kenneth Wilburn edited a volume that compared railway imperialisms.[43] Whereas T. G. Otte and Keith Neilson published an excellent volume exposing the intimacy between railways and international politics.[44] Although these studies examining foreign powers and individual railways are useful as sources of information, they do not form a coherent body of scholarship, lacking systematic analysis and the insights a comparative study could offer.

Elisabeth Köll accurately identifies that 'the history of China's railroad development is as much a global story as it is a Chinese story'.[45] As this book reveals, railway imperialism in China gave birth to railway nationalism in the 1900s where many Chinese considered the development of railways as being tied to the survival of the country. Many construction initiatives were led by overseas Chinese, provincial governments and local gentry/merchants. Yet some of the most important projects, such as the Chaozhou to Shantou railway built by Indonesian Chinese Zhang Yinan, the Xinhui to Xinning line constructed by Chinese American Chen Yixi, and the Beijing to Zhangjiakou railroad erected by the Zhili regional government, have not generated much academic interest.[46]

However, these projects did encourage others to invest locally to prevent railways from falling into the hands of foreign powers. Historians have labelled this the 'Railway Rights Recovery Movement'. En-han Lee has written about this 'quest for railway autonomy'.[47] Madeleine Chi studied the case of the Shanghai–Hangzhou–Ningbo line. Mary B. Rankin examined the rhetoric and practice of the national movement, whereas Elisabeth Kaske has highlighted gentry power in the making of the Chuanhan railway.[48] This is significant as attempts of nationalising the Chuanhan directly led to the 1911 Nationalist Revolution. As Xiaowei Zheng's book on the genealogy of constitutionalism and transformation of elite consciousness indicates, scholarship on the revolution is growing.[49] Yet interest in the railway itself remains at a low ebb, despite the fact that mainland and Taiwan scholars have continued to compile sources on the Chuanhan.[50]

The railway-inspired revolution gave birth to the Republic of China (1912–49). Dr Sun Zhongshan (Yat-sen), its first president, was passionate about railways. Although there is no shortage of scholarship on Sun, no one has as yet discussed his railway vision elaborated in 'Strategies to Build a Nation'.[51] The ROC regime standardised railway gauges in areas under its control which has been assessed by Chang Jui-Te.[52] It also constructed new lines in preparation for Japanese invasion. Likewise, warlords like Zhang Zuolin and Yan Xishan built regional lines. Sadly, they remain largely unexplored, even though they are mentioned in the scholarship on these warlords.[53] More importantly, few if any have pointed out that the so-called warlord era was only made possible by railways, since the warlords, like the foreign powers, used rail lines to divide and rule.

China's experience in the first half of the twentieth century, characterised by warlord and world wars, cannot be understood without appreciating the important role railways played in these conflicts. As a strategic weapon, railways were crucial to war ever since its invention, as Christian Womar has advanced in *Engines of War*.[54] Scholarship on the Anti-Japanese War or WWII in China is growing. But only a few, Hans van de Ven and Stephen MacKinnon for example, have highlighted that the Japanese conquest of China followed the path wrought by railways; the Japanese Army depended on railways and was consequently constrained by them.[55] The same can be said of the Civil War between the Communists and the Nationalists that ensued in 1945. Harold Tanner and Victor Cheng are two of the few who have pointed out the role of railways in this final showdown.[56] Military campaigns have dominated the study of wars, yet the railways, so crucial to victory or defeat, have been consigned to the footnotes of history.

The communist revolution actually began on the railways in the early 1920s, but few have written about this. Railways helped the Red Army – People's Liberation Army to battle the Japanese and defeat the Nationalists. The CCP prioritised railway construction after taking power in 1949, this has generated a few surveys and articles.[57] Its 'railway diplomacy', the construction of the Tanzania–Zambia railway in the 1960s and 1970s, has interested a handful of scholars who sang the praises of the CCP's 'proletarian internationalism'.[58] Post-Mao economic reform has renewed interest in Chinese railways and generated a few surveys.[59] China's railway ventures into Southeast Asia, Africa, Latin America, Central Asia and Europe under the BRI have received much recognition. Some have called it 'railpolitics', whereas Doig prefers 'high speed empire'.[60]

Railways changed the way Chinese travelled and stations gave birth to cities such as Zhengzhou, transforming culture and landscape. Jiang Pei and Xiong Yaping have examined the transformation of these cities and their surrounding countryside.[61] Li Siyi has recently published an excellent book on the 'temporal–spatial experience' and the 'cultural imagination' wrought by railway travel in the late Qing and the ROC era.[62] James Zheng Gao has compared the social impact of railways on the Chinese and Zimbabweans at the turn of the previous century.[63] Railways inspired a new kind of literature and art during the Mao era. *The Legend of the Red Lantern* [红灯记], the lead Peking opera, glorified the CCP's underground activity along the SMR. They also included films and propaganda posters, recounting the CCP's heroic struggle against the Japanese and promoting socialism, which Emma Zhang has written about.[64] We have seen a large but extremely compartmentalised literature, with the railway playing a marginal or incidental role in, rather than the focal point of, many of the works. They clearly expose problems.

Railways as research problematic

Railway Imperialism in China: A Political Biography picks up these intertwined challenges. First, it is very clear the railways that defined foreign encroachment and the late Qing's resistance to it, imperialism and Chinese nationalism, and which underpinned revolution, wars and nation making have often been sidelined and reduced to a footnote in history. What do we learn when we place the railway at the centre stage of modern Chinese history? Second, the literature is compartmentalised, devoted to individual lines. These individual lines deserve attention, but more attention must be paid to what the lines have in common and what their similarities and differences can reveal to us. Third, this dispersed literature does not establish a debate, nor do they converse with each other. Many lines, the Tangxu railway and the CER for examples, are, in fact, intimately related. Nonetheless, their relation to each other, and a discussion of the light they can shed on late Qing reform and geopolitics of railways, is completely missing. These problems have led to gaps in our knowledge.

Railways are the main protagonist of this story. However, tying together the vast and scattered histories poses challenges. Arjun Appadurai and Igor Kopytoff have offered new perspectives in the study of commodities which Appadurai defined as 'objects of economic value'.[65] Railways are sizeable objects of enormous economic value. Appadurai has maintained that 'commodities, like persons, have social lives'.[66] This volume extends this line of thought to the political arena due to the significance of politics on international, domestic and even personal levels that underpinned its introduction and development. Kopytoff has postulated that: 'biographies of things can make salient what might otherwise remain obscure'.[67] This can address the problem with existing scholarship head-on as they are scattered, dealing with different railways built by different regimes, leaving behind a trail of gaps in our knowledge. Kopytoff uses an example to explain his approach: 'the biography of a car in Africa would reveal a wealth of cultural data'.[68] The biography of railways in China, likewise, would expose a wealth of political data that are missing in the history of modern China which has led to misunderstandings. This genealogical method of historical inquiry is an appropriate theoretical framework because it enables me to connect the three phases of railway's political life: (1) as nemesis of reform when it was first introduced or born, to be discussed in Chapters 1 through 3, (2) as 'tools of empire' for European powers and Japan as it grew up, to be analysed in Chapters 4 through 7 and (3) as an instrument of nation building for Chinese regimes when it came of age, to be examined in Chapters 8 through 10.

The genealogical framework helps chart the political biography of railways; in doing so, it exposes the geopolitical circumstances under which they were introduced into China and how they galvanised domestic politics and shaped its development. Delving deep into the hunt for railway concession, finance, construction and administration, the political biography teases out the pattern and variations of railway imperialism. Moreover, it investigates how the scramble for railways fuelled railway nationalism and how the Chinese themselves took the technology to unprecedented heights. Elisabeth Köll is correct to argue that as 'a vehicle to convey both the history of modern China and the complex forces still driving the nation's economic success, rail has no equal'.[69] Köll has tackled the history of railways by discussing three of the lines from the perspective of a business institution. This volume will extend this important work from a political perspective as well as broadening its scope to include all major lines. Bringing these individual lines together, it investigates the transformation of railways from a bête noire within discussion about reform to the emblematic 'engines for empire' and an instrument of nation making. It does so by examining three overarching issues: first, why the late Qing were slow to embrace railways, even amidst intensified efforts at reform; second, the exact ways in which foreign powers used railways to carve out their spheres of influence and control; and, third, how railways became a catalyst for revolution and a tool of region or nation making.

Existing scholarship tends to highlight the Qing's hostility to railways, both reifying and dismissing it as a defining example of Chinese conservatism. Such an account is oversimplified. Why did reformers like Shen Baozhen and Li Hongzhang resist the development of railways despite launching and maintaining other modernisation projects? We cannot understand the Qing's resistance to railways without understanding the geopolitics that accompanied their introduction and the unique characteristics of this technology. A careful examination of previously overlooked material and exploration of new archival sources point us toward the pressing perturbation that underlay Qing policy and drove fierce internal debate. Having defeated the Qing during the Opium Wars and procured treaties, foreign powers were eager grow their trade and expand into the interior. Qing officials understood that railways could benefit reform, but they were also concerned that this would come at the cost of China's sovereignty, territorial integrity, the stability of an economy and society that was not ready for railways, and the environment – a prescient fear that inadvertently made them proto-environmentalists. The fears would materialise with the onset of railway imperialism. It is time we analyse the concerns of Qing officials, rather than continue to belittle them.

The existing literature on railway imperialism has never asked these two crucial questions: why did railways become 'engines for empire' in the 1890s, and how did the scramble for China change from seeking settlements in Treaty-ports to the hunt for railway concessions? If there was a pattern to the behaviour of railway imperialists, it has yet to be explored. This book advances, for the first time, a model for the study of railway imperialism. Railways were, without a doubt, a game changer. Treaties before the mid-1880s confined foreign powers to their 'settlement' in Treaty-port cities, but railways would almost certainly allow them to penetrate into the interior and expand, as the 'alienation zone' entitled them to claim new territory, to station troops, to extract resources and to construct stations which could grow into colonies. Ronald Robinson argued that 'lines extending from Berlin, Vienna, and Constantinople into the Balkans did much to bring about partition there'; the same might be said of 1890s China, which foreign powers divided 'with exclusive railroad rights'.[70] Railways help us enumerate the scramble for China and pinpoint the Qing's final days.

Existing scholarship has left an enormous gap between the late Qing and post-Mao era, overlooking the legacies of railway imperialism. How did they inform China's successive leaders, be they regional or national, and shape their endeavours? Railways drew the lines by which foreign powers carved out their spheres of influence and control, and railways also dictated the manner in which the Qing disintegrated, as warlord regimes staked out their territories by railways – railways thus prefigured the ROC era. This volume adds railway nationalism to the theories fashioned by Benedict Andersen and Anthony Reid. But it also challenges them, as railway nationalism often contended with a form of railway regionalism, which I propose more accurately reflects the subjectivities and intentions of regional engagement. The first half of the twentieth century was marred by world and civil wars in China – few scholars have acknowledged that they were fought over, on and along railways; railways have been sidelined by the weight of armed conflicts. What can we learn when we situate war in the context of railways? The CCP began by practising communism on railways; they depended on it to win the battle for China, and they used it to construct a new nation and assert itself on the global stage. The CCP's ongoing undertaking seems in keeping with an ancient Chinese saying: 'a gentleman's way to deal with someone is do unto him as he did to you'.

This book exploits brand new sources from the Rothschild Archive in London, the Sir Robert Hart Archive at the Queen's University Belfast, the Jardine Matheson Archive in Cambridge, the Sheng Xuanhuai Archive in Hong Kong and the Second Historical Archive in Nanjing. It presents materials from the National Archives at Kew, the Institute of Modern History Archive in Taipei, and several provincial and municipal archives in China, including Helongjiang, Jilin, Liaoning, Qingdao, Sichuan and Yunnan. It features material from the Library of Congress, the Woodrow Wilson Center, the Harry S. Truman Library and the Office of the Historian at the US Department of State. It employs a wide range of old newspapers, from *Shen Bao* and *North China Herald* to *La Dépêche Coloniale*, and *The Chinese Students' Monthly*, a rare student journal published between 1908 and 1929. The 'unequal treaties' are frequently mentioned in scholarship on Western and Japanese aggression in China; this book analyses them in detail. This book makes use of letters, diaries and memoirs in different languages, from officials like Weng Tonghe, to foreigners like Captain Golitsyn, merchant Archibald Little, geologist Ferdinand von Richthofen, researcher Ito Takeo, engineer Pierre Marbotte, painter Katherine A. Carl, to name a few. The book also showcases rare images from private collectors and the *Historical Photographs of China*.

Chapter outline

Composed of ten chapters, this book charts the political biography of railways from the late Qing to the post-Mao era in three parts, with each outlining one of the three phases of railway's political life. Part I focuses on railways as the nemesis of reform when its birth, that is its introduction by British merchants, galvanised domestic politics in the mid-nineteenth century. Chapter 1, 'Railway calling: British attempts', traces efforts made by the British to introduce the Shanghai to Suzhou line in the 1860s and the Wusong railway in the 1870s. Chinese resistance seems to justify the claim that the Qing rejected Western technology. However, this chapter challenges this interpretation; many officials were not against railways with their concerns lying in the destruction of existing socio-economic system and the environment. Letting foreign powers build railways meant foreign penetration into the interior, which would compromise the Qing's sovereignty and territorial integrity.

Li Hongzhang, the Qing's most important statesman and reformer, made his first attempt at railway development in the early 1880s. Chapter 2 'A journey of a thousand miles begins with a single step' investigates the circumstances under which the Tangxu was constructed and grew into the first important network in north China, which has been completely ignored by existing historical research. Despite opposition, Li pushed ahead. The Sino-French War in the mid-1880s exposed the urgent need for a Navy supported by railways, aiding Li's effort. The Tangxu was extended south to Tianjin. Reformers like Li knew that western technologies like railways could strengthen China and that foreign powers were simply the wrong messengers.

To secure the backing of Empress Dowager Cixi for his railway projects, Li gifted her with a 2-mile railroad for her birthday in 1888. Chapter 3, 'Converting the oriental despot' investigates the Xiyuan and other lines built for the exclusive pleasure of Cixi, all at the expense of reform. Li's gamble worked, and the Qing court finally gave its blessings to the Inside and Outside the Barrier railway, from Beijing to Fengtian in Manchuria. Once they discovered the benefit of railways, the Manchus ensured that they would be the first to benefit from it. Ironically, it would turn Manchuria into a 'cradle for conflict' for seven decades to come.[71]

The Qing's slow but steady progress at railway development took a dramatic turn after the First Sino-Japanese War in 1895. Part II is devoted to railway imperialism when railways became, what Daniel Headrick calls, the 'tools of empire', as European powers and Japan used this technology to carve out their sphere of interest and control.[72] Chapter 4, 'Engines for empire: Russia sets the standard' focuses on Russian railway imperialism as they used military alliances to impose the Chinese Eastern Railway, that cut through Manchuria on its way to Vladivostok and Liaodong Peninsula. Russia conquered the Qing without conquest. This whetted the appetite of the British, the Germans, the Japanese and the Americans, turning Manchuria into a 'bridgehead'.[73] The CER justified land grabs, a military presence, the exploitation of human and natural resources, and colonisation through station towns – a model to be perfected by the Germans and the Japanese.

Germany also had its eye on Manchuria, but the death of two missionaries in Shandong delivered an opportunity for them to gain a foothold there. Chapter 5, 'By order of the Kaiser: Germany outmanoeuvres', focuses on German railway imperialism, which has hitherto been overshadowed by the colony Qingdao. The construction of the Jiaoji railway allowed the Germans to seize land and to station troops, to exploit Shandong's rich natural and human resources, and to Germanise station towns. The Jiaoji would turn Shandong into another 'bridgehead' and 'cradle of conflict' in the period following the outbreak of WWI in 1914. Where there were railways, there were inevitably rivalries and conflicts.

While the Navy steered German railway imperialism, it was the Army that would drive Japanese railway imperialism, which is the focus of Chapter 6, 'Military preparedness in civilian garb: Japan catches up'. The Japanese conquest of China began with the South Manchurian Railway as the Army used it to acquire more territory, to not just station but also to build up a military force for the conquest of China Proper; to drain Manchuria's rich human and natural resources, sustaining four decades of aggression in and beyond Asia Pacific; and to Japanise through migration in order to dilute the Manchurian population. Railways were the most effective weapon to maintain colonial rule and sustain war.

While the Russians, the Germans and the Japanese battled it out in Manchuria and north China, the French and the British were scrambling for railways in southwest China. Chapter 7, 'Civilisation follows the locomotive: French variation', examines the French-constructed Vietnam to Yunnan or the Dianyue railway. The French were in direct competition with the British as the latter struggled to build the Burma railway into Yunnan. Unlike other powers, the French forced the Chinese to provide land and protection, which wrought havoc to Yunnan, as well as exploiting the region's rich resources. But Gallicisation was not as successful as Russification, Germanisation and Japanisation. The Dianyue railway illustrates the significance of variations of experiences.

Railway imperialism educated regional leaders and the ordinary Chinese people. They reacted by financing and building their own railways in the

1900s. Part III dissects railway nationalism as it transformed into an instrument of nation building under very different regimes and a 'new weapon of war' in the long twentieth century. Chapter 8, 'The quest for railway autonomy: China mobilises', explores the Railway Rights Recovery Movement in Sichuan that culminated in the Nationalist Revolution and gave birth to the ROC. This chapter challenges existing scholarship on the movement, which commonly highlights the role of nationalism and sidelines the part played by the Chuanhan railway. The real catalyst to revolution was anger over the substantial loss in shares and compensation as a result of nationalisation. Understanding this leads us to a different interpretation of the Nationalist Revolution – rather than acting as a vehicle for nationalism, the Railway Rights Recovery Movement was hijacked by Dr Sun Zhongshan and his United League in their effort to overthrow the Qing dynasty.

While imperialists carved up China with railways, so too did the warlords. No scholar has pointed out that it was railways that predestined the warlord-divided and war-torn ROC era. Chapter 9, 'He who controls railway controls China', discusses how Dr Sun used railways to imagine a new China, while warlords used them to divide and rule. As the chapter shows, railways played an important part in not just warlord conflicts but also world wars, providing battlegrounds and pretexts for the fighting. The existing scholarship has reduced railways to a footnote in these wars, but this chapter resituates them in the context of railways, arguing that whoever controlled the railway, controlled China.

A railway itself is a neutral agent, as it carries both friends and foes. But it becomes politicised and acquires new meaning in the hands of its financiers, builders, owners and inheritors. Chapter 10, 'High speed empire: China dreams', traces the endeavours of the CCP. Learning from the imperialists, Dr Sun and the warlords, the Communist regime embarked on a massive construction campaign during the Mao era. China's global ambitions can be seen in its use of railway diplomacy in Africa during the Cold War. This has expanded to Southeast Asia, Latin America, Central Asia and Eastern Europe under the umbrella of the Belt and Road Initiative (BRI) during the post-Mao era. Late Qing China may have fallen at the hands of railway imperialism, but post-Mao China is outshining old imperialists and conquering the world through railways. From the manifestation of and resistance to Western encroachment to the 'engine for empire', from catalyst for revolution and now instrument for global ambition, this volume ties individual railways together to illustrate the political journey of railways from late Qing to contemporary China. It highlights the fundamental significance of railways and sheds new light on the history and politics of modern China.

Notes

1 Lilit Marcus, 'China debuts world's fastest train', *CNN Travel*, 21 July 2021, at https://edition.cnn.com/travel/article/china-fastest-maglev-train-intl-hnk/index.html, accessed 27 December 2021.
2 Will Doig, *High speed empire: Chinese expansion and the future of Southeast Asia* (New York: Columbia Global Reports, Columbia University, 2018).
3 Edward M. Spiers, *Engines for empire: the Victorian army and its use of railways* (Manchester: Manchester University Press, 2015).
4 Arjun Appadurai, 'Introduction, commodities and the politics of value', and Igor Kopytoff, 'The cultural biography of things: commoditisation as process', in *The social life of things: commodities in cultural perspective*, ed. Anjun Appadurai (Cambridge: Cambridge University Press, 1986), 1–62 and 64–91.
5 Percy H. Kent, *Railway enterprise in China: an account of its origin and development* (London: Edward Arnold, 1907).
6 Mongton Chih Hsu, *Railway problems in China* (New York: Columbia University, 1915).
7 Lin Cheng, *The Chinese railways: a historical survey* (Shanghai: China United Press, 1935); Chang Kia-Ngau, *China's struggle for railroad development* (Baltimore: John Day, 1943).
8 Zeng Kunhua, *Zhongguo tielushi* (Taibei: Wenhai chubanshe, 1962); Mi Rucheng, *Zhongguo jindai tielushi ziliao: 1863–1911* (3 vols, Beijing: Zhonghua shuju, 1963). See Jin Xibian, Li Zhancai, Yang Yonggang, Huang Huaping, Kou Xingjun and Ma Xuebing for more.
9 Harold Stringer, *The Chinese railway system* (Shanghai: Kelly and Walsh, 1922); Colin Garrett, *China's railways: steaming into a new age* (Wellingborough: Stephens, 1988).
10 E-tu Zen Sun, *Chinese railways and British interests 1898–1911* (New York: King's Crown Press, Columbia University, 1954), 1.
11 Wayne Altree, 'A half-century of the administration of the state railways of China', *Papers on China*, vol 3 (1949): 83–121; Chi-ming Hou, *Foreign investment and economic development in China, 1840–1937* (Cambridge: Harvard University Press, 1965); John E. Schrecker, *Imperialism and Chinese nationalism: Germany in Shantung* (Cambridge: Harvard University Press, 1971); E-tu Zen Sun, 'The pattern of railway development in China', *The Far Eastern Quarterly* (pre-1986), vol 14, 2 (February 1955): 179–99; Joseph Esherick, 'Harvard on China: the apologetics of imperialism', *Bulletin of Concerned Asian Scholars*, vol 4, 4 (1972): 9–16; Conrad Thomason, 'Ambiguities of modernisation: the China case', *The China Geographer 5* (Fall 1976): 13–27.
12 Ernest P. Liang, *China: railways and agricultural development, 1875 to 1935* (Chicago: University of Chicago Department of Geography, 1982).
13 Ralph W. Huenemann, *The dragon and the iron horse: the economics of railroads in China, 1876–1937* (Cambridge: Council on East Asian Studies, Harvard University, 1984).

14 Elisabeth Köll, *Railroads and transformation of China* (Cambridge: Harvard University Press, 2019).
15 Edward Le Fevour, *Western enterprise in late Ch'ing China: a selective survey of Jardine, Matheson and Company's operations, 1842–1895* (Cambridge: Harvard University Press, 1968), 107–29.
16 David Pong, 'Confucian patriotism and the destruction of the Wusong railway, 1877', *Modern Asian Studies*, vol 7, 4 (1973): 647–76; Sandra Sturdevant, 'Imperialism, sovereignty, and self-strengthening: a reassessment of the 1870s', in *Reform in nineteenth century China*, eds Paul Cohen and John Schrecker (Cambridge: Harvard East Asia Center, 1976), 63–70.
17 Hsien-chun Wang, 'Merchants, mandarins, and the railway: institutional failure and the Wusong railway, 1874–1877', *International Journal of Asian Studies*, vol 12, 1 (January 2015): 31–53.
18 Peter A. Crush, *Woosung road: the story of China's first railway* (Hong Kong: Railway Tavern, 1999).
19 David Pletcher, *The diplomacy of involvement: American expansion across the Pacific 1784 to 1900* (Columbia: University of Missouri Press, 2001).
20 Arthur L. Rosenbaum, 'Railway enterprise and economic development: the case of the imperial railways of north China', *Modern China*, vol 2, 2 (April 1976): 227–72 and 'The Manchuria bridgehead: Anglo-Russian rivalry and the Imperial Railway of North China, 1897–1902', *Modern Asian Studies*, vol 10, 1 (February 1976): 41–64; Peter A. Crush, *The Imperial railways of north China* (Beijing: Xinhua chubanshe, 2013).
21 Peter A. Crush, 'Early Chinese narrow gauge railways part 2: the emperor's palace railway', at www.researchgate.net/publication/353767552_EARLY_CHINESE_NARROW_GAUGE_RAILWAYS_-_Part_2_-_The_Emperor's_Palace_Railway_aka_Violet_Tower_Railway_or_Xiyuan_Railway, accessed 23 September 2021.
22 Fisher Y. C. Yu, *The Chinese Eastern Railway: its redemption by the Chinese government and the financial involvements* (Cambridge: Harvard University, 1925); Ching Chun Wang, *The Chinese Eastern Railway* (Peking: Peking Leader Press, 1926); *The sale of the Chinese Eastern Railway* (New York: Council on Foreign Relations, 1933); George E. Sokolsky, *The story of the Chinese Eastern Railway* (Shanghai: North China Herald, 1929); Lien-en Tsao, *The Chinese Eastern Railway: an analytical study* (Shanghai: The Bureau of Industrial and Commercial Information, 1930).
23 Tan Guiluan, *Zhongdong tielude xiuzhu yu jinying: Er'guo zaihua shilide fazhan* (Taipei: Lianjin chubanshe, 2016). See also the works of Li Guoqi, Ma Weiyun and Wu Wenxian.
24 Tao Shing Chang, *International controversies of the Chinese Eastern Railway* (Taipei: Chengwen, 1971); Sow-Theng Leong, *Sino-Soviet diplomatic relations, 1917–1926* (Canberra: ANU Press, 1976); Blaine Chiasson, *Administering the colonizer: the Chinese administrative takeover of the Chinese Eastern Railway concession, 1918–29* (Vancouver: UBC Press, 2014); Victor Zatsepine, *Beyond the Amur: frontier encounters between China and Russia, 1850–1930*

(Vancouver: UBC Press, 2017); Chris Miller, *We shall be masters: Russian pivots to East Asia from Peter the Great to Putin* (Cambridge: Harvard University Press, 2021); Paul Dukes, *Russia in Manchuria: a problem of empire* (London: Routledge, 2022).

25 Nikita Vul, 'He, who has sown the wind: Karakhan, the Sino-Soviet conflict over the Chinese Eastern Railway, 1925–26 and the failure of Soviet policy in northeast China', *Modern Asian Studies*, vol 48, 6 (November 2014): 1670–94; Michael M. Walker, *The 1929 Sino-Soviet War: the war nobody knew* (Lawrence: University Press of Kansas, 2017); Bruce A. Elleman, 'The Soviet Union's secret diplomacy concerning the Chinese Eastern Railway 1924–1925', *The Journal of Asian Studies*, vol 53, 2 (May 1994): 459–86; Bruce Elleman, *Diplomacy and deception: the secret history of Sino-Soviet diplomatic relations 1917–1927* (Armonk: M. E. Sharpe, 1997); Bruce A. Elleman and Stephen Kotkin, eds, *Manchurian railways and the opening of China: an international history* (New York: Routledge, 2010).

26 Olga Bakich, 'Origins of the Russian community on the Chinese Eastern Railway', *Canadian Slavonic Papers*, vol 27, 1 (March 1985): 1–14; Nikita Vul, 'Chinese Eastern Railway and Russian minority groups in Manchuria', *Modern History of Russia*, vol 11, 1 (2021): 89–102; Mark Gamsa, *Harbin: a cross-cultural biography* (Toronto: University of Toronto Press, 2020).

27 David Wolff, *To the Harbin station: the liberal alternative in Russian Manchuria, 1898–1914* (Stanford: Stanford University Press, 1999); Olga Bakich, 'Russian city in China: Harbin before 1917', *Canadian Slavonic Papers*, vol 28, 2 (June 1986): 129–48; Thomas Lahusen, 'A place called Harbin: reflections on a centennial', *The China Quarterly*, vol 154, June (1998): 400–10.

28 Smith D. Woodruff, *The German colonial empire* (Chapel Hill: University of North Carolina Press, 1978); Klaus Mühlhahn, ed., *The cultural legacy of German colonial rule* (Berlin: Walter de Gruytar, 2017); Hartmut Berghoff *et al.* eds, *Exploration and entanglement: Germans in the Pacific world* (Oxford: Berghahn Books, 2018).

29 George Steinmetz, *The devil's handwriting: precoloniality and the German colonial state in Qingdao, Samoa, and Southwest Africa* (Chicago: The University of Chicago Press, 2007); Klaus Mühlhahn, 'Negotiating the nation: German colonialism and Chinese nationalism in Qingdao: 1897–1914', in *Twentieth century colonialism and China: localities, everyday and the world*, eds Bryna and David S. G. Goodman (London: Routledge, 2012), 37–56.

30 Elisabeth Köll, 'Chinese railroads, local society and foreign presence: the Tianjin-Pukou line in pre-1949 Shandong', in *Manchurian railways*, 106–23.

31 Peter Duus and Ramon H. Myers eds, *The Japanese informal empire in China, 1805–1937* (Princeton: Princeton University Press, 1991); Leonard A. Humphreys, *The way of the heavenly sword: the Japanese army in the 1920s* (Stanford: Stanford University Press, 1995); Louise Young, *Japan's total empire: Manchuria and the culture of wartime imperialism* (Berkeley: University of California Press, 1998); Emer S. O'Dwyer, *Significant soil: settler colonialism and Japan's urban empire in Manchuria* (Cambridge: Harvard University Asia

Center, 2015); Bill Sewell, *Constructing empire: the Japanese in Changchun, 1905–1945* (Vancouver: UBC Press, 2019).

32 Ramon H. Myers, 'Japanese imperialism in Manchuria: the South Manchuria Railway Company, 1906–1933', in *Japanese informal empire*, 101–32; S. C. M. Paine, *Imperial rivals: China, Russia and their disputed frontier* (Armonk: M. E. Sharpe, 1996), 175–90; Yoshihisa Tak Matsusaka, *The making of Japanese Manchuria, 1904–1932* (Cambridge: Harvard University Asia Center, 2003).

33 C. Walter Young, *Japanese jurisdiction in the South Manchuria Railway areas* (Baltimore: The Johns Hopkins Press, 1931). See also the works of Yoshii Kenich, Wada Keiko and Wada Hirofumi.

34 John Young, *The research activities of the South Manchuria Railway company 1907–45: a history and bibliography* (New York: Columbia University Press, 1966); Ito Takeo, *Life along the South Manchurian Railway: the memoirs of Ito Takeo* (Armonk: M. E. Sharpe, 1988).

35 John F. Cady, *The roots of French imperialism in Eastern Asia* (Ithaca: Cornell University Press, 1954); James R. Fichter, *British and French colonialism in Africa, Asia and the Middle East* (Cham: Springer, 2019).

36 Virginia Thompson, *French Indo-China* (London: George Allen and Unwin, 1937); Robert Lee, *France and the exploitation of China 1885–1901* (Hong Kong: Oxford University Press, 1989).

37 Amaury Lorin, '"La civilisation suit la locomotive": le *credo* ferroviaire de Paul Doumer, gouverneur général de l'Indochine (1897–1902)', *Revue d'histoire des chemins de fer*, vol 35, 1 November (2006): 41–54; Vatthana Pholsena, 'Technology and empire: a colonial narrative of the construction of the Tonkin-Yunnan railway', *Critical Asian Studies*, vol 47, 4 (2015–10): 537–57; Selda Altan, 'Politics of life and labor: French colonialism in China and Chinese coolie labor during the construction of the Yunnan-Indochina railway 1898–1910', *International Labor and Working Class History*, vol 101 (2022): 77–99.

38 Jean-François Rousseau, 'An imperial railway failure: the Indochina-Yunnan railway, 1898–1941', *The Journal of Transport History*, vol 35, 1 (2014): 1–17; Florence Bretelle-Establet, 'Resistence and receptivity: French colonial medicine in southwest China, 1898–1930', *Modern China*, vol 25, 2 (April 1999): 171–203.

39 J. Chandran, *The Burma–Yunnan railway: Anglo-French rivalry in mainland Southeast Asia and south China* (Athens: Ohio University Center for International Studies, 1971); B. E. Kennedy, 'Anglo-French rivalry in Southeast Asia 1763–93: some repercussions', *Journal of Southeast Asian Studies*, vol 4, 2 (September 1973): 199–215; John Keay, 'The Mekong exploration commission, 1866–68: Anglo-French rivalry in South East Asia', *Asian Affairs*, vol 36, 3 (November 2005): 289–312.

40 Zen Sun, *British interests*; D. McLean, 'Chinese railways and the Townley Agreement of 1903', *Modern Asian Studies*, vol 7, 2 (March 1973): 145–64; Philip Joseph, *Foreign diplomacy in China 1894–1900* (London: G. Allen and Unwin, 1928); Nathan A. Pelcovits, *Old China hand and the foreign office* (New York: King's Crown Press, 1948).

41 Ian P. Austin, *Ulyssess S. Grant and Meiji Japan 1869–1885* (London: Routledge, 2019).
42 Gregory Moore, *Defining and defending the open door policy: Theodore Roosevelt and China, 1901–1909* (Lanham: Lexington Books, 2015), 111–37; Maury Klein, *The life and legend of E. H. Harriman* (Chapel Hill: University of North Carolina Press, 2011); William R. Braisted, 'The United States and the American China development company', *The Far Eastern Quarterly*, vol 11, 2 (1952): 147–65.
43 Clarence B. Davis and Kenneth E. Wilburn, Jr eds, *Railway imperialism* (Westport: Greenwood Press, 1991).
44 T. G. Otte and Keith Neilson, eds, *Railways and international politics: paths of empire, 1848–1945* (London: Routledge, 2006).
45 Köll, *Transformation*, 297.
46 Yen Ching-hwang, 'Chang Yu-nan and the Chaochow railway (1904–08): a case study of overseas Chinese involvement in China's modern enterprise', *Modern Asian Studies*, vol 18, 1 (February 1984): 119–35.
47 En-han Lee, *China's quest for railway autonomy, 1904–1911: a study of the Chinese railway rights recovery movement* (Singapore: Singapore University Press, 1997); Kia-ngau Chang, *China's struggle for railway development* (New York: Da Capo Press, 1943).
48 Madeleine Chi, 'Shanghai-Hangchow-Ningbo railway loan: a case study of the rights recovery movement', *Modern Asian Studies*, vol 7, 1 (January 1973): 85–106; Mary B. Rankin, 'Nationalistic contestation and mobilisation politics: practice and rhetoric of the railway-rights recovery at the end of the Qing', *Modern China*, vol 28, 3 (July 2002): 315–61; Elisabeth Kaske, 'Sichuan as a pivot: provincial politics and gentry power in late Qing railway projects in southwestern China', in *Southwest China in a regional and global perspective (c. 1600–1911): metals, transport, trade and society*, eds Ulrich Theobald and Jin Cao (Leiden: Brill, 2018), 379–423.
49 Xiaowei Zheng, *The politics of rights and the 1911 revolution in China* (Stanford: Stanford University Press, 2018).
50 See the works of Wang Ermin, Dai Zhili, Wu Lunni and Xian Yuhao among others.
51 Sun Zhongshan, *Jianguo fanglue* (Beijing: Changan chubanshe, 2011) and *The international development of China and fundamentals of national reconstruction* (Taipei: China Cultural Service, 1953).
52 Chang Jui-Te, 'Technology transfer in modern China: the case of railway enterprise (1876–1937)', *Modern Asian Studies*, vol 27, 2 (May 1993): 281–96.
53 Gavan McCormack, *Chang Tso-lin in northeast China, 1911–1928: China, Japan and the Manchurian idea* (Stanford: Stanford University Press, 1977); Chi Man Kwong, *War and geopolitics in interwar Manchuria: Zhang Zuolin and the Fengtian clique during the northern expedition* (Leiden: Brill, 2017); Donald G. Gillin, *Warlord Yen Hsi-shan in Shanxi province 1911–49* (Princeton: Princeton University Press, 1967).

54 Christian Wolmar, *Engines of war: how wars were won and lost on the railways* (New York: Public Affairs, 2010).
55 Hans van de Ven, *China at war: triumph and tragedy in the emergence of the new China* (Cambridge: Harvard University Press, 2018); Stephen R. MacKinnon, *Wuhan, 1938: war, refugees, and the making of modern China* (Berkeley: University of California Press, 2008); Odd Arne Westad, *Cold War and revolution: Sino-Soviet rivalry and the origins of the Chinese civil war, 1944–46* (New York: Columbia University Press, 1993).
56 Harold M. Tanner, 'Railways in communist strategy and operations in Manchuria, 1945–48', in *Manchurian railways*, 149–70 and *The battle for Manchuria and the fate of China, Siping 1946* (Bloomington: Indiana University Press, 2013); Victor Cheng, *China between peace and war: Mao, Chiang and the Americans, 1945–1947* (Canberra: ANU Press, 2023).
57 Chao Yung Seen, *Railways in communist China* (Hong Kong: Union Research Institute, 1955); Chi-Keung Leung, *China: railway patterns and national goals* (Chicago: University of Chicago Department of Geography, 1980); Kenneth Cantlie, *The railways of China* (London: China Society, 1981); Barry Naughton, 'The Third Front: defence industrialisation in the Chinese interior', *The China Quarterly*, vol 115 (September 1988): 351–86; Covell F. Meyskens, 'Third Front railroads and industrial modernity in late Maoist China', *Twentieth-Century China*, vol 40, 3 (October 2015): 238–60.
58 Richard Hall and Hugh Peyman, *The great uhuru railway: China's showpiece in Africa* (London: Gollancz, 1976), introduction; Kasuka S. Mutukwa, *Politics of the Tanzania-Zambia railway project: a study of the Tanzania-China-Zambia relations* (Washington DC: University Press of America, 1979); Jamie Monson, *Africa's freedom railway: how a Chinese development project changed lives and livelihoods in Tanzania* (Bloomington: Indiana University Press, 2009).
59 Jian Hong Wu and Chris Nash, *Railway reform in China* (Leeds: Institute for Transport Studies, 1988); Elspeth Thomson, *The reform and re-structuring of China's railways* (Singapore: East Asia Institute, National University of Singapore, 2002); Linda Yin-Nor Tjia, *Explaining railway reform in China: a train of property rights rearrangements* (New York: Routledge, 2016).
60 Shang-Su Wu and Alan Chong, 'Developmental railpolitics: the political economy of China's high speed rail projects in Thailand and Indonesia', *Contemporary Southeast Asia*, vol 40, 3 (December 2018): 503–26.
61 Jiang Pei has authored many articles; Xiong Yaping, *Tielu yu huabei xiangcun shehui bianqian, 1880–1937* (Beijing: Renmin chubanshe, 2011).
62 Li Siyi, *Tielu xiandai xin* (Taibei: Shibao wenhua chubanshe, 2020).
63 James Zheng Gao, *Meeting technology's advance: social change in China and Zimbabwe in the railway age* (Westport: Greenwood Press, 1997).
64 Emma Yu Zhang, 'Socialist builders on the rails and road: industrialization, social engineering, and national imagination in Chinese socialist films, 1949–1965', *Twentieth-Century China*, vol 42, 3 (2017): 255–73.
65 Appadurai, 'Politics of value', 3.

66 Appadurai, 'Politics of value', 3.
67 Kopytoff, 'Cultural biography', 67.
68 Kopytoff, 'Cultural biography', 67.
69 Köll, *Transformation*, jacket.
70 Ronald E. Robinson, 'Introduction', in *Railway imperialism*, 2.
71 Owen Lattimore, *Manchuria: cradle of conflict* (New York: Macmillan, 1935).
72 Daniel Headrick, *The tools of empire: technology and European imperialism in the nineteenth century* (New York: Oxford University Press, 1981).
73 Rosenbaum, 'Manchuria bridgehead'.

Part I

Nemesis of reform

1

Railway calling: British attempts

> We should promote a transportation system; the more convenient, lighter, faster, and better, the more beneficial to us. We should make railways like those foreign countries, travelling at 7000 to 8000 *li* per day and night. Let it make profit for us and let us limit others from copying it.[1]

Spoken by Hong Rengan, cousin of Hong Xiuquan, the leader of the Taiping Rebellion (1850–64), these words demonstrate faith in the power and potential of railways. This technology would aid their economy and military campaign – they understood what railways could do for the new Christian kingdom they established. The Taiping rebels were open-minded and forward-looking. We can only imagine what might have happened if the Taiping had succeeded in overthrowing the Qing. Their vision of the railway is important because it would inspire fellow Cantonese, Dr Sun Zhongshan. While the Hong cousins dreamed of railways, the refugees their rebellion created left China for the United States where the *Sacramento Daily Union* declared that 'Chinese laborers can be profitably employed in grading railroads in California'.[2] They toiled and died building railroads across North America while the story of the railway in their own homeland took a completely different turn. This chapter revisits the controversial Wusong [吴淞] railway. It challenges existing scholarship on Chinese resistance to the introduction of railways, which generations of historians have dismissed as anti-modern conservatism. It exposes the positive reception of the Wusong, which has been ignored in the effort to emphasise the Qing's hostility to this foreign technology. It offers a new interpretation on the politics of railways that shaped not just the fate of the Wusong but also late Qing reform.

'When they felt the need of them'[3]

Nearly all general histories of railways in China begin with the Wusong railway constructed in 1876. Its importance was not limited to the fact that it was the first railway built on Chinese soil but because it was dismantled within a year of its completion. These histories have ignored the Shanghai–Suzhou line proposed a decade before. The political genealogy of railways must begin with this early attempt as its rejection will help us comprehend not just the dismantling of the Wusong but, more importantly, the politics of railways.

The end of the First Opium War in 1842 saw a growing number of foreigners descend upon Shanghai to conduct business. Many of them saw the potential of railway development in such a vast country: 'the major mercantile ambition was to see the inauguration of the railway age in China, partly as a field for investment and partly as a mean of penetrating internal markets'.[4] Some thought the task an easy one: 'Get a map of China, put some tracing paper over it, and having ascertained where the provincial capitals are situated, trace out a network of railways, and when finished exclaim – "there is what the Chinese government required to regenerate itself".'[5] Rowland MacDonald Stephenson, whom Danial Thorner called 'outstanding pioneer of railways in India', was brought in by Jardine Matheson to undertake surveys, recommending multiple lines.[6] One proposed line fanned out from Hankou, travelling eastward to Shanghai, westward to Sichuan, southwestward to Yunnan and India, and southward to Guangdong (Canton). Another centred in Zhenjiang (Nanjing), travelling northward to Beijing via Tianjin and southward to Shanghai. Jardine developed an interest in railways as early as 1845.[7] An old China hand, Jardine made its fortune in the opium trade and was instrumental in the opening of China. It had an incentive to invest because the opium trade was growing rapidly, and Shanghai allowed Jardine to diversify and venture into many other sectors.

'A large number of merchants', according to British diplomat William Frederick Mayers, 'requested the concession of authority to build a line from Shanghai to Soochow [Suzhou] in 1863'.[8] They included Americans.[9] The Shanghai–Suzhou railway proposal was presented to Li Hongzhang, then Governor of Jiangsu province.[10] The choice of Suzhou was obvious, as it was the centre of silk production and the closest city on the Grand Canal that could provide access to the interior and north China. The British and Americans pinned their hopes on Li who had cooperated with them during the suppression of the Taiping. Open-minded and pragmatic, Li seemed to be someone they could do business with, but these foreigners did not understand that the push to reform had been triggered by the need to suppress the Taiping. It was an open question whether the Qing court would continue with reform after the rebellion was repressed in 1864. Li expressed his opinion in early 1865:

> As for the two matters of establishing telegraphs and constructing railways, every foreign country would want to do it, especially the British because most

> of the ships coming to China are theirs. This is something that is most beneficial to merchants. When Suzhou was first recovered [from the Taiping], its Consul General had asked to build a line from Shanghai to Suzhou, this was not permitted. ... Constructing railways is a gigantic affair that costs money, it involves moving mountains and valleys. They also know that this is impossible. We should refuse them outright.[11]

The call for a Shanghai–Suzhou line was rejected. The *North China Herald* (*Herald* hereafter) summed up foreign disappointment: 'To us, railways mean free intercourse, enlightenment, commerce and wealth; to the mandarins, they suggest rowdyism, the overthrow of time-honoured custom and tradition, disturbance and ruin.'[12] Instead of trying to understand Li's rationale, this line of thought has been perpetuated by generations of historians who blamed the lack of enthusiasm for reform on Confucianism. This is best seen through Mary Wright's influential verdict on the Tongzhi Restoration (1862–74) that 'even in the most favourable circumstances there is no way in which an effective modern state can be grafted onto a Confucian society'.[13] This assessment has been reinforced by Albert Feuerwerker in his book on China's early industrialisation and by John Rawlinson in his work on the late Qing failure to establish a navy.[14] Ralph Heunemann described the Chinese resistance as a 'policy conundrum'.[15] As we can see from Li's response, he expressed legitimate concerns in rejecting the call for a line from Shanghai to Suzhou; he did not reject railways outright. His concerns did not have much to do with Confucianism but the practicalities of managing and protecting an agriculture-based economy and society. Li's words deserve our attention, not continued criticism.

Li's rationale was threefold. First, railways would only benefit foreigners, the British in particular, as 'most of the ships are theirs'. The introduction of railroads would allow foreigners to penetrate further into the interior, which the Qing were anxious to avoid. Britain had humiliated the Qing in the Opium Wars and imposed unwanted treaties that granted them exclusive privileges; nonetheless, they continued to push for more access, to the detriment and growing concern of the Qing. Li was in no mood to grant the British further advantages. Second, Li believed that there was no Chinese demand for a railway. Existing means of transportation, by mule/horse wagons and via rivers/canals, were cheap and suitably efficient. Though railways might have aided local businesses to grow, by and large ordinary Chinese people did not need railways. In short, there was not yet a business case to be made in its favour. Third, railway construction required the destruction of great swathes of the natural habitat by 'moving mountains and valleys', taking land away from peasants and forcing some to re-allocate burial grounds, a serious offence to the ancestor-worshipping Chinese people. This had the potential, as was the case of the Taiping and other rebellions, to turn the landless poor into dangerous vagrants and rebels. Li's colleague, Shen Baozhen, the Governor of Jiangxi province, also weighed in: 'As for railway, there are many concerns and problems. Erasing mountains and rivers would create enormous disaster for the future. This would damage burial sites in the countryside, exposing us to conflicts.'[16]

Li and Shen did not share the impulses and notions of the modernising West, which sought to conquer nature. For them, the gargantuan project of 'erasing mountains and rivers' for railways would only serve foreign interests. Their concern for the destruction of the environment, which previous scholarship has not paid attention to, deserves more than just a paragraph here. The Chinese are a people deeply attached to the land from which they derive their existence and where their ancestors are buried. The Confucian worldview teaches its adherents to live in harmony with nature. Railway construction forcibly uproots people from their land, deprives them of their livelihood and damages the environment – this put humans in conflict with nature. Seen from the hindsight of today's concerns for the environment, Li and Shen had plenty to legitimately worry about. Late Qing conservatism had much in common with today's progressive environmentalism – a fascinating parallel that deserves more studies.

Li and Shen were joined in their concerns by their superiors, Zeng Guofan, the governor-general of Liangjiang, Mao Hongbing, governor-general of Liangguang, and Zuo Zongtang, another reformer and then governor-general of Shaangan – all of whom were towering figures in late Qing officialdom. Mao raised other important issues: 'Chinese horse wagons cannot compete with [the railway]' and 'there are many bandits that have not been suppressed, they are waiting for opportunities to make trouble, when they see such things as railways, they would use it to mobilise and rise up again'.[17] Fresh from the wounds inflicted by the Taiping Rebellion and the Second Opium War, Qing China was hungry not for change but stability. The British, however, were not disheartened by the rejection of the Shanghai–Suzhou line. Robert Hart, director-general of the Imperial Maritime Customs, urged the Qing to build railways in 1865; as did diplomat Thomas Wade in 1866.[18] Some called for a Guangzhou–Foshan line, while others for a Shanghai–Hangzhou line; all expected Sir Rutherford Alcock, British Minister to China (1865–69), to press the Qing court.[19]

David Pong has argued that Shen's attitude towards railways had changed by 1867.[20] Pong's argument is further verified when considering Shen's own words, stating, 'The Qin [dynasty 221–206 BC] built the Great Wall. It was a misfortune at the time but later dynasties relied on it. The same can be said of telegraph and railway; if built, it would benefit us in the future.'[21] Though the 'misfortune' he cited referred to the loss of tens of thousands of lives in the construction of the Great Wall, Shen's words reveal that, despite his concerns, he was not opposed to the introduction of railways. Li shared his thoughts on the subject a month later: 'When it comes to telegraph and railway, they greatly benefit foreigners but they harm us; more so with railway than with telegraphs.'[22] For Shen and Li, the issue was clear: benefits to China were indispensable in decision-making – a point that has been overlooked. Knight Biggerstaff's research reveals that of the seventeen ranking officials that the *Zongli Yamen*, the Qing's new Foreign Office, consulted in the late 1860s, only Shen and Li believed that 'such modern improvements might be beneficial to China', [and that] 'when the necessary capital could be raised the Chinese construct railway and telegraph lines themselves, making use of foreign engineers and imported equipment'.[23] Li and Shen were not against railways; they merely trod carefully around an issue many considered unsuitable to the Chinese economy and society at the time.

Li was accurate in surmising that railways would primarily serve to expand foreign interests and influence in China. The case of colonial India offers a useful point of comparison and provides insights to the rationale of Qing officials. Tara Sethia has questioned why investment in Indian railroads did not stimulate economic and industrial growth in Hyderabad. Her research points to the fact that the introduction of railways 'was not a response to the local politics or needs but originated from the needs and interests of several groups in England favouring railway construction in India', and they were used 'as instruments' to accomplish British political and economic interests 'at the expense of Hyderabad'.[24] This was the same with Argentina where railways were built and governed by the British, with what Winthrop Wright called a 'calculated disregard for Argentine interests in general'.[25] Qing China faced a similar situation, albeit with some important differences: in the Indian case, coercion played an important role in British strategy. But this did not work with the late Qing. Shen was a pioneer of reform who knew that railways could propel the Qing onto the path of modernisation, but he would not be coerced. Li felt similarly and insisted that he 'would not be hurried'.[26] Percy Kent seems to sympathise with them: 'the country as a whole was not conscious of the need of improved communication, and the officials were determined that railways, if introduced at all, should be introduced by the Chinese themselves when the need arose'.[27]

Shen and Li were prominent participants in this debate, and their actions would soon decide the fate of the first railway built on Chinese soil. As early as 1872, *Shen Bao*, the widely read newspaper published in Shanghai, began to carry news about the railways in Britain and discuss their potential in China.[28] It seems that public opinion was changing. To charm the Chinese public, British manufacturers and merchants erected a sample railway in Tianjin's British Settlement.[29] Under such headlines as 'Tianjin Tries Railroad', *Shen Bao* reported that local officials were invited to ride in the upper-class carriages with the British Consul in September 1872.[30] Many Chinese enjoyed free rides on the trains while spectators sang its praises. *The London and China Telegraph* reported that:

> Another movement has been organised under influential auspices for establishing a railway system in China. The supporters of the present scheme include the Duke of Sutherland, the Marquis of Westminster, the Marquis of Salisbury, Sir Rutherford Alcock, and other noblemen and gentlemen, and it is proposed that a short line of railway with rolling-stock and all accessories should be constructed for presentation to the Emperor of China, with the object of obtaining his sanction to the establishment of a complete railway system throughout the country.[31]

Like the Jesuit missionaries before them, the British were convinced that converting the emperor would aid their mission. They were now joined by a growing number of Americans in Shanghai. The Huangpu River was shallow, making it difficult for large ships to enter the city, a problem compounded by sediment, leading to serious problems. These issues led Oliver B. Bradford, the American Deputy Consul in Shanghai, to propose a narrow-gauge railroad from the International Settlement to the anchorage at Wusong.[32] We now know from Robert Hart's diary that as early as 1865 British merchants were manoeuvring for a line to Wusong. The entry for 25 February 1865 reads: 'Yesterday a letter from Mossman arrived, forwarding a printed prospectus of a railway line from S'hae to Woo Sun.'[33] A short railroad would not only cut distance and cost; it would also accelerate transportation and facilitate better movement of goods. But Bradford failed to attract enough interest and capital from the United States, and the Americans were forced to resort to 'admitting a preponderating British interest'.[34] Jardine was this dominating interest. They approached the Shanghai *Daotai*, the region's highest-ranking official in 1872, and purchased a strip of land from the Settlement to the anchorage. As Kent argues, despite their intention to build a railway, the British and Americans told the *Daotai* that they were going to build a *ma lu*, a 'horse road'.[35]

Herein lies the beginning of the problematic interpretations of the Wusong – which this chapter aims to correct. Kent's 1907 book has influenced many scholars as general histories of railways in China often emphasise the fact that the foreigners misled the Shanghai authority. While there is some validity in this assessment, it does not represent the whole story. *Shen Bao* announced that 'Railway will be built between Shanghai and Wusong' in May 1873.[36] By August, further news emerged that the British planned to present this railway as a gift to China.[37] This kind of reporting continued throughout the year 1874 as construction began in the summer.[38] It seems that *Shen Bao* was spreading the gospel of locomotion and preparing the locals for the arrival of the railway. The plan for and construction of the Wusong were public knowledge. Qing officials in Shanghai, especially the *Daotai* or what Yuen-sang Leung called the 'linkage man', whose job was to keep an eye on and liaise with the foreigners, could not have failed to notice this.[39]

The construction of the Wusong was not the only reason 1874 would become a significant year. In May, more than 3,000 Japanese soldiers landed in Taiwan.[40] This kind of threat from the seas and the subsequent debate on the need to establish a navy is crucial to our understanding of the Chinese perceptions of railways – a detail which existing scholarship has not considered. It was naval construction that would change the fortune of railroads. The Japanese came to seek revenge for fifty shipwrecked Liuqiu (Ryukyu) people who had been killed by indigenous Taiwanese in 1871. Liuqiu had been an independent kingdom and an important trade link between China and Japan. However, Japan considered them 'Japanese subjects', and their rationale was clear.[41] The 'Taiwan expedition' marked the beginning of Japanese overseas expansion.[42] Robert Eskildsen has denoted the episode the 'birth of Japanese imperialism'.[43] In 1879 Japan annexed the littoral kingdom and it became the Okinawa prefecture.[44] The Qing court reacted to the Japanese landing by sending Shen Baozhen as Imperial commissioner to Taiwan. This experience convinced Shen to adopt a series of measures to modernise Taiwan's defence.

Japan's surprise attack, although thwarted this time, exposed its designs on China and alarmed many officials. The debate around establishing a navy, which had first emerged in the 1840s, was soon revived. Ding Richang, then Governor of Jiangsu province, proposed three naval forces: the North Navy for the Zhili-Shandong coasts, the East Navy for the Jiangsu–Zhejiang provinces and the South Navy for the Taiwan–Fujian–Guangdong region.[45] As the debate about China's naval defence intensified, Li Hongzhang pushed the discussion further. In December 1874, he argued that naval construction would benefit from railways and a telegraph system:

> Things happen unexpectedly and military intelligence changes constantly. If we have telegraphs like western countries, then information can travel a thousand *li* at an instant to reach anywhere along the coast. If we have railways built with soldiers stationed nearby, upon hearing emergencies, they can rush and travel thousands of *li* a day so the commander won't make mistakes. I worry because we don't have them.[46]

Matters of national security decidedly changed the minds not just of Li, but many other officials on the railways. A consensus for naval construction had been reached and the Qing court reacted in 1875. Shen Baozhen was appointed Nanyang, or South Ocean commissioner, in charge of naval defence in the East and South China Seas. Li Hongzhang became Beiyang, or North Ocean commissioner, in charge of affairs in the Bohai-Yellow Seas. These appointments signalled a real change. The two opium wars had not instigated the same level of urgent mobilisation; nothing had made the Qing court change its mind until the Japanese landed in Taiwan. It appears that the Qing court had come to terms with naval construction facilitated by railways that could quickly move troops and weapons. The railway had become fundamental in the construction of a navy, a matter of national security. Like Taiping-suppression induced reform, this was born out of urgency, thus reactive and impromptu. But building up a modern navy supported by railways demanded leadership, money and expertise, all of which the late Qing lacked. This, nevertheless, appeared to be good news for the Wusong under construction. Shen Baozhen, the governor-general of the region with jurisdiction over Shanghai, may prove more amenable to the railway or even accept it as a *fait accompli*.

What is done can be undone

Ralph Heunemann began his general economic history of railroads in China with a lament: 'the story of how railroads came to China is a familiar and unhappy one'.[47] His use of the word 'familiar' indicates the usual Chinese resistance to reform; whereas 'unhappy' most likely referred to the fate of the Wusong railway. But this was not exactly the case. The construction of the Wusong has been documented by scholars like Percy Kent, which I will not repeat. Using previously overlooked sources, such as *Shen Bao*, Shanghai's predominant daily that reflected public opinion, this section revisits the perception and dismantling of the Wusong to address some of the misinterpretations.[48]

Hope was in the air for those who invested in the Wusong by 1875. In February, *Shen Bao* reported that 'about three-quarters of a mile has been laid, and the 'Pioneer' has performed the first railway run ever made in China' (see Figure 1.1), and further speculated that the 'Wusong railway will soon be completed'.[49] Shanghai officials could not have missed this. The tracks, steam engine and test run confirmed *Shen Bao* reports that the foreigners were not building a 'horse road'. Hence, it is inaccurate to claim that Shanghai officials had only just discovered the deception.[50] After all, local newspapers had been reporting the construction and appearance of the Wusong railway since at least May 1873. David Pong and Hsien-Chun Wang provided accounts of the diplomatic wrangling from

Figure 1.1 The 'Pioneer' (Peter Crush private collection. Courtesy of Peter Crush)

March 1876. It began in Shanghai between the *Daodai*, Feng Junguang, who asked the British to halt construction, and the British Consul, Walter Henry Medhurst. It soon spilt over to Beijing where Prince Gong, head of the *Zongli Yamen*, and the British Ambassador, Thomas Wade, were engaged in a tense negotiation over the Margary Affair, when a second steam engine named the 'Celestial Empire' arrived in May 1876.[51] In June, Wade even went so far as to summon British warships to Shanghai, just outside of Wusong, in a show of force meant to intimidate the Qing. Wang aptly described this crisis as 'the lowest point in Sino-British relations since the Second Opium War'.[52] As the row continued, the first half of the Wusong, the segment between the Settlement and Jiangwan, was close to completion.

Shen Bao fuelled the anticipation surrounding the arrival of the Wusong, reporting several times in June 1876 that 'The railway will be completed'.[53] This anticipation was finally met in the beginning of July when the local Chinese were invited to test the new machinery.[54] Richard Rapier, whose company supplied and supervised the construction detailed this:

> On Saturday the 1st July, the Chinese were invited to travel free on the railways – great, indeed, was the jostling, pushing, and crowding to secure places in so small a means of conveyance. All, however, passed off well and without accident, and to the great delight of the people. On the 3rd July traffic commenced, and the receipts were at once of the most satisfactory character.[55]

Many Chinese were clearly excited about the arrival of the railway in Shanghai. *Shen Bao* advertised the service, schedules and prices as well as reporting great enthusiasm among Chinese passengers.[56] So too did the *Herald*:

> The absence of popular ill-will to the undertaking, and the delighted eagerness with which the natives hastened to take advantage of the opportunity afforded to them by the proprietors on the day following its opening to have a first ride in the train, are pretty plain proofs that the Chinese people are not hostile to the introduction of railways and other inventions.[57]

Elisabeth Köll also highlights that Chinese publications at the time expressed no 'hostility or fear in relation to this new-fangled Western invention'.[58] Ordinary Chinese were receptive to the railway and even Chinese officials went for a ride, despite the fact that a man was accidently killed by a passing train.[59] Li Hongzhang and Thomas Wade met in Yantai, Shandong province, in August 1876 to settle the Margary Affair, a British official killed in Yunnan which produced yet another treaty, the *Chefoo Convention*.[60] Li and Wade also discussed the matter of the Wusong railway. They instructed their subordinates to continue negotiations in Shanghai.[61] While Mayers continued his role on behalf of Britain, the Chinese side was now represented by Sheng Xuanhuai. A comprador who shuttled between foreign companies and the Qing, Sheng was Li's protégé. They reached an agreement titled 'Articles of Agreement Relating to the Woosun Railway' in October 1876: the Chinese would purchase the railway at the cost of 285,000 taels of silver, once it was completed.[62] This deal benefited the investors, who not only made a profit with the sale but could still use the railway after the Chinese takeover.

Construction of the second segment continued. In the meantime, the railway had become a local attraction even before its completion:

> The newly constructed railway has reached the Xu Family Garden area and so many men and women went to see it; the crowd never stopped. They include those from Shanghai and also those who have travelled dozens or even hundreds of miles – they all rush to see it. Everyday there are nearly thousands of visitors, some rode on their horse-driven carriage, some on sedan chairs, some on rickshaws, some on Japanese carriages and others on two-wheel small carts, etc. There are also those who sell fruits and snacks. The company is still moving materials and laying the tracks. It is not yet completed. Can you imagine what a sight it would be once it is finished and start to carry passengers?[63]

The unfinished Wusong had become a tourist attraction; some even travelled from afar just to see it, or better yet, to take a curiosity ride. They welcomed the coming of railways, as *Shen Bao* continued to report on its progress.[64] By early December 1876, the second half of the line, from Jiangwan to Wusong, was completed. The entire line, about nine and half miles, began to operate.[65] This moment was witnessed by many, including women and children, who rushed to experience the spectacle. Reactions amounted to nothing short of a fanfare. Local scholars wrote about it, and one verse was simply titled 'Song to the Fire Wheel Vehicle':

> Wheels follow the iron road, turning and swirling
> It flies to Wusong with passengers feeling like immortals
> Other provinces don't know such ingenious machines
> They only hear boats and ships can now travel on land.[66]

These words indicate the strong and positive impression left by the railway on the Chinese. The 'fire' denoted the smoke emitted from coal burning; implied in the word 'immortals' was the idea that riding in the machine felt like flying, which was, of course, an experience only possible in fantasies at the time. There were many other similar descriptions.[67] The Shanghainese appeared to enjoy and marvel at the railway. Indeed, the imagery in the most popular pictorial, *Dianshizhai Pictorial*, is testament to this enthusiasm (see Figure 1.2).

The six-car train was divided into three classes; it ran six round trips every day, between Shanghai and Wusong, which took about 40 minutes. By February 1877, business was thriving and three more cars were added.[68] By May, rumours circulated of an extension to the line, which was to go as far as Suzhou.[69] The railway was making a profit, amounting to £27 per miles per week, 'thus nearly equalling the average receipts from passenger traffic on English railways'.[70] It was little wonder that Jardine's Shanghai office called the operations 'highly satisfactory'.[71] By September 1877, the line had carried more than 161,331 passengers, most if not all Chinese.[72] Rapier was hopeful about the future of the Chinese operation:

> Not only were the people anxious for employment on the railway, but they manifested very great aptitude. The platelayers soon became quite equal to English workman, and as breaksmen and firemen the Chinese were all that could be desired. Some of them were also very passable as fitter and turners, and in course of time there is no doubt they would make excellent drivers.[73]

Although the Chinese took ownership of the Wusong in October 1877, the battle for the railway had only just begun. The question of who owned the railway remained unclear, as was who would take responsibility for its management. Did it belong to the central government, as a result of the involvement of the *Zongli Yamen* and Li Hongzhang, or to the regional government headed by Shen Baozhen, since the railway lay within his sphere of administration? Should the railway be managed by

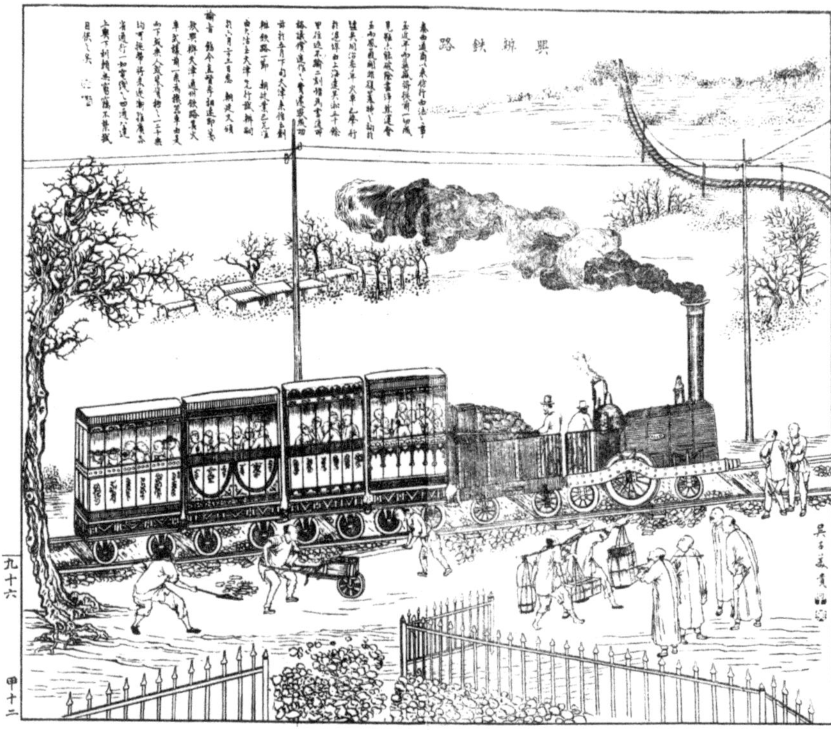

Figure 1.2 The 'Fire wheel vehicle' (*Dianshizhai huabao* (10 vols, Guangzhou: Guangdong renmin chubanshe, 1983), vol 1, 96)

the government, by private enterprise or by government supervised commercial companies? Late Qing China did not have the expertise needed to drive and repair trains and railroads. This meant they needed to hire and work with foreign conductors and engineers. Li Hongzhang proposed that the railway be sold in shares and managed by Chinese merchants with government supervision. But it appears that *Shen Bao* caught word of possible demolition and stirred up public opinion: 'How could something perfect have suddenly become something problematic, is it so hard to decide what to do?'[74] After describing how railways in India helped in famine relief, *Shen Bao* reminded its readers why so many people died in Shanxi province:

> without all three: telegraph, railway and steamship, how can we quickly help those who suffer from disasters, that is why they all died. This is enough for us to see the benefits of railway. But does such a thing exist that, because it is a foreign invention, we should not see its value, and even consider it unimportant and unnecessary?[75]

Many Chinese had come to see the benefits of railways to the country beyond Shanghai; the fact that it was a foreign invention did not matter. Clearly, they were expressing their concern over rumours of possible demolition, but they had no say over its future. As the debate raged on and to the dismay of many, Shen Baozhen dismantled the Wusong in November 1877.[76] He also threatened to uproot the telegraph line along the route.[77] The *Herald* detailed the journey of the 'last train', as hundreds of Chinese queued to take the final ride'.[78] *Shen Bao* also lamented the disappearance of the Wusong.[79] The decision to dismantle the railway raises many questions, not least because Shen was a reformer. Two of the most powerful officials disagreed about the fate of the Wusong, but one of them was able to outmanoeuvre the other. Why was Shen so eager to dismantle the railway, given that he had championed reform and endorsed the introduction of railways in the 1860s debate where Li had been more cautious? The following section offers a new interpretation on politics of late Qing reform that contributed to the destruction of the Wusong.

Personalities and reform rivalries

The Wusong railway has been made infamous, sensationalised to be more precise, in accounts of China's modernisation. However, the academic debate it has generated is limited to a handful of journal articles. David Pong concluded in his 1973 article that Confucian patriotism had driven Shen's behaviour, and he advanced four rationales. First, a 'strong reaction against foreign encroachment on Chinese territorial and administrative integrity'; second, a 'strong distaste for the corrupting elements'; third, a 'genuine concern for the well-being of the poorer sections of the Chinese society'; and lastly, a 'concern for the development and independence of Chinese economic interests'.[80] Sandra Sturdevant challenged Pong's argument in 1976, stating that 'Chinese foreign policy was defensive. It aimed at containing Westerners, at preventing them gaining more "rights" than they already enjoyed. Ultimate control was key.'[81] Using Foreign Office papers and the Jardine Matheson archive, Hsien-chun Wang picked up the Wusong controversy in 2015, placing the blame firmly on 'institutional failure' and the lack of a railway policy, arguing that the 'unreformed fiscal system was ill-suited to support modern large-scale engineering projects, such as the railway'.[82] Maintaining the line would have been costly and entailed the hiring of foreigners. Pong, Sturdevant and Wang have helped us understand the dismantling of the Wusong from different perspectives.

But what was the fate of the dismantled railway? Ralph Huenemann wrote in his 1984 book: 'As is well known, Shen bought out the foreign owners, ordered the track dismantled, and had the equipment shipped to Taiwan, where it was abandoned to the elements.'[83] This represents a standard interpretation, perpetuated from earlier works, such as Albert Feuerwerker's 1958 book, which David Pong also referred to.[84] But if the intention was simply to dispose of the equipment, Shen could have left it to the elements in Shanghai rather than taking the trouble to ship it to Taiwan. Does the standard interpretation stand to scrutiny if we include the fact that Li and Shen fought over the dismantled railway? Li had wanted to put it to use in the Kaiping coal mine, which will be discussed in the next chapter, but to his frustration, Shen had already taken it to Taiwan.[85] Both had a vested interest in the Wusong. Li and Shen were each granted mining permissions in 1875. Shen's project, the Jilong coal mine in Taiwan, began to produce coal in 1876, while Li's venture, the Kaiping coal mine, was less successful as he was confronted with the problem of transportation.[86] Shen was eager to develop Taiwan.[87] Is it possible that the 'destruction' served an ulterior motive, or perhaps that there was more to the destruction of Wusong than mere Confucian patriotism and institutional failure? Historians of China have catalogued the achievements of both Li and Shen; but few looked into their rivalries. A comparison of their personalities and competition in acquiring modernisation projects would offer an insight into not just what shaped the fate of the Wusong but also what undermined late Qing reform.

After passing the *jinshi* degree in 1847 and working in the Hanlin Academy for a few years, Li Hongzhang (b. 1823) joined the staff of Zeng Guofan, who was at that time leading the campaign against the Taiping. The end of the Second Opium War in 1860 made it possible for the Qing to turn its full attention to the Taiping rebels. The desperate Taiping forces decided to assault Shanghai; Zeng tasked Li with the job of defending. Following his superior's example, Li quickly put together the Huai Army in his native Anhui province. Deploying his troops to Shanghai proved an immense challenge having to pass the Taiping-held Nanjing on the Yangzi River. The Qing had no modern ships of its own to aid such a manoeuvre. Li's Huai Army was eventually shipped to the city, hidden at the bottom of British merchant ships. This was Li's first encounter with the West, and it profoundly shaped his outlook. During this journey, he witnessed and experienced, for the first time, the efficiency of steam engines. Upon his arrival in Shanghai, Li visited foreign battleships anchored outside Wusong. Enormously impressed, he knew that it was this advanced technology that made the West powerful, and that if he could use them, he would defeat the Taiping rebels. He worked with the Ever-Victorious Army, a mercenary force funded by merchants and led first by Frederick Ward and then Charles Gordon. Together they defended Shanghai and captured rebel-held

Suzhou.[88] Li used this foreign army and its advanced weapons but was also keen to prevent them from becoming too powerful. He cut their expenses, which prompted Gordon to resign in frustration, providing the ideal pretext for Li to take over. After the Taiping was crushed in 1864, Li disbanded the Army and took all its weapons. In light of these manoeuvres, it is little wonder John Rawlinson described Li as 'an opportunist rather than a strategist'.[89]

Li's flexibility, pragmatism and willingness to learn was key to his success then and throughout his career, as has been noted by many historians.[90] He was also adept at utilising the resources of others to procure what the Qing could not produce itself. He put into practice the old Chinese proverb: 'borrow hens to lay eggs'. The experience taught him that working with foreigners did not have to be confrontational, cultivating a friendship with Ward, and employing Samuel Halliday Macartney as army doctor. He also hired John Fryer, a missionary, to translate scientific works for Jiangnan Arsenal, China's first manufacturer of Western weapons/ammunitions. He would hire many more foreigners to undertake his reform projects and train Chinese soldiers.[91] In 1872, Li acquired another reform project, the China Merchants' Steam Navigation Company. He even persuaded the Qing court to send young boys to study in the United States between 1872 and 1875, an extraordinary step which the *Herald* hailed as 'a prophecy of great progress to the oldest and largest nation on the globe'.[92] Among them was a boy of twelve called Zhan Tianyou (Jeme Tien Yow) who would become the 'father of Chinese railways'. Li dared to experiment, quietly undertaking reform, and was unconcerned about the influence of foreigners, as long as they did not constitute a political threat. Soft-spoken and amicable, many foreigners found Li an approachable figure.

Moreover, Li's readiness to learn from and use foreigners did not come at the expense of compromising principles, as evidenced in his rejection of the Shanghai–Suzhou proposal. Li became the most important official in matters of foreign affairs after Zeng Guofan died in 1872. During his negotiations with Wade, he registered his protest that the British had violated the Qing's authority, but he also tried to find a way out of the predicament; he knew that the Wusong would also benefit China. To solve the problem, he approached the *Zongli Yamen* in the hope that they would support the purchase of the railway. He urged them to work with Shen Baozhen because Shen was the Nanyang commissioner and 'matters in Shanghai should consult him'.[93] Li also wrote to Shen Baozhen in whose hands the future of Wusong lay, articulating that the continued stalemate with the British would be detrimental and advised him to accept a Chinese purchase. Li did not hide his dissatisfaction when Shen dismantled the line, writing to a friend that 'Shen is not well exposed, [he is] narrow-minded and too proud.'[94] Although they were both 'learning from the West', they differed greatly in their perception of Westerners and in the ways in which they dealt with them.

If Li was mainly concerned with the result, then the means to that end probably mattered more to Shen Baozhen. Uncompromising and confrontational, Shen was Li's antithesis. Born in 1820, Shen also obtained his *jinshi* degree in 1847. Such timing meant that he and Li were in Beijing for the final examination and entered the Qing bureaucracy at the same time. This situation was just as likely to forge competition as it was close friendship. Like Li, Shen also rose to political prominence during the suppression of the Taiping and reform; David Pong called him an 'extraordinary able official'.[95] The court awarded him the governorship of Jiangxi province. When Zuo Zongtang, another prominent reformer, was called to pacify the Dungan Revolt in 1866, Shen Baozhen succeeded him in overseeing the Fuzhou Shipyard and Naval Academy. In 1875 Shen succeeded Li Hongzhang as governor-general of Liangjiang, while Li became the governor-general of Zhili. Despite also employing foreign experts, such as Prosper Giquel to provide expertise in naval construction, Shen was less trustful of them, more protective and often took a hard-line on issues that arose.[96] In this regard, he resembled his uncle, Commissioner Lin Zexu, who was unyielding in his battle against the British during the First Opium War.

His principle-based moralistic stance can be seen from the way in which he dealt with the Nanchang Anti-missionary Incident, Shen's 'first encounters with foreigners', which is very different from Li's experience.[97] Pong puts the blame on Shen, arguing that his position as governor meant that he could have: 'taken measures to avert mob violence', but Shen did not intervene.[98] Pong understands why: 'For Shen who had become governor two days earlier, an unduly lavish reception of the missionary would have gravely undermined his authority and prestige and set a dangerous precedent.'[99] Shen's hostility to foreigners is also evident in his letter to Feng Junguang, who was negotiating with Mayers over the Wusong in May 1876: 'we step back, they step forward until they pushed us to the side of the well and they are still angry if we don't back down further.'[100] Shen's distrust of the British can be seen in his expressed opinions surrounding the Margary Affair and the Jiangwan incident which saw a Chinese man lose his life on the Wusong. 'As for Jiangwan incident, the damage is not that serious and can be settled easily. But Wade's intention lies not at the Margary Affair but in commerce, and everyone knows it.'[101] This distrust was further reinforced in a letter to another colleague where Shen wrote that, 'Wade is using the Margary Affair to coerce us.'[102]

Shen Baozhen was, in fact, correct. While doing away with the expenses of an all-out war, the Margary Affair helped Britain force the *Chefoo*

Convention upon the Qing with an indemnity of 200,000 taels.[103] With the death of an official as leverage in the negotiations, Britain procured four new Treaty-ports and the privilege to operate at several non-Treaty cities. It also secured rights in Yunnan. Had Shen been sent to negotiate with Wade in Li's place, an agreement may not have been reached, in which case Wade may have found a use for the gunboats he summoned. Under these circumstances Shen could not imagine allowing the British to also have a railway at their disposal – a line that would certainly permit them to have more control of the Qing's economic life and move troops in times of war. Li Hongzhang had perhaps forgotten about the Lay–Osborn Affair but Shen Baozhen certainly had not.[104] The Qing court decided to purchase gunboats to move troops for the final assault on the Taiping rebels in 1861. Prince Gong tasked Horatio Nelson Lay, then director-general of the Imperial Maritime Customs, with the job.[105] Lay returned from England in 1863 with seven cruisers as well as a commander, Captain Sherard Osborn. However, it soon became apparent that they would not take orders from anyone but the emperor. Lay also demanded separate funds, entitlements such as a princely palace and an official sedan chair.[106] The Qing had spent 920,000 taels on a flotilla they could not command; they had to annul the contract. Consequently, Lay was replaced by Robert Hart.

The Wusong railway, in some sense, was no different from the Lay–Osborn flotilla: it was British made and controlled, and the British had shown themselves willing to compromise the Qing's sovereignty and undermine its integrity. To compound his sense of righteousness, the Americans and the British had made a mockery of the bureaucracy, and had openly flouted Chinese authority. They had managed to obtain land and operate under a false pretence, suggesting a familiarity with the system, as well as an ability to manipulate it, that must have been galling to Shen. If this affront was tolerated, it would encourage further encroachment by others seeking to advance their own interests. Shen reiterated this point again in his letter to Feng: 'Foreign merchants undertook construction work without asking Chinese officials for permission. If we do not punish this dictatorial behaviour ... it seriously damages our sovereignty and injures our honour.'[107] Dismantling the railway sent a clear message, not unlike the message conveyed by terminating the Lay–Osborn flotilla. It asserted Shen's authority by showing that such behaviour would not be tolerated – a perspective that was made very clear in another letter after the demolition:

> When you set out to build it, you did not ask for our permission. The fact that we did not use force to stop you shows our peaceful intention and friendliness. We cannot give you a license to continue. The other day, *Daotai* Feng told me that your intention was to let all Chinese see the benefit of railways. I dismantled it ... everyone can see that.[108]

Although we do not know what Shen meant by 'force', his actions made clear that this was his territory, and it was he who decided the fate of Wusong. Dismantling the line was also an indirect response to Li. As governor-general of Liangjiang and Nanyang commissioner, Shanghai fell under Shen's jurisdiction. However, it seems that Li held a considerable degree of power over the region, presiding over the negotiations with the British and sending his protégé Sheng Xuanhuai to nail down the agreement. Although both Shen and Li worked closely with the *Zongli Yamen*, it would seem that Li wielded more influence than Shen and was able to control affairs outside of his administrative power. In his discussion of the enduring relationship between Li Hongzhang and Shanghai, Yuen-sang Leung argued that Li maintained substantial informal influence over the various affairs of the city after he was succeeded by Shen Baozhen.[109] Although he was the governor-general of Zhili, far away in north China, Li was able to cast his net into Shen's territory and meddle in local affairs. This situation would surely have continued if the railway was allowed to operate, as Li suggested. David Pong mentioned that the Wusong railway would have posed a threat to the China Merchants' Steam Navigation Company, but he did not elaborate. Founded by Li and managed by his protégés, the company was a profitable enterprise, able to compete with foreign ships. But if Shen had allowed the Wusong to continue its operations, the railway could have succoured the British to stifle their Chinese competitor. It could also strengthen Li's position, as he would have acquired another reform project through his confidant Sheng Xuanhuai who was involved in the negotiations and purchase of the Wusong. Shen had no incentive to aid the British; nor did he want to enable Li. The destruction of the railway consequently killed two birds with one stone. Moreover, he had plans to put the technology to use in Taiwan.

Shen and Li's differing attitudes towards foreigners were compounded by the rivalry that had grown between them, a common occurrence in Qing officialdom. The conditions for such rivalries intensified during the late Qing as the issue of reform became a political platform where battle lines were drawn. Some of these lines were ideological, as many were against the push to modernisation emanating from the West. Some were ethnic, as the Manchu and Han divide deepened. The most vicious was factional, the battle between the North and South parties was a defining example of that. Hans van de Van has labelled this 'intra-elite factionalism', as different cliques fought to secure positions and interests for their own members.[110] Others, however, were personal, stemming from prejudice and jealousy. Li was Zeng Guofan's protégé and successor, but Shen worked closely with their rival, Zuo Zongtang. Zuo and Li fought openly for government resources. Zuo argued for a focus on 'frontier defence,' that is, directing resources to the battleground he oversaw against Russian incursions and Central Asian rebels in Xinjiang. Li, in contrast, demanded greater investment in 'naval defence' to combat the more menacing enemies from the sea. Some mainland historians believe that the destruction of the Wusong saw Shen and Li part company.[111] But their rivalry went far beyond as they competed and kept up with each other when it came to acquiring reform projects. It would soon come to a climax and, in a very literal sense, a dead end.

Both Shen and Li began to import foreign gunboats to strengthen the naval defence after they were appointed as commissioners in 1875. Hans Van de Ven's research sheds light on the Qing's naval race with Japan since then.[112] As John Rawlinson details, the main drive of this race was the internal rivalry between Shen and Li. They were constantly fighting for the Qing's Sea Defence Fund and competing with each other to build up their fleets. When Li bought two British ships in 1872, Shen soon bought two from America. When Li placed an order for four Armstrong gunboats through Robert Hart in 1876, Shen immediately bought two of the same vessels with the help of Prosper Giquel. Li intended to buy more ships while trying to discourage Shen from doing so.[113] Such was the scale and intensity of their competition. This was how this round of competition ended:

> An ingenious swap made by Li in 1879 reveals this rivalry. The four improved 'epsilon' gunboats, ordered by the southern commissioner [Shen] in 1875, had arrived. Through Hart, Li arranged to have them delivered to Tientsin. There he wanted to keep them. After all, Li argued to Shen, Li's responsibility embraced the stormy northern seas, while Shen's responsibility for the quiet Yangtzi was not so demanding. Therefore, Li's original 'gammas', although pitted, barnacled, and in need of a refit at Kiangnan yard, would be adequate for Shen.[114]

Li hijacked Shen's orders and left his defunct vessels to Shen. We can only imagine Shen's outrage. Li might have had his revenge for the loss of the Wusong; he might have also expected retribution, but Shen died suddenly in December 1879 – leaving his ships to Li and the Wusong railway to the elements in Taiwan. Did this kind of fierce competition contribute to Shen's premature death? Rawlinson's conclusion is to the point: 'in his relations to Shen Pao-chen, Li was flexible. He disagreed with Shen's desire to fight Japan in 1879, and in matter of ship buying, evidently regarded the southern commissioner as a rival.'[115]

Rivalry among Qing officials galvanised railway politics and dictated the fate of the first line built on Chinese soil. It also exposed how much control the late Qing court had on the far-flung empire and how rivalry among officials undermined reform. Would reform have translated into better results without this kind of sustained, all-compassing cut-throat competition? This question deserves more unpacking than the few paragraphs here. If the foreigners' strategy was 'raw rice already cooked', or what is done is done, the Chinese, to strain the metaphor further, stubbornly refused to eat the cooked rice despite being hungry. The fallout of the Wusong misadventure was threefold: for the Qing court and officials, the railway became a taboo subject; furthermore, as Kent has rightly observed, 'some years now elapsed before foreigners again became actively interested in the railway question'.[116] The British learnt a hard lesson. They soon opted to finance, rather than build and manage railways. The introduction of railway in China was inevitably entangled with issues of geopolitics, the politics of reform and rivalry among officials within the Qing bureaucracy. This kind of politicking would continue into the late 1870s and 1880s, when a new urgency to introduce railways emerged.

Notes

1 Hong Rengan, *Hong Rengan xuanji* (Beijing: Zhonghua shuju, 1978), 13 (1 *li* equals 500 metres).

2 'News of the morning', *Sacramento Daily Union*, 15 June, 1858: 2; Manu Karuka, *Empire's tracks: indigenous nations, Chinese workers, and the Transcontinental railroad* (Oakland: University of California Press, 2019), 82–101.

3 Kent, *Enterprise*, 15.

4 Pelcovits, *China hand*, 109.

5 Anonymous, 'Coal and railways at the north', in *Papers on China* (Hong Kong: no publisher or date of publication, donated by Thomas Wade to Cambridge University library), 35–37.

6 Daniel Thorner, *Investment in empire: British railway and steam shipping enterprise in India 1825–1848* (Philadelphia: University of Pennsylvania Press, 1950), 45; Rowland M. Stephenson, *Report upon the feasibility and most effectual means of introducing railway communication into the empire of China* (London: J. E. Adlard, 1864), 4–13.

7 Le Fevour, *Western enterprise*, 107.

8 William F. Mayers, *The treaty ports of China and Japan* (London: Trubner, 1867), 412.

9 Tyler Dennett, *Americans in eastern Asia: a critical study of the United States' policy in the Far East in the nineteenth century* (New York: Barnes and Noble, 1922), 594.

10 Mi Rucheng, *Zhongguo jindai tielushi ziliao: 1863–1911* (3 vols, Beijing: Zhonghua shuju, 1963), vol 1, 4–5.

11 'Jiangsu xunfu Li Hongzhang zhi zongshu', in *Tielushi ziliao*, vol 1, 20.

12 'Editorial selections', *North China Herald* (*Herald* hereafter), 22 April, 1867: 27.

13 Mary C. Wright, *The last stand of Chinese conservatism: the T'ung-chih restoration 1862–1874* (Stanford: Stanford University Press, 1962), 300.

14 Albert Feuerwerker, *China's early industrialization: Sheng Hsuan-Huai (1844–1916) and Mandarin enterprise* (Cambridge: Harvard University Press, 1958),

141; John L. Rawlinson, *China's struggle for naval development 1839–1895* (Cambridge: Harvard University Press, 1967), 201.

15 Huenemann, *Economics*, 1–2.

16 'Jiangxi xunfu Shen Baozhen zhi zongshu', in *Tielushi ziliao*, vol 1, 20.

17 'Liangguang zongdu Mao Hongbing zhi zongshu', in *Tielushi ziliao*, vol 1, 20.

18 Wenxiang *et al.*, *Chouban yiwu shimo: Tongzhichao* (100 vols, Beijing: Zhonghua Shuju, 1960), vol 40, 20 and 31.

19 Mayers, *Treaty ports*, 113 and 328.

20 David Pong, *Shen Baozhen and China's modernization in the nineteenth century* (Cambridge: Cambridge University Press, 1994), 294.

21 'Zongli chuanzheng Shen Baozhen tiao', in *Tielushi ziliao*, vol 1, 23.

22 'Huguang zongdu Li Hongzhang zouzhe', in *Tielushi ziliao*, vol 1, 25.

23 Knight Biggerstaff, 'The secret correspondence of 1867–68: view of leading Chinese statesmen regarding further opening of China to Western influence', *Journal of Modern History*, vol 22, 2 (June 1950): 122–36.

24 Tara Sethia, 'Railways, Raj, and the Indian states: policy of collaboration and coercion in Hyderabad', in *Railway imperialism*, 103–20.

25 Winthrop R. Wright, *British-owned railways in Argentina: their effect on the growth of economic nationalism 1854–1948* (Austin: University of Texas Press, 2021), 5.

26 Kent, *Enterprise*, 15.

27 Kent, *Enterprise*, 7.

28 'Yingguo boyu dalue', *Shen Bao*, 13 May 1872: 3 and 'Lun neidi jianggoushe huolunche lu', 182, 28 November 1872: 1.

29 'Presentation of an address to Thomas Wade', *Herald*, 1 December 1876: 543 and 'The Hong Kong Chamber of Commerce', 21 March 1879: 282.

30 'Tianjin shixin tulu huoche', *Shen Bao*, 30 September 1872: 3.

31 'Railways in China', *The London and China Telegraph*, Monday 14 July, 1873: 465 and 'The Nuptial locomotive', Monday 28 July, 1873: 496.

32 Dennett, *Americans*, pp. 595–96; Pletcher, *Diplomacy of involvement*, 139.

33 Robert Hart, *Diary*, vol 7 May 1865 to March 1866, MS 15.1.7.034–35, Sir Robert Hart project at Queen's University Belfast, https://digital-library.qub.ac.uk/digital/collection/p15979coll3/id/1350, accessed 21 September 2022.

34 Dennett, *Americans*, 596.

35 Kent, *Enterprise*, 9.

36 'Shanghai zhi Wusong jiangzao huolunche tielu', *Shen Bao*, 6 May 1873: 1.

37 'Yingren xianyi huoche tielu', *Shen Bao*, 7 August 1873: 2.

38 'Wusongkou jianzao huoche tielu' *Shen Bao*, 26 February 1874: 1, 'Huolunche wei fuguo zhiju', 15 July 1874: 1, 'Wusong jiangxing huoche', 4 August 1874: 3 and 'Shanghai zhi Wusong zhu huochelu', 11 November 1874: 2.

39 Yuen-sang Leung, *The Shanghai taotai: the linkage man in a changing society 1843–90* (Singapore: Singapore University Press, 1990).

40 'Liuqiu shangren wei Taiwan shengfan shahai', *Shen Bao*, 15 November 1872: 4, 'Dongyang kanglun Liuqiu shi', 28 July 1873: 2 and 'Fanyi Dongyang hengbin xizibao', 8 May 1874: 2; Liu Yan'ni and Chen Sheng, *Chuanzheng yu jindai zhongguo haijun* (Fuzhou: Fujian renmin chubanshe, 2016), 66.

41 William F. Mayers, *Treaties between the empire of China and foreign powers* (Shanghai: North China Herald Office, 1877), 187.

42 Matthew Fraleigh, 'Transplanting the flower of civilization: the "peony girl" and Japan's 1874 expedition to Taiwan', *International Journal of Asian Studies*, vol 9, 2 (July 2012): 177–209; Danny Orbach, 'By not stopping: the first Taiwan expedition (1874) and the roots of Japanese military disobedience', *The Journal of Japanese Studies*, vol 42, 1 (2016): 29–55.

43 Robert Eskildsen, *Transforming empire in Japan and East Asia: the Taiwan expedition and the birth of Japanese imperialism* (Basingstoke: Palgrave Macmillan, 2019).

44 Ulysses S. Grant tried to mediate, see Ian P. Austin, *Ulyssess S. Grant and Meiji Japan 1869–1885* (London: Routledge, 2019), 91–92.

45 Ding Richang, *Fuwu gongdu* (50 vols, Taibei: Huawen shuju, 1968), vol 25, 15.

46 Li Hongzhang, *Li Wenzhonggong zougao* (80 vols, Shanghai: Guji chubanshe, 1997), vol 24, 621–29.

47 Huenemann, *Economics*, 1.

48 Weipin Tsai, *Reading Shen Bao: nationalism, consumerism and individuality in China 1919–1937* (London: Palgrave Macmillan, 2010).

49 'Wusong tielu jiangcheng', *Shen Bao*, 17 February 1876: 2 and 'Huolunche kao', 19 February 1876: 3.

50 Kent, *Enterprise*, 10.

51 'The Woosung railway', *Herald*, 6 May 1876: 424.

52 Wang, 'Merchants, mandarins'.

53 'Tielu jiangcheng', *Shen Bao*, 8 June 1876: 2, 'Tielu gaocheng', 14 June 1876: 1 and 'Tielu jinkuang', 20 June 1876: 2.

54 'Tielu luocheng', *Shen Bao*, 30 June 1876: 2 and 'Huoche kaixing', 1 July 1876: 3.

55 Richard C. Rapier, *Remunerative railways for new countries with an account of the first railway in China* (London: E. & F. N. Spon, 1878), 102 and 'Brief account of the Wusong railway', in *Minutes of the proceedings of the Institution of Civil Engineers*, 59 (1880): 274–77.

56 'Huolun chelu gaobai', *Shen Bao*, 1 July 1876: 6, 'Huoche kaishi', 4 July 1876: 2, 'Ji huake chucheng huoche', 3 July 1876: 2 and 'Genggai huolunche wanglai geshi', 22 July 1876: 6.

57 'The opening of the Woosung railway to Kungwang', *Herald*, 1 July 1876: 2 and 'The Woosung railway', 22 July 1876: 83.

58 Köll, *Transformation*, 23.

59 'Xie Cishi xunyue tielu', *Shen Bao*, 31 July 1876: 1 and 'Huoche shangren', 4 August 1876: 3.

60 Mayers, *Treaties*, 44–48; 'The Chefoo Convention', *Herald*, 16 September 1876: 272 and 'Chefoo Convention supplementary clause', 5 October 1876: 351.

61 'Bingshang tielu', *Shen Bao*, 19 October 1876: 1.

62 'Articles of agreement relating to the Woosung railway', *Herald*, 23 November 1876: 513.

63 'Jikan tielu', *Shen Bao*, 31 March 1876: 1.
64 'Lun yong tielu huoche', *Shen Bao*, 8 November 1876: 1 and 'Tielu fuxing', 29 November 1876: 1.
65 'Huolunche gongsi gaobai', *Shen Bao*, 2 December 1876: 6.
66 'Yong huolunche', *Shen Bao*, 18 December 1876: 2.
67 'Huolunche shi yuanyun', *Shen Bao*, 1 January 1877: 4 and 'Chen huolunche zhi Wusong zuoge', 20 October 1877: 2.
68 'Huoche shengyi renao', *Shen Bao*, 22 February 1877: 3.
69 'Xukai huochelu zhi Suzhou', *Shen Bao*, 28 May 1877: 1.
70 'Xukai huochelu zhi Suzhou', *Shen Bao*, 28 May 1877: 110; Le Fevour, *Western enterprise*, 108.
71 Shanghai to W. Keswick in Hong Kong dated 29 July and 5 August 1876, Private Letter Book (PLB), Jardine Matheson Archive.
72 'Lun Wusong tielu', *Shen Bao*, 18 September 1877: 1.
73 Rapier, *Remunerative railways*, 109.
74 'Lun Wusong tielu', *Shen Bao*, 18 September 1877: 1.
75 'Lun Wusong tielu', *Shen Bao*, 18 September 1877: 1.
76 'The Woosung railways', *Herald*, 1 November 1877: 394.
77 'Shang hui diangan', *Shen Bao*, 4 January 1878: 2.
78 'The Shanghai-Woosung railway', *Herald*, 25 October 1877: 367.
79 'Tielu jianji', *Shen Bao*, 28 November 1878: 3.
80 Pong, 'Confucian patriotism'.
81 Sturdevant, 'Imperialism, sovereignty', in *Reform in nineteenth century*, 63–70 (65).
82 Wang, 'Merchants, mandarins'.
83 Heunemann, *Economics*, 3.
84 Feuerwerker, *Industrialization*, 269, note 28; Pong, 'Confucian patriotism', footnote 15.
85 'Yizao tielu', *Shen Bao*, 9 August 1878: 2.
86 Ellsworth Carlson, *The Kaiping mines 1877–1912* (Cambridge: Harvard University Press, 1971), 9.
87 Zhu Congbing, *Li Hongzhang yu zhongguo tielu* (Beijing: Qunyan chubanshe, 2006), 104–11.
88 See the works of Andrew Wilson, Charles H. Allen, A. Egmont Hake and Stanley Spector, *Li Hung-chang and the Huai army: a study in nineteenth century Chinese regionalism* (Seattle: University of Washington Press, 1964).
89 Rawlinson, *Naval*, 69.
90 Samuel C. Chu and Kwang-ching Liu, eds, *Li Hung Chang and China's early modernization* (Armonk: M. E. Sharpe, 1994).
91 Richard J. Smith, 'Foreign-training and China's self-strengthening: the case of Feng-huang-shan, 1864–1873', *Modern Asian Studies*, vol 10, 2 (1976): 195–223.
92 'Shu chuyang xuetong fu meiguo', *Shen Bao*, 28 October 1873: 2; 'Chinese educational mission', *Herald*, 19 December 1872: 534.
93 Zhu Congbing, *Li Hongzhang*, 94.

94 Li Hongzhang, *Li Hongzhang quanji* (37 vols, Hefei: Anhui jiaoyu chubanshe, 2008), vol 32, 233.
95 Pong, *Shen Baozhen*, 1.
96 Steven A. Leibo, *Transfering technology to China: Prosper Giquel and the self-strengthening movement* (Berkeley: Institute of East Asian Studies, University of California, 2005).
97 Pong, *Shen Baozhen*, 87.
98 Pong, *Shen Baozhen*, 91.
99 Pong, *Shen Baozhen*, 91.
100 Lin Qingyuan and Wang Daocheng, *Shen Baozhen xinza kaozhu* (Chengdu: Bashu chubanshe, 2013), 445.
101 Lin and Wang, *Shen Baozhen*, 461.
102 Lin and Wang, *Shen Baozhen*, 478.
103 Mayers, *Treaties*, 44–48.
104 Pong, *Shen Baozhen*, 106.
105 Jack J. Gerson, *Horatio Nelson Lay and Sino-British relations, 1854 to 1864* (Cambridge: Harvard University Press, 1972).
106 Hans van de Ven, *Breaking with the past: the Maritime Customs Service and the global origins of modernity in China* (New York; Columbia University Press, 2014), 40–43.
107 Lin and Wang, *Shen Baozhen*, 481.
108 Lin and Wang, *Shen Baozhen*, 487.
109 Leung, *Taotai*, 111–16.
110 Van de Ven, *Breaking*, 121.
111 Ouyang Yuefeng, 'Cong tongchou dikai dao fendao yangbiao', *Wuyi Daxue Xuebao* 10, 3 (2008): 42–45.
112 Van de Ven, *Breaking*, 107–12.
113 Rawlinson, *Naval*, 70–74.
114 Rawlinson, *Naval*, 74.
115 Rawlinson, *Naval*.
116 Kent, *Enterprise*, 15.

2

A journey of a thousand miles begins with a single step

> Electricity, telegraphs and railways are what western nations consider vital to a country. But China believes that it does not need these technologies, nor does it hurry to acquire them. Additionally, China firmly believes that it should not have them. Some say they will mislead people and others believe they disturb *fengshui*. Still others say they will lead to endless legal cases. ... Alas, how can they say that China must not have them when the Chinese people have not seen them with their own eyes![1]

The words of Wang Tao, an influential Shanghai scholar, indicate how contentious the question of railways had become by the late 1870s. Labelled the 'fugitive in paradise', Wang had seen with his own eyes the advantage of modern inventions in Britain and was one of an increasing number of intellectuals and officials calling for the introduction of railways.[2] Others included diplomats like Guo Songtao, Xue Fucheng, Ma Jianzhong, Li Gui and entrepreneurs like Zheng Guanyin, to name a few.[3] Their writings exposed the ordinary people to the potential of railways and shaped public opinion.[4] *Shen Bao* also played its role, carrying news about railways in Japan and Vietnam to promote the argument for their introduction.[5] There was now a strong current of public opinion in favour of railways. Li Hongzhang once compared the Qing to a broken old house and himself to a 'plasterer' who was called upon to repair it.[6] In the early 1880s, while the 'cracks' in the house were growing, Li set about building China's first railroad – a feat few historians have researched, despite being mentioned in books on Li and the Kaiping coal mine.[7] This chapter picks up the theme from the previous chapter but looks at how and why railways were introduced by the Chinese themselves. Although barely eight miles in length, the Tangxu [唐胥] railway is extremely significant, not just because it is the mother of all lines in north China but because it would come to shape geopolitics of China well into the mid-twentieth century.

'A railway built by stealth'[8]

Labelled 'China's first modern corporation', the China Merchants' Steam Navigation Company (CMSNC) was initiated by Li Hongzhang. A number of scholars have written about its achievement.[9] A defining example of China's success at modernisation, the company's vessels traversed the ports of Asia and competed with foreign ships.[10] Headed by comprador-turned-entrepreneurs like Tang Tingqu and Sheng Xuanhuai, it operated on the premise of 'government supervised commercial management', an old mechanism that enabled the government to maintain control of and profit from the company's ventures by licensing and overseeing merchants who were barred from monopolising strategic industries.[11] Albert Feuerwerker referred to this arrangement as 'Mandarin enterprise'.[12] This mechanism fed into corruption, since merchants had to cajole and bribe officials in order to gain access to profitable projects, and officials expected 'returns' for the licences they issued on behalf of the government. This situation often led to confusion as the issue of ownership was not always a legally clear one.

A more pressing problem was that the CMSNC had to import coal from Japan, which was prohibitively expensive.[13] It was not the only enterprise that needed fuel. The Jiangnan Arsenal also demanded coal, as did other modernisation projects, including the Fuzhou Shipyard and the foreign gunboats bought by the various regional leaders. A long-term solution was called for. China had been mining coal for over 2,000 years.[14] Kenneth Pomeranz argues that 'Chinese ironmakers, for instance, seem to have known things about the creation and use of coke (purified coal) that would not be discovered elsewhere for centuries', but the mines they worked in were 'mostly small and poorly positioned to take advantage of China's richest and most-hungry fuel market' by the eighteenth century.[15] The coal mines in Tangshan were good examples of this: they required new mining technology and modern transportation. Li Hongzhang and Shen Baozhen received permissions in 1875 to establish modern coal mines: Li in Tangshan, north of Tianjin and Shen in Jilong (Keelong), Taiwan.[16] While the Jilong coal mine began to operate in 1876, Li's project, the Kaiping coal mine, ran into a serious problem – how to transport coal from the colliery to Tianjin and the coast (see Figure 2.1).[17]

Indeed, Tangshan had previously been surveyed in 1868, but the project was abandoned due to the difficulties posed by transportation. Li and his protégé, Tang Tingqu, who headed the Kaiping mine, needed to find a solution, but they faced reprisal from anti-railway and anti-mining officials. The 'government supervised commercial management' model was perfect for Li. While he manoeuvred from behind the scenes, Tang could act as his mouth piece, proposing and implementing his plans. Tang initially suggested a

Figure 2.1 Tangshan and Tianjin in north China (Baidu public domain)

railway from the coal mine to a port near Tianjin, but this was rejected. Tang then proposed a shorter line from the colliery to Lutai, a district in Tianjin that sits on an interlacing river/canal system which would facilitate further transportation. He hired foreign experts in 1878: R. R. Burnett, who later sank the first shaft, J. M. Molesworth and the railway engineer Claude William Kinder. They were followed by six others, making a total of nine English engineers and foremen on Kaiping's payroll by the end of the year.[18]

This was further proof that Li was not reluctant in using foreigners for his projects and was more interested in achieving results.[19] However, as Claude Kinder recalled, 'before any effective start could be made, orders came to stop all work, the Imperial sanction for a railway having been withdrawn' in February 1879.[20]

It seems the introduction of railways was still a taboo subject in the aftermath of the Wusong debacle despite the Kaiping coal mine being a Chinese company; Li and Tang had to be careful not to ruin the hard-won project. To avoid agitating their adversaries, Tang proposed an even shorter, barely 8-mile tramway from the colliery in Tangshan to the village of Xugezhuang, hence the Tangxu [唐胥], a portmanteau of the first syllable of each name, and was to be followed by a 'coal river', a canal to be precise, from Xugezhuang to Lutai. The equipment from the dismantled Wusong would have been well-suited for the Tangxu. This had not escaped the notice of Claude Kinder, who was hired to survey and construct railways for the coal mine; nor did it bypass *Shen Bao*, which reported that Li had intended to use the dismantled Wusong for a line in Kaiping – a detail that appears to have escaped the attention of many scholars.[21] Although the construction of the Tangxu tramway was allowed, it came with a caveat, as Kinder remembered clearly: 'in granting permission for the line it was expressly stipulated that no locomotives were to be used'.[22] This meant the project would have to rely on slow-moving mule driven carts, proving itself no real solution to the problem with transportation that had plagued the coal mine.[23]

At this crucial moment, Liu Mingchuan, a retired Huai Army general who had helped to crush the Taiping, suddenly fired the first shot across the bows in support of railway construction in early 1880. Liu reminded his colleagues that the Qing faced ongoing threats from Russia and Japan: 'As for the ways to strengthen ourselves, to train soldiers and make weapons are important but the most urgent is the construction of railways. The benefits of railways lay in grain supply, disaster relief, commerce, mining, tax collection and transportation.'[24]

Liu listed the benefits of railways and considered the matter urgent. To the surprise of many, the court praised Liu's desire to strengthen the Qing and even asked the Beiyang commissioner, Li Hongzhang, and the Nanyang commissioner, Liu Kunyi, to 'carefully consult, discuss and propose' policies with regard to railways.[25] *Shen Bao* did its bit to keep the debate alive, running editorials about the benefits of railway.[26] Thus, the official attitude, as well as public opinion, was changing. Why did Liu openly call for the construction of railways at this crucial moment and why did the words of a retired military general carry such high esteem when the court had ignored earlier calls? The answer lies first and foremost in Li Hongzhang's manoeuvring, which took advantage of the Russian annexation of western

Xinjiang – a crisis so severe that Liu had to be recalled from his comfortable retirement to lead the campaign against Russian advance.[27] Indeed, Liu's memorial was the result of Li's manipulation from behind the scenes. On his way to Beijing to take up his new position, Liu passed through Tianjin where he visited his old patron. As one of Li's protégés, who rose through the ranks of the Huai Army founded and led by Li, Liu's proposal was used by Li to test the court's attitude towards railways.[28] The court's instruction to 'carefully consult, discuss and propose' was not exactly a green light but rhetoric to appease Liu. However, it was, at least, not an outright rejection.

Wasting no time, Li submitted a memorial immediately, arguing that the railway would bring nine benefits to the country. He elaborated in great detail:

> The region north of *Jianghuai* depends on land and roads, unlike southern provinces where there are many rivers which contribute to the smooth flow of goods. That is why of the 20–30,000,000 levies on foreign goods and *lijin* taxes, we received nine tenths from the southern provinces but only one tenth from the north. If we build railways, make them connect like rivers and enable people to migrate, then the northerners might change their sluggish ways and become hardworking, land and labour can be fully utilised, towns might emerge and commerce may thrive. At important railway stations, we can impose *lijin* and levies, the north would generate the same amount of revenue like the south. This serves our national economy. This is its first benefit.
>
> Forces are strong when combined and weak when scattered. The coastline of our country is more than ten thousand *li* long, we cannot set up defences everywhere because we don't have the resources nor the methods. But if we have railways, they would make transportation easier, far places like Yunnan, Guizhou, Gansu and Shaanxi could be reached within ten days, the defenders of our eighteen provinces can be a guerrilla force. We reduce the army and military spending in the future, they can be an effective force, available when called upon because we are connected. One soldier can be used like ten soldiers. This benefits our military. This is the second benefit.[29]

Li began by addressing two of the most urgent concerns of the Qing – revenue and national security. This was only the beginning. The third benefit he outlined addressed the court's ability to control its far-flung empire with a mobile military force that could be summoned to protect the capital quickly. The fourth revolved around disasters, since the Qing suffered from calamities of all kinds and relief became routine for the government. Railways could speed up the transportation of grains to reduce the effects of famine. Li's fifth point argued that years of neglect and banditry had made the rivers and canals of southern China – the country's 'rice basket' – unsafe. The Qing had been forced to use the seas to transport foodstuff, tax revenue, military staples and weapons. Railways would, in other words, help 'maintain public order' and 'respond to domestic emergencies', as Edward Spiers illustrates in the case of the Victorian Army.[30] Li continued to enumerate the benefits, highlighting how railways could enhance the Qing's postal system to facilitate better communication between the court and its officials scattered across the country, how it could coordinate with the river/canal system and open travel to remote areas, enhancing the flow of goods, and also help the regime to monitor and nip rebellions in the bud. Li ended the memorial by articulating: 'If China builds railways, we can mine our own coal to prevent the rich profit from leaking to foreign countries.' The foreigners who observed this debate closely seemed to have finally understood that 'the apparent reluctance of the Chinese government to sanction the introduction of railway into the empire is not due to the fact they were ignorant of the benefits that would be derived from the innovation'.[31]

Li Hongzhang relentlessly expounded the benefits of railways. Nonetheless, he was rebuffed by Liu Xihong in early 1881. A ranking official who had served in the Qing's embassy in Britain and Germany, Liu argued that railways were unfeasible from eight perspectives and that it would harm the country in nine ways – a direct counter to Li's nine benefits.[32] Liu explained why the Qing should not undertake railway construction:

> A few years ago, I was an envoy to Europe. Before departure, I met with Thomas Wade who said China could only become rich by mining coal and could only save the military by building railways. ... When I arrived in England, many tried to persuade us to build railways and were eager to lend us money. In my opinion, this kind of enthusiasm cannot be anything good. ... Railways are one way for them to coerce us and bring misfortune, we must not listen to them.[33]

For Liu, railway was a tool for Britain to coerce China into further concessions. The more the British tried to persuade him, the more he was suspicious of their intentions, in spite of knowing that railways could help strengthen the country – but this process could not be controlled by foreign powers. Liu was followed by others. Xu Zhixiang's memorial 'On the evils of railway' touched on all the issues that mattered to the Qing's economy and society:

> If we use railways for transportation, the construction will take two or three years, no matter how slow or quick, it will lead to decline and unemployment in the boat industry, hunger and poverty will create thieves and rebels – this is its first harm.
>
> The Yellow River floods every year ... Tens of thousands [in money] would be spent, not on river management but to make things convenient for foreigners. People will complain and be disheartened – this is its second harm.

> ... Once a railway was built, foreigners would build more homes, warehouses, and churches, they become important places and we have to share these places with them – this is its third harm.
>
> Foreigners have wanted to build railways for more than ten years but they have not found a reason and don't know where to start. If we advocate it, they will benefit, we would open door to intruders and if we refuse them, they would wage wars – this is its fourth harm.
>
> What we rely on to fend off foreigners is our dangerous terrain, ... wherever boats could not travel to, trains can reach, where is Chinese sovereignty? – this is its fifth harm.
>
> ... If we use soldiers to guard railways that stretch thousands of *li*. How can we make sure they are protected everywhere? If this is not done properly, military foodstuff, equipment and weapons would be left to enemies – this is its sixth harm.
>
> On the matter of post communication, ... All the postal stations will disappear, vehicles and horses will be idle. Those unemployed will be angry, horses will die of hunger, this goes beyond the people but also includes goods – this is its seventh harm.
>
> ... Machines produced by foreigners don't last long, cannot be repaired if broken ... after ten years, all the money made would go to the foreigners ... we have to invest again, when is this going to end? – this is its eighth harm.[34]

Xu catalogued the destructive effects of railways on the economy and society, unlike Li Hongzhang, for whom benefits outweighed the detriments. The time and costs involved, the greater influx of foreigners, the difficulties to maintain security, the destruction of indigenous businesses, the resultant chaos of unemployed vagrants and rebels, and China's inability to repair equipment were some of the factors that occupied the minds of railway sceptic officials, and they are legitimate concerns. Railway would destroy the existing economic system, disrupt social stability and lead to reliance on foreigners; it would compound the problems that the Qing had been battling with. While Li saw what railways could do to aid the Qing's quest for wealth and strength, Liu and Xu elucidated what they would destroy at the cost of the Qing's political and socio-economic stability. Their concerns are strikingly similar to the slogan of the current Stop HS2 campaign in the UK: No business case. No environmental case. No money to pay for it. Like a monster, the railway tears through the country and leaves a trail of destruction in its wake.[35]

Ralph Heunemann discusses this kind of anti-railway rhetoric in his book, dismissing its proponents as 'downright silly'.[36] Late Qing officials like Liu and Xu were certainly not silly. On the contrary, they expressed grave concerns because they had foreseen the most devastating consequence – entrapment by foreign funding and expertise. Xu concluded his memorial:

> We borrow from foreigners, that only increases their profits. We use their methods, and then we have to follow their scheme. When we borrow from them, we rely on them for everything from skilled artisans to tools. These foreigners only see profits and benefits, they will not share with us their best gadgets. This is very obvious.[37]

Xu was absolutely right to assert these issues. Late Qing China did not have the funds, expertise and resources to build and maintain railways. They would have to borrow from and rely on the foreigners, who were only interested in their own gains. What Xu identified as a threat was a debt trap. He was farsighted enough to have forecasted the types of problems underdeveloped countries would continue to experience in the twenty-first century. Many have accused the post-Mao regime of 'debt-trap diplomacy' in Southeast Asia and Africa. However, they did not invent the mechanism. With all the important issues at stake, it often seems that Qing officials spent more time debating than undertaking the practical work. Yet they were right to be wary; their fears were verbalised succinctly by another official, Yan Mao:

> When we see the powerful ships and weapons of foreigners, we admire them, and we fall into their trap. ... when we build, we must hire foreigners, we must buy materials from them; when we repair, we must use foreigners too. The issue with railways is that the right and benefits all go to foreigners and all the potential problems stay with us.[38]

He concluded that the admiration and acquisition of foreign machines lead to dependency. Such an interpretation precedes the 'dependency theory' fashioned in the mid-twentieth century in the study of developmental economics. Yan's statement also confirms that late Qing officials were aware of what was required to strengthen their country, but it would come at a price they could not accept. Railways would make Qing China dependent on foreigners – that constituted the loss of sovereignty. Mary Wright's famous verdict of the Tongzhi Restoration between 1862 and 1874 as 'the last stand of Chinese conservatism' seems a fitting description for this anti-railway stance.[39] However, David Platcher, a historian of railways and diplomacy in the American context, offers an insightful assessment of the rationale of late Qing officials: 'Behind this rationalisation officials concealed a variety of more sophisticated arguments against railroads. They would never be profitable, for the Chinese were too poor to buy tickets and would never generate enough freight, so the railroad would go into debt and have to be rescued.'[40]

Late Qing officials were certainly not 'downright silly'. They had embraced European clocks and watches in the eighteenth century. They quickly accepted Western armaments during the suppression of the Taiping Rebellion. They appreciated and imported many Western gunboats. So why did they reject railways? The key issue was a concern for autonomy. European clocks/watches, weapons and gunboats, once purchased, belonged to Chinese individuals and regional leaders. But railways would tie the Qing to Westerners who would impose their will and advance their own goals. The Qing and its officials were keen to maintain control as evidenced by their reform projects: from the Jiangnan Arsenal and Fuzhou Shipyard to the CMSNC and the Kaiping coal mine. Initiated and supervised by Zeng Guofan, Zuo Zongtang, Shen Baozhen and Li Hongzhang, these projects were managed by their protégés, even though foreigners were employed to provide expertise. It was vital for the Chinese authorities to undertake reform on their own clock and in their own terms, free of foreign interference. It was this desire that politicised the introduction of railways. The Qing were able to hold off railway encroachment for now, but the arrival of railway imperialism in the mid-1890s would undermine the Qing's sovereignty and territorial integrity, the focuses of Chapters 4 through 7. Hindsight would lend force to the resistance and arguments of anti-railways officials.

As this debate raged on, 'the canal to Lutai with its connecting railway to the colliery were laid out in 1880 and completed in 1881'.[41] Although Kinder knew of the prohibition placed on the use of locomotives, he prepared for the coming railway:

> During the winter of 1880–81 a locomotive of extraordinary design was quietly built in the shops from various odds and ends procurable without attracting much attention. ... Before this engine was finished its preparation became known and strict orders were issued that it should be stopped at once.[42]

Kinder was caught in the act. To not further agitate his anti-railway colleagues, Li memorialised the court in the early summer of 1881, explaining:

> To travel from Tangshan to Tianjin, one has to pass Lutai where road transportation becomes difficult, especially during the summer-autumn flooding season, and leads to congestion; we cannot even use horse wagons. As a result of this, we have decided to construct a water system, a canal of about 70 *li* from Lutai to Xugezhuang for the purpose of transporting coal, connected by a horse road of 15 *li* to the colliery.[43]

Li stated very clearly that the coal from Tangshan would travel on a 'horse road' to Xugezhuang, where it would be offloaded to the canal. There is no reference to railways or locomotives, even though their preparation continued in secret. It appears that Li was doing what Jardine had done with the Wusong. No wonder that the *New York Times* called the Tangxu 'a railway built by stealth'.[44] Li manoeuvred hard behind the scene as Kinder's own article reveals that 'the monster was allowed to be completed. On the 9th June, the first spike was driven by Mrs. R. R. Burnett'.[45] This old boiler-turned steam engine was named the 'Rocket of China' (see Figure 2.2), a reference to 'George Stephenson's "Rocket" in the Rainhill trials of 1829'.[46] Kinder was instrumental in the adoption of the standard gauge of 4 feet 8.5 inches or 1.435mm.

In his book on the Kaiping coal mine, Ellsworth Carlson illuminates that 'Li seems to have gambled on the successful operation of the line justifying his action' and he planned to 'use its success as justification for further building'.[47] Li was not disappointed. The coal mine looked promising, and its share price shot up in October 1881.[48] The *Herald* predicted that the Kaiping mine 'will be able shortly to supply the whole of north China with coal at a cheap rate' by January 1882.[49] But bad news came in March 1882 when 'an order has been issued stopping work at the Kai'ping mine' because 'the foreign methods of mining go down into the vitals of this great earth-Dragon – Chinese methods only tickle his hide – hence he is uneasy, in fact angry'.[50] Anti-mining sceptics hoped to obstruct the progress of the coal mine, arguing that the commotion generated disturbed the Eastern Tombs

Figure 2.2 The 'Rocket of China' (Image courtesy of the Peter Crush Chinese railway collection and special collections, University of Bristol library, Historical Photographs of China (HPC hereafter), at https://hpcbristol.net/visual/PC27-296, Identifier PC27-296)

in Zunhua, about 50 miles to the northwest, where Manchu emperors were buried.[51] Though largely sympathetic to Chinese sensibilities, Kent finally scoffed at the idea of *fengshui*, stating that, 'The professors of this science are both active and numerous, and for them the occasion was a great one.'[52] Kent did not know that 'fengshui was interwoven with Qing governance, especially its legal system' and the Qing 'incorporated fengshui into its administration of gravesites, forests, mining'.[53] The *Herald* reported that 'the railway engines to be used at the Kai'ping mines have been suppressed', which led to 'a considerable quantity of railroad rolling-stock, in the way of trucks, platform, etc' to be left on the Bund.[54]

The *Herald* also predicted that 'the delay may only be temporary'. They were soon proved correct through Li's efforts behind the scene.[55] To persuade sceptical officials, Tang Tingqu invited them to visit Kaiping and ride on the railway in the hope that they would recognise the difference between mule and steam power. Kinder believed that this offer was a 'great success' because 'nothing more was said about mules and the stables built for their use were quickly demolished'.[56] In October 1882 two new engines were imported from England. By November, Tangshan's coal was transported regularly from the colliery on the Tangxu railway.[57] By January 1883, the Kaiping authority imported a few more coaches and began offering a passenger service between Tangshan and Xugezhuang. David Pletcher correctly referred to the Tangxu as the 'first successful line', the barely 8-mile line was built after many memorials and a debate that ran from 1878 to 1882.[58] For Li, too much time and energy had been spent on battling the resistance that greeted every effort he made. It would be an overstatement to say that this was the first government-sanctioned railway, since funding and resources came from the CMSNC and the Kaiping coal mine. Li's resourcefulness can be seen in his use of one reform project to fund another. He relied on his own connections and means, rather than handouts from the government, to see his project through.

As the Chinese proverb states, 'a journey of a thousand miles begins with a single step', and indeed the short Tangxu would become the mother of all railways in north China, shaping the history and politics of modern China for seven decades to come.[59] But the Tangxu was only a tiny step; as the coal had to be offloaded to barges after a journey of less than 8 miles and travel on the 'coal river', a canal that required constant dredging on its journey of roughly 80–90 miles to Dagu port and Tianjin. *Shen Bao* kept up the pressure, highlighting the slow pace of river transportation as a contrast to the rapidity of railways.[60] It also published the profits of the Kaiping coal mine and reminded its readers of the difficulty of transportation to Dagu, and thus the need to extend the railway.[61] But to continue his railway enterprise, Li Hongzhang would need more than just newspaper headlines – he needed a breakthrough. The prediction made in his memorial, that the Qing needed a mobile military force which could be called upon and despatched quickly to wherever necessary in times of threat to national security, materialised in early 1883.

'Chinese railways made by the navy [中国铁路海军造]'[62]

France had annexed south Vietnam in the early 1860s. By the early 1870s, French forces had advanced into north Vietnam, reaching the Chinese border by the early 1880s when it brought to the Qing court a cry for help from the Vietnamese. In the summer of 1883, a local Chinese militia called the 'Black Flags' or Guangxi Braves, whom Bradley Davis characterised as 'imperial bandits', was deployed to battle the French.[63] Lewis Chere has referred to this Sino-French conflict as an 'undeclared war'.[64] The first series of battles took place in northern Vietnam, where the Black Flags managed to repel French forces by the spring of 1884.[65] The conflict reached a stalemate by the summer: neither side could claim overwhelming victory. Unable to bring the Chinese to their knees on land, the French instead took the war to China's long coastline in August 1884. The French Navy attacked ports along the Taiwan, Fujian and Zhejiang seaboard, and up to Shanghai. They landed on north Taiwan, using the island to launch a naval embargo.[66] They destroyed the Fujian Naval Force, which had been built up by Shen Baozhen with the help of French naval officer Prosper Giquel. The French drew a war zone that stretched from the southwest land frontier with Vietnam through the South China Sea and Taiwan Strait to the East China Sea, an area encompassing six provinces. This long war zone made it almost impossible for the Qing to anticipate where the enemy might strike next, and it made the movement of troops and logistics extremely onerous.

Despite being overstretched, the Chinese forces, under the leadership of the aforementioned General Liu Mingchuan, fought hard in Taiwan and held the French at bay in a small area on the island. The Chinese also managed to fend off the enemy on land despite the lack of modern weapons. This was an astonishing victory for the Qing. While the Chinese were unable to defeat the French, they also managed not to lose the war – a situation that resulted in a deadlock, which arguably constituted a win for China, albeit a precarious one.[67] The Chinese might well have won a more decisive victory had they been able to move troops and supplies quickly – a feat that railways would have facilitated. Empress Dowager Cixi was eager to end the war, not because of the not-disastrous result, but because she was obsessed with celebrating her 50th birthday. Li Hongzhang was sent to negotiate. What the French did not win on the battlefield, they procured in the Treaty of

Tientsin, also known as the Li–Patenotre Treaty, signed in June 1885.[68] The provisions stipulated that Vietnam was to become a French Protectorate, Qing China was to open the Sino-Vietnam border for commerce, and several Treaty-port cities were awarded to France. The French also asked for the right to build railways in China. This was the first instance of a foreign power demanding railways in a treaty, a signal that, by the mid-1880s, the game of empire was changing.[69]

The French had opened a new frontier, their own exclusive passageway to penetrate the Qing empire. In the chaos of the Sino-French War, Japan made an attempt to conquer Korea in December 1884, while the British Navy landed on Hamilton Island at the southern tip of the Korean Peninsula in April 1885. These threats to national security came from the seas, and they once again revived the debate around the construction of a national navy. The Qing's naval construction was extemporaneous and disorganised as regional leaders bought, and attempted to make, their own gunboats. In his excellent work on the Qing's naval development, John Rawlinson counted 'four distinct fleets', one of them being the aforementioned Fujian Naval Force. Rawlinson pointed out that there was no 'national navy' and that 'loyalties were provincial or personal', which is apparent in the case of Li Hongzhang and Shen Baozhen.[70] This led to confusion and rivalries:

> China was not too poor to afford a modern navy. But there was no budget, and the revenues were divided by quota in such a way as to encourage rivalry between the central government and the provinces, and among the provinces, as well as among the parts of each provincial administration.[71]

This was exactly the case with railway development. The Qing court reacted to French aggression by establishing a Department of Naval Affairs, what foreigners called the Admiralty, tasked with the job of forming and managing a national navy.[72] Like Japan's first attempt to invade Taiwan, which had prompted the Qing to appoint Beiyang and Nanyang Commissioners, the Sino-French War stimulated progress on naval construction. At the height of the Sino-French conflict, many Chinese newspapers, especially *Shen Bao*, ran headlines like 'Railways are imperative'.[73] The *Herald* even reported in June 1884 that the court had, in principle, agreed to the construction of railways.[74] Li must have welcomed the news as he planned to extend the Tangxu.[75] Prince Chun, father of the Guangxu Emperor, was put in charge of the Admiralty.[76]

Li Hongzhang grabbed this opportunity. He ingratiated himself with Prince Chun, inviting him to visit the Beiyang Navy – again, we see the 'provincial' and 'personal' aspect of the Qing's naval construction.[77] Prince Chun spent May and June 1886 inspecting the various battleships on the Bohai Sea, as well as the facilities scattered in Tianjin, Dalian, Weihai and

Yantai. Li paraded his Beiyang fleet, bought from Europe, and several drills were conducted.[78] The Prince was impressed and much relieved that the Qing finally had its own 'national navy', but Li took every opportunity to remind him that to keep this navy operational, coal was an absolute necessity – and coal was very slow to arrive from Tangshan. Li made sure Prince Chun understood the vital importance of railways to the Beiyang Navy in transporting this precious resource, as well as in moving troops and weapon supplies. Much as Li expected he would, Prince Chun, as head of the Admiralty, wrote to the Empress Dowager. His memorial was titled 'The Admiralty asks to be allowed to build the Jingu railway':

> I know the railway's reputation has been negative in past debate and there were different opinions. I heard them all and also held prejudice against it until I visited the Beiyang Navy. Now I really know the difference between outsiders' empty talk and insiders' reality. ... Zeng Jize spent eight years in Europe and saw with his own eyes that all European countries use railways to move troops, transport weapons, facilitate commerce and benefit people. There are many benefits, from frontier defence to the livelihood of the ordinary people – there is really no harm at all.[79]

Prince Chun raised the issue of railway to the level of national security, which Kent referred to as the 'banner of military expediency'.[80] Major newspapers like *Shen Bao* and the *Herald* devoted plenty of space to the news.[81] When the father of the Guangxu Emperor, brother-in-law of the Empress Dowager, and respected Manchu noble proposed a railway, there were few who dared to challenge it. The Admiralty was soon put in charge of railway construction, giving credence to the common phrase 'Chinese railways made by the Navy'. It appears that Li's manoeuvrings had worked – he had perfected the game of pulling the strings from behind the scene.[82] The proposed line ran from Tianjin to the port of Dagu, which gave the line the portmanteau name of Jingu [津沽]. But this project could not proceed without first extending the Tangxu. Prince Chun continued:

> Five hundred *li* north of Dagu and Beitang, there is hardly any defence. In the case of an emergency and if we have a railway, we can send troops in the morning and they will arrive in the evening. In other words, we can station an army in one area but can use it for several areas. This cuts the budget needed to maintain the military. The Kaiping coal mine built a 20 *li* railway in Guangxu's seventh year [1881], but it was not enough to quickly transport coal to the ships, hence another 65 *li* would need to be constructed in order to connect with *Yanzhuang* on the River Ji. This is the central part that leads to the Mountain-Sea Barrier, the only passageway for moving troops. If we extend south to Dagu and north to the Barrier, the ten-thousand troops commanded by General Zhou Shengbo can dash and defend all places in between, like tens of thousands of troops.[83]

We can see why the Prince was keen. Five hundred *li* north of Tianjin sits the Mountain–Sea Barrier, the shortest and most convenient gateway to and from Manchuria. Defending his homeland clearly mattered to the Prince and the Qing court. The '20 *li* railway' referenced in his memorial was the Tangxu. He pointed out its inadequacies and called for the Tangxu to be extended to Yanzhuang in Lutai. He gave specific instructions as to how the line could be stretched even further:

> If this is a big project that is difficult to raise funds for, we can first build the 80 *li* segment between Yanzhuang and Dagu. Then we can gradually connect the 100 or so *li* of segment between Dagu with Tianjin. After the Jingu railway is completed, we can extend north to the Mountain-Sea Barrier.[84]

Prince Chun favoured two major extensions from Lutai: south to Dagu, then Tianjin and north to the Barrier. It seems that the Prince had become a passionate advocate for railways, but in reality his memorial was Li Hongzhang's words uttered through his mouth and pen – the Prince had never visited the places he mentioned, nor could he have been familiar with the distances between them, but he did understand the importance of coal to the Navy and the urgency to build railways. Li had successfully manipulated the Prince from behind the scenes. But the Prince's plea and the court's sanction were little more than words. Li was left to raise the funds for the project alone. Undaunted, he leapt into action, preparing for his future railway enterprise and commissioned the Kaiping Railway Company. As China's first railway corporation, the Company bought the Tangxu and undertook the extensions. Li tasked Wu Tingfang with the job. Born in Malacca and educated in Britain as a lawyer, Wu would become one of the most paramount officials and diplomats for the late Qing and later Republic of China.[85] This appointment again demonstrates Li's open-mindedness and pragmatism, as Wu was not a Chinese degree holder, thus an outsider to Confucian officialdom. Li was not averse to seeking talent where it could be found, and using it.

Preparatory work for the Tangxu to Lutai extension began in the autumn of 1886.[86] These developments excited foreigners – German, French, American and British. Tyler Dennett wrote 'in 1886 there were gathered in Tientsin and Peking the largest number of "concession hunters" yet seen in China.'[87] William Keswick of Jardine Matheson even predicted an 'imminent China railroad boom' in 1887.[88] Jardine was the agent for Decauville, a French manufacturer of narrow-gauge railways that was eager to do business in China. To interest the Chinese public, Decauville shipped a few miles of moveable railway to Tianjin and laid it at the back of the British Settlement – a repeat of what the British did in 1872.[89] All Chinese could take a free ride; and many did, with *Shen Bao* reporting the excitement.[90] But these foreign hopefuls were disappointed when nothing happened after this publicity stunt, and no contract was awarded to any potential investor. Dennett understood the situation better than others; the Chinese 'were now clearly determined to retain control not merely of the ownership but of the construction of whatever railways might be built'.[91]

The desire to retain Chinese autonomy is made clear in the activity of the Kaiping Railway Company. The Company raised money from Chinese merchants through the issuing of shares.[92] The Lutai extension, roughly 20 miles, was completed in 1887. The extension from Lutai to Dagu, the port which the British demolished during the Second Opium War, began immediately.[93] Despite resistance from peasants who refused to sell land or move burial sites, the Dagu extension, about 24 miles, was completed by April 1888 and Tianjin, about 31 miles, was reached in August.[94] This completion meant that the collieries in Tangshan were now linked with the port city of Tianjin.[95] Kinder recalled that the railway was 'kept so quiet … for five years very few high officials had dared to visit it or extol its merits'.[96] This remained the case until Li's visit to the railway that he had schemed so hard to build in October 1888; he rode from Tianjin to Tangshan, the stations which gave the Jintang [津唐] railway its name (see Figure 2.3).[97] By November, trains between Tangshan and Tianjin were running daily.[98] The humble Tangxu had come a long way, and it would soon grow into the Inside and Outside the Barrier Railway.

Figure 2.3 Li Hongzhang inspecting the Jintang railway (Baidu public domain)

It seems that the prospects looked good in 1888. Greatly encouraged, Li Hongzhang pushed on with a strategic extension of the Jintang westward to Tongzhou. The *Herald* simply called this line the 'Peking railway,' as Tongzhou was a borough of Beijing.[99] The borough was also the terminus of the Grand Canal, a depot for goods from around the country. The idea of this line excited many Chinese merchants, who even promised to contribute 10 per cent of their annual profit to naval expenditure – music to Li's ears.[100] Kinder, the chief engineer, began a field survey in January 1889.[101] But it seems that Li's luck soon ran out: by February the Tongzhou proposal had wrought the wrath of his opponents.[102] As Kent put it: 'the reactionaries had for the time become too strong for the progressive party at Court'.[103] They attacked Li; even his reform-minded colleagues did not come to his aid. It seems that the more reform projects Li acquired, the more rivals he generated. One of them was Zhang Zhidong, whose role in this story once again illustrates the problem of competition that politicised the life of railways.

Zhang served in various civil jobs after obtaining his *jinshi* degree in 1863. A reformer, he also initiated and led modernisation projects, the Hanyang Iron Works being prominent among them. He was the fourth and last most important late Qing official after Zeng Guofan, Zuo Zongtang and Li Hongzhang. Instead of opposing Li's Tongzhou scheme, Zhang simply proposed another line, the Luhan [芦汉] railway, from Lugouqiao, another borough in Beijing, to Hankou – the heart of south-central China on the Yangzi River or what Kent called the 'great emporium of central China'.[104] Clearly, Zhang was competing with Li and directing attention to southern China, where he first served as governor-general of Liangguang (Guangdong and Guangxi provinces) from 1883, and then Huguang (Hubei and Hunan provinces) from 1889.[105] Zhang may have killed Li's Tongzhou initiative for now, but Li would make it difficult for the Luhan to start as he manoeuvred behind the scene, diverting funds for the Luhan to his project – the Inside and Outside the Barrier railway.[106] It would take a decade of politicking and foreign scramble for the Luhan to start in 1898 as the vertical line encroached on the interests of several European powers. Such is the complexity of railway finance and construction in the late Qing. Like Shen Baozhen, Zhang rose to challenge Li. Their advocacy for reform brought them together, but competition over railways would keep them apart. Zhang's career would also come to be closely associated with railways.[107]

The development of railways suffered from the same problems that plagued naval development. Lacking a national strategy, its construction was reactionary, shambolic and provincial. We can see this from the Wusong, which was at the mercy of Shen Baozhen; the Tangxu and the Jintang, which were controlled by Li Hongzhang; and now the Luhan, that would 'belong' to Zhang Zhidong. These issues reflect fundamental flaws with late Qing reform. Impromptu and piecemeal, reform lacked a national approach and, as a result, neither did the development of railways and the navy. We see the court react time and again when its security was challenged. When they did so by turning on the green light on a case-by-case basis, specifics were left in the hands of regional leaders who turned reform into personal projects with which they could grow their power and compete with others. This practice politicised the process of naval development, as John Rawlinson has brilliantly argued. It also politicised the life of railways. Some historians believe that late Qing reform had failed, and they put the blame on the strictures of Confucianism. But as we can see from railway and naval development, it was the lack of national vision and leadership, compounded by rivalries among officials, that prevented reform from achieving its best results. This was clear during the Sino-French War, when Li Hongzhang refused to aid the Fujian Naval Force. As fate would have it no one came to his aid a decade later during the First Sino-Japanese War, which led to the annihilation of his Beiyang Navy.[108] The Admiralty failed to fulfil its mission to create and lead a unified national navy.

Shrewd and persistent, Li Hongzhang racked his brains for ways to promote railways. His resourcefulness has been clearly showcased in his effort to use one reform project to generate another, his use of Tang Tingqu to propose different routes, his manoeuvring of Liu Mingchuan's timely call for railways and his courting of Prince Chun's stamp of approval in his campaign to build a navy supported by railways. Li fought openly, but also laboured secretly, segment by segment, for the extension of the Tangxu to Lutai followed by Dagu and Tianjin. No wonder Stanley Spector called Li a 'man of action'.[109] The banner of national security changed the fate of the Tangxu, but it did not fundamentally change the fortune of railways in China. Although more fortunate than the Wusong, the Tangxu's success was, sadly, a direct result of French aggression. Arguably, what Li had won thus far could hardly be called a victory: Prince Chun's proposals were met with no opposition, but Li's Tongzhou extension brought down the wrath of anti-railway proponents and his rival. Undaunted, Li turned his attention northward, to the 'five hundred *li* north of Dagu'. The battle for the railway was far from won and Li would have to up his game in order to initiate another railway to 'plaster' the old, crumbling house.

Notes

1 Wang Tao, *Taoyuan wenlu waibian* (12 vols, Shanghai: Guji chubanshe, 2010), vol 3, 25.

2 Elisabeth Sinn, 'The fugitive in paradise: Wang Tao and cultural transformation in late nineteenth-century Hong Kong', *Late Imperial China*, vol 35, 1 (1998): 56–81.

3 Guo Songtao, *Zuiyan cunlue* (Shanghai: Shanghai shudian, 1994), 20; Xue Fucheng, *Yongan wenbian* (Taibei: Wenhai chubanshe, 1973), vol 2, 17–18; Zheng Guanying, *Shengshi weiyan* (Beijing: Zhonghua shuju, 2013), 421–29; Ma Jianzhong, *Shikezhai jiyan* (Beijing: Zhonghua shuju, 1959), vol 1, 10–18; Li Gui, *Huanyou diqiu xinlu* (Changsha: Hunan renmin, 1980) and Charles A. Desnoyers, *A journey to the East: Li Gui's A new account of a trip around the globe* (Ann Arbor: University of Michigan Press, 2004); Xiong Yuezhi, *Xifeng dongjian yu jindai shehui* (Shanghai: Jiaoyu chubanshe, 2019).

4 Jenny Huangfu Day, *Qing travellers to the Far West: diplomacy and the information order in late imperial China* (Cambridge: Cambridge University Press, 2018); 'Searching for the roots of western wealth and power: Guo Songtao in Victorian England', *Late Imperial China*, vol 35, 1 (June 2014): 1–37.

5 'Shidong shilu', *Shen Bao*, 4 April 1879: 3, 'Xigong yikai tielu', 5 May 1880: 1, 'Riben xifa siyong', 18 July 1880: 4 and 'Yuenan jinshi', 7 October 1880: 2.

6 Wu Yong and Liu Zhixiang, *Gengzi xishou congtan* (Shanghai: Shanghai shuju, 1996), 129. Walls and windows were covered by paper. When worn, more would be plastered upon, the tradesmen were called *biaohu jiang* [裱糊匠]; the closest translation is plasterer.

7 Feuerwerker, *Industrialisation*, 13–14; Spector, *Li Hung-Chang*, 234–35; Carlson, *Kaiping*, 18–20.

8 'A railway built by stealth', *Times*, 2 November 1884: 14.

9 Feuerwerker, *Industrialization*, 150–88; see also works by Stephen Halsey and Anne Reinhardt.

10 'Prospectus of the China Merchants' Steam Navigation Company', *Herald*, 21 April 1877: 400, 'Shanghai Steam Navigation Company', 18 January 1877: 57 and 'China Merchants' Steam Navigation Company annual report', 27 September 1881: 320.

11 Feuerwerker translated it as 'official supervision and merchant management', see *Industrialization*, 1.

12 *Industrialization*, Feuerwerker book title.

13 'Yunmei shuo', *Shen Bao*, 4 August 1873: 1 and 'Changqi meikuang chushou', 8 December 1874: 2.

14 Zhongguo gudai meikuang kaifashi bianxiezu, *Zhongguo gudai meikuang kaifashi* (Beijing: Meitan gongye chubanshe, 1986).

15 Kenneth Pomeranz, *The great divergence: China, Europe and the making of the modern world economy* (Princeton: Princeton University Press, 2021), 62 and 64.

16 Li Hongzhang, *Zougao*, vol 40, 41–43.

17 'Kaiping jiayin', *Shen Bao*, 11 December 1878: 2; Carlson, *Kaiping*, 18–21.

18 Carlson, *Kaiping*, 12–13.

19 Kent, *Enterprise*, 23; 'The miscellaneous: Kai-Ping coal mines', *Herald*, 24 January 1882: 105 and C. W. Kinder, 'China railway company', 24 February 1888: 220.

20 Claude W. Kinder, *Railways and collieries of north China* (London: The Institution of Civil Engineers, 1891), 4; 'Tielu tingzhu', *Shen Bao*, 8 February

1879: 2, 'Tancha kuangmiao', 29 May 1879: 2 and 'Kaiping meikuang qingxing', 25 January 1881: 2.

21 Kinder, *Railways*, 4; 'Yizhao tielu', *Shen Bao*, 9 August 1878: 2.

22 Kinder, *Railways*, 12.

23 David Gwyn, *The coming of railways: a new global history 1750–1850* (New Haven: Yale University Press, 2023), 81–82.

24 Liu Mingchuan, *Liu Mingchuan wenji* (Heifei: Huangshan shushe, 1997), 44.

25 Liu, *Wenji*, 46.

26 'Lun Kaiping chuankai tielushi', *Shen Bao*, 4 May 1882: 1, 'Yingguo tielu shumu', 14 April 1882: 1 and 'Dagu jinwen', 9 June 1882: 1.

27 'Jingbao quanlu', *Shen Bao*, 15 December 1882: 3–4; Spector, *Li Hung-chang*, 253.

28 Weng Fei, *Li Hongzhang yu huaijunde chuangjian* (Hefei: Huangshan shushe, 2012).

29 Li Hongzhang, *Zougao*, vol 39, 344–47.

30 Spiers, *Engines*, 15 and 21.

31 'Li Hung-Chang on railway', *Herald*, 14 July 1882: 36.

32 Liu Xihong, *Qingchao xuwenxian tongkao* (400 vols, Taibei: Xinxing shuju, 1963), vol 362, 11061–70; 'Jingbao quanlu', *Shen Bao*, 6 April 1880: 4, 'Gonglu yuzhi', 21 January 1881: 2 and 'Gonglu yuzhi', 30 March 1881: 1.

33 Liu Xiheng, 'Michen buke jiekuan zaolu pian', in *Tielushi ziliao*, vol 1, 102.

34 Xu Zhixiang, 'Lun tielu zhihai zhe', in *Yangwu yundong wenxian huibian* (8 vols, Taibei: Shujie shuju, 1963), vol 6, 167–68; 'Susheng fuyuan shiyi', *Shen Bao*, 18 March 1874: 5, 'Yuzhi', 4 March 1879: 2 and 'Benguan jie Fengdian', 21 September 1885: 1.

35 Stop HS2, at http://stophs2.org/, accessed 2 September 2020.

36 Huenemann, *Economics*, 38–40 (38).

37 Xu Zhixiang, 'Bakai tielu jixiu hegong zhe', in *Tielushi ziliao*, vol 1, 104–5.

38 Yan Mao, 'Zou tielu buke kaiban zhe', in *Tielushi ziliao*, vol 1, 107–8.

39 Wright, *Last stand*.

40 Pletcher, *Diplomacy of involvement*, 138–39.

41 Kinder, *Railways*, 11.

42 Kinder, *Railways*, 12.

43 Li, *Quanji: zouyi*, 339–40.

44 'A railway built by stealth', *Times*, 2 November 1884: 14; Herr Kiepert, 'A Chinese railway', *The Times* (London), 14 October 1884: 5.

45 Kinder, *Railways*, 12.

46 Kinder, *Railways*, 27 and Spiers, *Engines*, 1.

47 Carlson, *Kaiping*, 21.

48 'Kuangwu jinwen' *Shen Bao*, 11 October 1881: 1 and 'Gufen zhangjia', 1 November 1881: 1.

49 'Miscellaneous: the Kai-Ping coal mines', *Herald*, 24 January 1882: 105 and 'Kaiping', 9 June 1882: 625.

50 'Tientsin', *Herald*, 29 March 1882: 342; 'Meiwu jinkuang', *Shen Bao*, 5 March 1882: 1 and 'Meikuang xiaoxi', 7 April 1882: 2.
51 Mi, *Tielushi ziliao*, vol 1, 121.
52 Kent, *Enterprise*, 24.
53 Tristan Brown, *Laws of the land: fengshui and the state in Qing China* (Princeton: Princeton University Press, 2023), 1 and 193–222.
54 'Tientsin', *Herald*, 4 April 1882: 366 and 22 April 1882: 421.
55 Tientsin', *Herald*, 22 April 1882: 421.
56 Kinder, *Railways*, 12.
57 'The Kaiping railway', *Herald*, 15 November 1882: 531.
58 Pletcher, *Diplomacy of involvement*, 140.
59 'The imperial Chinese railway of north China', *Herald*, 16 October 1899: 745.
60 'Lun Kaiping chuangkai tielushi', *Shen Bao*, 4 May 1883: 1.
61 'Meiwu jinwen', *Shen Bao*, 5 March 1882: 1, 'Dagu jinwen', 9 June 1882: 1, 'Kaikuang yijianchou yundaolun', 19 December 1882: 1, 'Kuangwu shengse', 1 May 1883: 1 and 'Kaiping jinxin', 5 December 1883: 2.
62 Lei Yi, *Li Hongzhang yu wanqing sishinian* (Taiyuan: Shanxi renmin chubanshe, 2013), 332.
63 'The Peking government and the Black Flags', *Herald*, 28 November 1883: 602; Bradley C. Davis, *Imperial bandits: outlaws and rebels in the China-Vietnam borderlines* (Seattle: University of Washington Press, 2017).
64 Lewis M. Chere, *The diplomacy of the Sino-French war (1883–1885): global complications of an undeclared war* (Notre Dame: Cross Culture Publications, 1998).
65 'Rout of French troops reported loss', *Herald*, 2 January 1884: 20, 'Canton', 14 August 1885: 184 and 'Attack on the French in Annam', 27 June 1884: 732; Rawlinson, *Naval*, 109–28.
66 'The Franco-Chinese imbroglio', *Herald*, 8 August 1884: 146 and 'Fighting between the French and Chinese fleets', 18 February 1885: 183.
67 Lloyd Eastman, *Throne and Mandarins: China's search for a policy during the Sino-French controversy 1880–1885* (Cambridge: Harvard University Press, 1967); Charles Fourniau, 'La genese et l'evolution de l'affaire du Tonkin', *Revue Historique*, vol 246, 2/500 (October 1971): 377–408.
68 'The signing of the Treaty', *Herald*, 19 June 1885: 700.
69 'The Franco-Chinese Treaty', *Herald*, 19 June 1885: 705.
70 Rawlinson, *Naval*, 65 and 201–2.
71 Rawlinson, *Naval*, 202.
72 'Dijing jishi', *Shen Bao*, 3 November 1885: 2, 'Haijun sheshu', 20 November 1885: 2 and 'Chucai yiduo youdijing', 25 December 1885: 1; Rawlinson, *Naval*, 63–95.
73 'Tielu buke bujikai', *Shen Bao*, 2 November 1883: 1, 'Xulun huoche tielu', 4 March 1884: 1 and 'Tielu jidianxian ercheng', 21 June 1884: 1.
74 'Railways in China' and 'The Tientsin railway', *Herald*, 20 June 1884: 696.
75 'Tielu jiangxing shi', *Shen Bao*, 8 January 1886: 1.

76 'The crisis in Peking', *Herald*, 16 April 1884: 443 and 'Prince Chun's dinner', 27 October 1886: 446.
77 Zhou Fu, *Chunqingwang xunyue beiyang haifang riji* (Beijing: Xueyuan chubanshe, 2006), 378–442.
78 'Lun haijun xiongsheng', *Shen Bao*, 10 May 1886: 1 and 'Dijia lijin xushu', 16 May 1886: 2; 'Prince Chun in Chefoo', *Herald*, 28 May 1886: 558, 'Prince Chun in Port Arthur', 4 June 1886: 592 and 'Prince Chun in Tientsin', 25 June 1886: 671.
79 Yi Xuan, *Xuwenxian tongkao*, vol 363, 11071–75.
80 Kent, *Enterprise*, 28.
81 'Railway memorial from the board of admiralty', *Herald*, 13 April 1887: 414; 'Zhongguo chuanshe haijunyi', *Shen Bao*, 12 January 1887: 11.
82 'Li fuxiang zhaogu kaitielu', *Shen Bao*, 31 May 1887: 1 and 'Tielu pishi', 1 June 1887: 2.
83 Yi Xuan, *Xuwenxian tongkao*.
84 Yi Xuan, *Xuwenxian tongkao*.
85 Linda Pomerantz-Zhang, *Wu Tingfang: Reform and modernization in modern Chinese history* (Hong Kong: Hong Kong University Press, 1992).
86 'Tientsin', *Herald*, 8 December 1886: 613.
87 Dennett, *Americans*, 597–98.
88 Yokohma to Jardine's, dated March 15, 1887, Unbound Correspondence (UC), Jardine Matheson Archive.
89 'Tianjin shixing tulu huoche', *Shen Bao*, 30 September 1872: 3.
90 'Tielu daoshen', *Shen Bao*, 30 June 1886: 2 and 'Xizuo huoche', 7 December 1886: 1.
91 Dennett, *Americans*, 597–98.
92 'Prospectus of the Kaiping Railway Company', *Herald*, 22 April 1887: 445.
93 'Kaiban tielu', *Shen Bao*, 23 April 1887: 1.
94 'Tientsin', *Herald*, 17 June 1887: 667; 'Tielu shuwen', *Shen Bao*, 29 June 1888: 2 and 'Tielu shixing', 29 August 1888: 2.
95 Feuerwerker, *Industrialization*, 13–14.
96 Kinder, *Railways*, 13.
97 'Shangxiang kangong', *Shen Bao*, 17 October 1888: 1.
98 'Tielu shuwen, *Shen Bao*, 10 November 1888: 2 and 'Huoche shuwen', 29 December 1888: 2.
99 'Tungchow: the Peking railway', *Herald*, 23 October 1896: 706.
100 'Tielu binggao', *Shen Bao*, 29 October 1888: 1.
101 Kent, *Enterprise*, 32.
102 'Railways', *Herald*, 6 July 1889: 19; Feuerwerker, *Industrialisation*, 14.
103 Kent, *Enterprise*, 33.
104 Zhang Zhidong, *Zhang Zhidong quanji* (12 vols, Wuhan: Wuhan chubanshe, 2008), vol 2, 267–69; 'Miscellaneous: railways', *Herald*, 9 November 1888: 529.
105 'Zhongli yanmen zoufu zunchou tielu zhegao', *Shen Bao*, 29 September 1889: 1 and 30 September 1889: 1; 'The Admiralty board on railways', *Herald*, 4 October 1889: 419.

106 'Tungchow: the Peking railway', *Herald*, 23 October 1896: 706; 'Weiyuan goudi', *Shen Bao*, 30 May 1897: 1, 'Jiulu chongkan', 27 December 1897: 2, 'Weiyuan kanlu', 20 May 1898: 1; He Hanwei, *Jinghan tielu chuqi shilue* (Hong Kong: Chinese University, 1979), 7–9; Qinhuangdao gangwuju, *Qinhuangdaogang tielu yunshu fazhanshi* (Beijing: Haiyang chubanshe, 1993), 18.

107 Knight Biggerstaff, *The earliest modern government schools in China* (Ithaca: Cornell University Press, 1961), 7.

108 Rawlinson, *Naval*, 109–14.

109 Spector, *Li Hung-Chang*, 235.

3

Converting the oriental despot[1]

> Palace slaves pull yellow ropes on left and right
> Train tracks are laid as far as the *Yingxiu* Garden
> Order for imperial lunch is placed at the North Sea
> The wheels roll straight through the *Fuhua* Gate.[2]

This poem paints a vivid picture of the Empress Dowager Cixi's lunch run. When the train passed through the *Fuhua* Gate, crowds gathered for a glimpse of the royal fanfare. The railway and the six-coach train were a birthday gift presented to her in 1888. The railway's arrival within the hallowed grounds of the imperial palace could be seen as a sign of change and progress. But who had sent the Empress Dowager this extravagant gift? Why was it being pulled by 'palace slaves', rather than a steam engine? Unknown to many scholars, this 1.8-mile Xiyuan [西苑] railway, and other lines built exclusively for Cixi, have not received any academic attention.[3] Cixi played a significant role in the politics of railways, not just because the barely 2-mile line was an amusement for the despot. Her conversion to the benefits of railways would further dramatise railway development. Cixi has fascinated many observers, but her popularity tends to be reflected in non-academic works, even as she is broached or discussed in scholarship as the archenemy of late Qing reform. This chapter focuses on her life with railways. It sheds new light on how late Qing reform was moulded to the whims of a despotic ruler. Cixi subjected the development of railways to her political needs and personal pleasure, oblivious to the scramble of imperialist powers that would soon usher in the age of railway imperialism.

'The last stand of Chinese conservatism'[4]

From her relatively humble beginning as a lower-level consort to the Xianfeng Emperor, Cixi was eventually elevated to the position of empress dowager when her six-year-old son was put on the throne after the death of Xianfeng in 1861. This position entitled her to 'reign from behind the curtain' and reform was launched, bearing the name of her son's regnal title, the 'Tongzhi restoration'. In January 1875, the Tongzhi Emperor suddenly died at the age of nineteen, leaving behind a pregnant empress who also perished in March under mysterious circumstances. A skilled politician by now, Cixi quickly filled the power vacuum by adopting her nephew, the son of Prince Chun and her own younger sister. Thus, as the 'mother' of the new four-year-old Guangxu Emperor, she began her second 'reign from behind the curtain' as Empress Dowager again. Unchallenged and acknowledged as a symbol of stability, Cixi was called the 'Old Buddha'. In November 1885, she celebrated her 50th birthday, an event which gave rise to a larger plan to renovate the royal complex with modern luxuries such as electric lights and boilers, and to restore the palaces and gardens damaged during the Second Opium War. Included in the plans was the redevelopment of Xiyuan, which translates to 'West Garden', situated to the west of the Forbidden City – the North, Central and South Sea pleasure complex.[5]

Cixi's vanity and appetite for pleasure presented an opportunity for Li Hongzhang. Audacious and shrewd, Li sought to change the fortune of railways by appealing to her taste for luxury. He informed the Empress Dowager that foreign merchants would pay to build a leisure railway within the royal compound for her exclusive pleasure. Once again, Li was pushing the boundaries of the permitted and making his schemes. Prince Chun's visit to the Beiyang Navy led to important policy changes regarding railways in 1886. This change led to the gathering of what Tyler Dennett termed 'concession-hunters' in Tianjin and Beijing. The Americans went out of their way to impress, and Dennet describes '[a] model of American railway track and a small train operated by clockwork [being] set up within the palace grounds for the amusement of the boy emperor and the court'.[6] But they had stiff competition from the French manufacturer Decauville, which had also set up a 3 km-long railway in Tianjin in December 1886. This was a successful advertainment, and the *Herald* reported that 'since the official trial of the Decauville railway last Sunday the Chinese have come in great crowds every day to see the locomotive and the carriages'.[7] Yet another testimony that the Chinese public welcomed these technological innovations.

Li Hongzhang surely calculated that all the excitement would result in these sample railways finding a permanent home on Chinese soil. Indeed, the Decauville engineer Gaston Galy was eager to impress the Qing court and ranking officials in return for favours. But their agent, Jardine's director William Keswick, was cautious when the railway 'boom' he had anticipated did not materialise into loans and contracts. The two men quarrelled over the issue and Keswick made his condition clear:

> The free presentation of the railroad could be made to tie in with something else – the concession for the construction of the Tientsin-Peking line? – a tangible quid-pro-quo, not a shadowy promise from the Marquis Tseng which, with the highest and most honourable intentions, the very nature of the business would prevent him carrying out.[8]

Jardine was eyeing the Tianjin–Tongzhou railway, proposed by Li Hongzhang but shot down by his rival Zhang Zhidong. Indeed, with this aim in mind, Keswick had approached Zeng Jize, son of Zeng Guofan and the Qing's Ambassador to Britain, France and Russia. However, Keswick would soon find out that Jardine had competition:

> The portable railway plant manufactured and sent by Decauville to this place for trial has been sold to the municipal authorities. It was once offered to H. E. Li Hung-chang at its cost price, which was about Tls. 20,000 but the negotiation failed through the competition of M. Thévenet who engaged to furnish a like plant for the modest sum of Tls. 4000. M. Thévenet's portable railway plant will be sent and used in the Imperial Gardens a few miles from Peking.[9]

As president of the Industrial Mission of France in China, M. Thévenet, understood how to navigate the competitive environment that the *Herald* called the 'war of syndicates' and 'battle of turrets'.[10] His syndicate was operating a successful business, building Lüshun or Port Arthur, and constructing an iron bridge in Tianjin.[11] He promised to make an exact replica of the Decauville at less than a quarter of the cost. The *Herald* stated clearly that the Decauville was bought by Tianjin's 'municipal authorities', which was under Li Hongzhang's administration. But which of the two railways was gifted to Cixi? Peter Crush, a railway enthusiast, believes that it was Thévenet's cost efficient imitation.[12] Beyond the question of which railway eventually became the Empress Dowager's birthday gift, the question of where the 20,000 or 4000 taels came from remains unanswered. Li's philosophy of 'borrowing hens to lay eggs' suggests it is unlikely that the money came from his governor-general's office. Was it paid out of the Admiralty, or one of his reform projects such as the Kaiping coal mine? Was it perhaps a bribe from Jardine–Decauville or Thévenet in return for contracts?

Li Hongzhang managed to turn a railway into a birthday gift for the 'Old Buddha' – he was doing everything possible to persuade Cixi to see the benefits of railways. His rationale was clear: if the Empress Dowager enjoyed riding the train, she might also change her attitude towards this Western technology and endorse his endeavours. Procuring the railway and the train were only the beginning of a complicated plan to fit them into the carefully designed and elaborately decorated environment of the Forbidden City with its adjacent 'three seas' separated by manmade islands. Although no railway engineer, Li Hongzhang personally assumed control and put a great deal of thought into route design.[13] Since Cixi normally received ranking officials and foreign dignitaries in *Ziguang Ge*, or the Purple-light Pavilion, in the South-central Sea, this was an ideal place to showcase her prestige and the Qing's modernisation. The 1.8-mile Xiyuan railway started here, thus it was also called the Purple-light Pavilion [紫光阁] railway.[14] Travelling north along the shore, the line passed the *Fuhua Men*, or Happy-glory Gate, which traversed a path that acted as the boundary between the South and North Seas where the public could catch a glimpse of the moveable tracks being laid and of the passing train (see Figure 3.1).

Construction on the first segment, from the Pavilion to the Gate, began in spring 1888, exactly the same time when the Tangxu railway was extended to Tianjin. Continuing north, the second segment terminated at *Jingqing Zhai*, or the Clear-mirror Hall, at the northern-most shore of the North Sea (Beihai Park today) where Cixi usually took her lunch.[15] Zigzagging along the shores of the two seas provided the longest route and the best views of the lakes and islands with the Forbidden City in the background (see Figure 3.2).

Cixi was so eager for the railway to be completed that she used the best equipped and most skilled Manchu fighting forces to construct the second segment.[16] These 'workers' laboured day and night, finishing the second segment before her birthday. A train station suited to the surrounding royal landscape was built at the northern end for her enjoyment. The imperial train was refurbished to royal standards. Peter Crush believes that it was even fitted with its own electric lights – an unattainable luxury at the time.[17] No expense was spared for Cixi's indulgence at a time when anti-Qing sentiment was gathering steam and foreign powers were escalating their scramble. Weng Tonghe, teacher of and adviser to the Guangxu Emperor, recorded the event in his diary (8 December 1888), where he utilised a common literary practice and identified Li Hongzhang with the capital of his home province, Hefei. Weng wrote 'Hefei presented a six-coach train, today they were presented to the Empress Dowager for her pleasure. The *Ziguang Ge* railway is completed.'[18] The gift train was made up of six coaches: one upper-class, two first-class built for twenty-eight passengers each, two middle-class (also twenty-eight seaters) and one luggage carrier.

Cixi liked everything about the railway and the train except the sound from the engine and its whistle-blow, which created such commotion and noise that it shook the royal ground and threatened to disturb imperial *feng-shui* – the same rationale used to ban drilling in Kaiping and locomotives on the Tangxu. Cixi was not alone in her distaste for the raucous noise; Dimitri

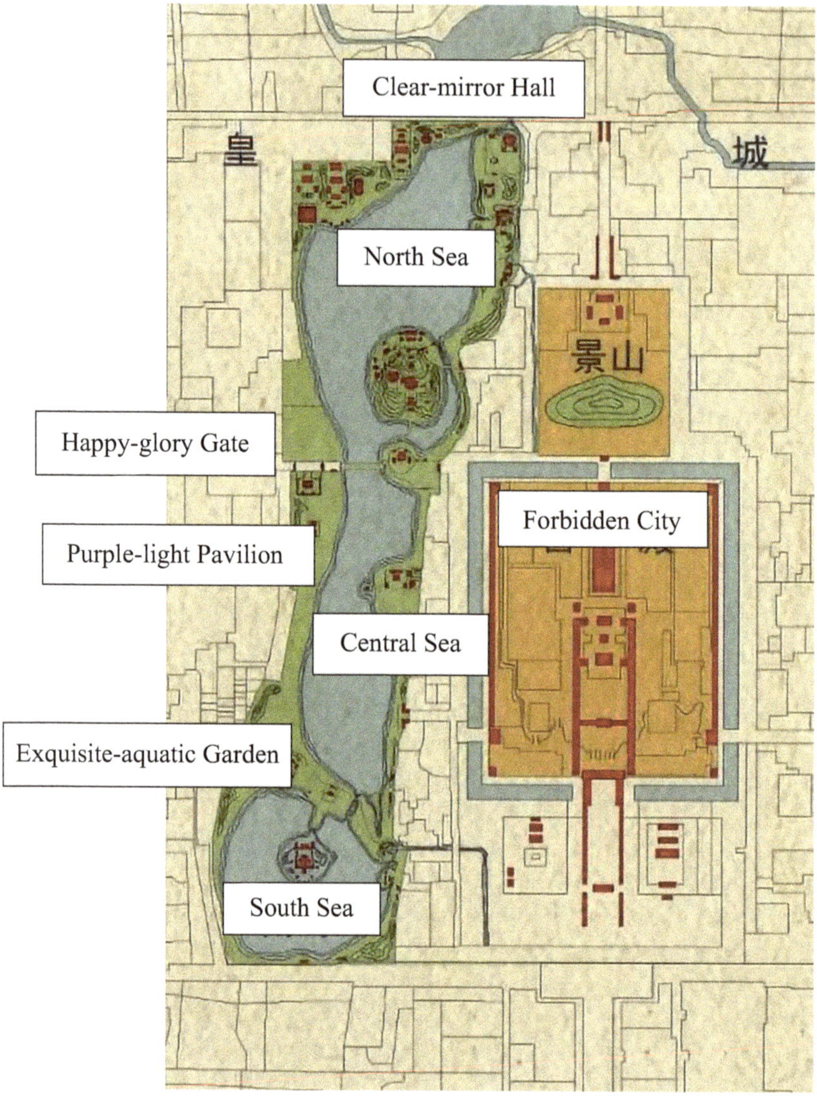

Figure 3.1 Xiyuan route along the 'three seas' (Wei Kaizhao *et al.*, *Beijing mingyuan qvtan* (Beijing: Jianzhu gongye chubanshe, 1994), 175)

Smirnov wrote about the sonic impact of Russian built railways on the peoples of Central Asia.[19] The only difference was that unlike the Central Asians who had to put up with Russian 'sonic occupation', the Empress Dowager had the power and resources to order the engine be replaced by eunuchs, who pulled the train using ropes. Eunuchs had become human

Figure 3.2 Foreign troops next to the Xiyuan railway track in 1901 (Image courtesy of the National Archives WO28-302 (113) and University of Bristol library, HPC, https://hpcbristol.net/visual/NA06-39, Identifier NA06-39)

steam engines![20] This spectacle of wealth and power was heightened by Royal Guards of Honour who preceded her journey with ceremonial pomp.

Cixi enjoyed her train ride so much that the line was extended further to traverse all three seas, crisscrossing south down to *Yingxiu Yuan*, or Exquisite-aquatic Garden, on the strip of land that separates the Central from the South Sea. From 1889 until its destruction by foreign soldiers during the suppression of the Boxer Rebellion in 1900, Cixi's lunch-time routine was not only transformed into a joy ride accompanied by Manchu princes and Chinese officials but also into an extraordinary spectacle for the ordinary people. A ride along this track between work and lunch allowed the Empress Dowager to show off to her guests, especially Westerners. Sarah Pike Conger, the wife of the American Ambassador, wrote about her visit to the Palace and her ride on the railway in December 1898:

> We were taken to another gate inside of which was standing a fine railroad coach presented to China by France. We entered this, and eunuchs dressed in black pushed and hauled it to another stopping place, where we were received by many officials and served with tea. The railroad passed through a beautiful city, clean and imperial.[21]

Mrs Conger's memoir seems to confirm that the railway and train were French-made and powered by eunuchs. Jardine termed this French success a 'lavish but judicious palm greasing', possibly a sign that Thévenet's imitation railway was the one presented to Cixi.[22] Li Hongzhang also took advantage of this 'palm greasing' and used this railway to convert Cixi. An official loyal to Li was appointed to manage the line exclusively.[23] Li ensured his control of the line, and thus his easy access to absolute power.

While Li and foreign merchants went to great lengths to cajole Cixi, her position is a defining example of what Karl A. Wittfogel called 'oriental despotism', as the 'common substance in the various Oriental societies [which] appeared most conspicuously in the despotic strength of their political authority'.[24] Although his work has rightly been criticised for its prejudices and inadequacies, the concepts Wittfogel offered are still useful. The prevalence of regimes ruling with 'despotic strength' has been a common feature of Chinese history. It thrived during the late Qing and continues to thrive in the post-Mao era after more than a century of modernisation and revolution. An important facet to Wittfogel's concept is the term 'hydraulic society', which denotes the 'agromanagerial and agrobureacratic characters' of oriental societies.[25] China was a defining example of agriculture-based economy and society – a hydraulic civilisation indeed. He believed that the absolutist regimes of hydraulic society are often headed by an autocratic individual, usually male, who has 'total power'.[26] The only difference now is that the autocrat was a female.

The toy railway provides a unique window onto late Qing reform. Cixi's 'total power' can be seen in her control of the Tongzhi and Guangxu Emperors, of reform down to individual railways and is most symbolically demonstrated in the eunuchs who powered her train. The totality of her power can also be seen in the endeavours of officials like Li Hongzhang and foreign companies such as Jardine Matheson and the Industrial Mission of France. Knowing the Chinese art of corruption, Jardine placed many agents in key positions to further its interests. One of these agents was Alexander Michie whose 'most significant activit[y] was his attempt, renewed over the course of several years, to gain a favoured place for Jardine, Matheson and Company at the imperial court'.[27] Cixi held the ultimate power, Li made every attempt to cultivate and convince her in his effort to promote his railway agenda. The fate of every railway, from the Wusong and the Tangxu to the Inside and Outside the Barrier, rested in her hands. So, too, did the fate of late Qing reform; efforts to save the Qing dynasty were subject to her political and personal whims. 'Total power' breeds corruption, the malaise that brought down all dynasties before – the Qing were no exception.

Wittfogel argued that 'total power' generated 'total terror' and 'total submission'.[28] It was this power that gave rise to hydraulic politics; the term 'hydraulic' denotes a liquid moving in a confined space under pressure. Chinese officials were similar to moving liquid within the pressurised space of the Qing bureaucracy. Those we have encountered: Li Hongzhang, Shen Baozhen, Liu Mingchuan and Zhang Zhidong are notable in the case of railways. Mavericks of their times, they were all capable, resourceful and accomplished men, but whatever their achievements, they all had to submit to the 'total power' in order to realise their ambitions. If Li was most tenacious, then Liu Mingchuan was the most daring, throwing his weight behind railways after the Wusong fiasco knowing that the regime counted on him to deter Russian advance. Zhang Zhidong, like Shen Baozhen, was competitive; he was also cunning, diverting attention from Li's success. These individuals were a few among hundreds who dreaded but also ingratiated themselves with the autocrat for their own ends. This had further politicised railway development as it was subjected to the needs of different characters and variables.

'The Manchurian bridgehead': genesis[29]

Cixi's appetite for power and glory was gratified by a toy railway, but did this audacious gift change her mind about railways, or help Li Hongzhang to build more of them? In May 1889, the court quietly issued an edict not just permitting, but almost encouraging, railways. The edict stated that they were 'important tools of self-strengthening and part of a larger national strategy. It benefits the country and does not harm the people. It should be a permanent policy and can be undertaken without seeking the court's permission.'[30] This encouragement was reinforced by another more specific edict in March 1890 permitting the construction of what was termed the Inside and Outside the Barrier [关内外] railway, when translated literally.[31] The 'barrier' was the Mountain–Sea Pass (see Figure 2.1). 'Inside' indicated what lay to its south, or north China, whereas 'Outside' denotes Manchuria. Proposed after the success of extending the Tangxu to Tianjin in 1888, this new railway was an ambitious line from Tongzhou (Beijing) to Tianjin, Tangshan and northward to Manchuria. Li's rival Zhang Zhidong had impeded the westward segment to Tongzhou by demanding the Luhan line. Li, therefore, turned his attention northward to Manchuria.[32]

But why did Li push for a railway to connect the capital with Manchuria, and why did the Qing court concede to this particular venture despite the fact that it would only traverse a sparsely populated area with little commerce?[33] It is paramount we understand the rationale because its construction would turn Manchuria into a 'bridgehead', which would come to shape the history and politics of not just railways, but also East Asia for more than six decades to come. The choice of Manchuria points to Li Hongzhang's concern about Russian ambitions for the region. Indeed, his suspicion regarding Russian aggression was well founded.[34] The Russian empire had primarily been centred around its European lands until the end of the sixteenth century when it expanded eastwards into Siberia.[35] Their southward march towards China was thwarted by the Kangxi Emperor and led to the Treaty of Nerchinsk in 1689 which clarified the border between the two empires.[36] Russia took the opportunity of the Qing's preoccupation with the Taiping Rebellion and Second Opium War to annex more than 900,000 km^2 of land north of the Amur river in 1858 and east of the Ussuri river in 1860 (see Figure 3.3).[37] They returned during the chaos of the Hui or Chinese Muslim Rebellion (1862–73), taking roughly 430,000 km^2 of land from Lake Balkhash to the Ili region in the Tianshan Mountain valley.[38]

As early as 1874 *Shen Bao* reported that 'Russia will build railways to reach the Chinese border'.[39] Li and many officials were aware of the plans for the Trans-Siberia railway, a development that would undoubtedly make Manchuria more vulnerable to Russian incursion.[40] Liu Mingchuan, who came out as an early supporter of railways in 1880, began his memorial with a warning about Russia:

> Russia is building a railway from Europe and is approaching the Khanate of Kokand. It is also constructing from Vladivostok to Hunchun. They have not yet sent in troops, not because they fear us but because the railways have not been completed. Disaster will strike within ten years.[41]

Liu was right; Russian railways were built beyond Kokand to Andijian (Uzbekistan), close to the Qing border.[42] His prediction of a Russian return would materialise in the 1890s. Russia would use the Trans-Siberian railway to hem Qing China in from the west on the border with Xinjiang, in the north via Mongolia and in the east through Manchuria. Li Hongzhang believed that the Russians would strike Manchuria again. He also believed that any Russian gains would serve to whet the Japanese appetite – hindsight proves that Li was correct. Despite a futile attempt to take Taiwan in 1874, Japan made a similar attempt on Korea in 1884 while China was preoccupied with the Sino-French War. Japanese designs for Korea also put Manchuria at risk, as it was sparsely populated and unfortified. Li's concerns were legitimate. The strategic importance of Manchuria meant that it would play the most important role in the history and politics of railways. The threat was profoundly felt, and in 1889 *Shen Bao* kicked off a lengthy national debate on the importance of railways for Manchuria.[43] They made the public aware of the danger and elaborated on the necessity of building railways in order to combat the threat of the Trans-Siberian line which would encircle China from three directions.

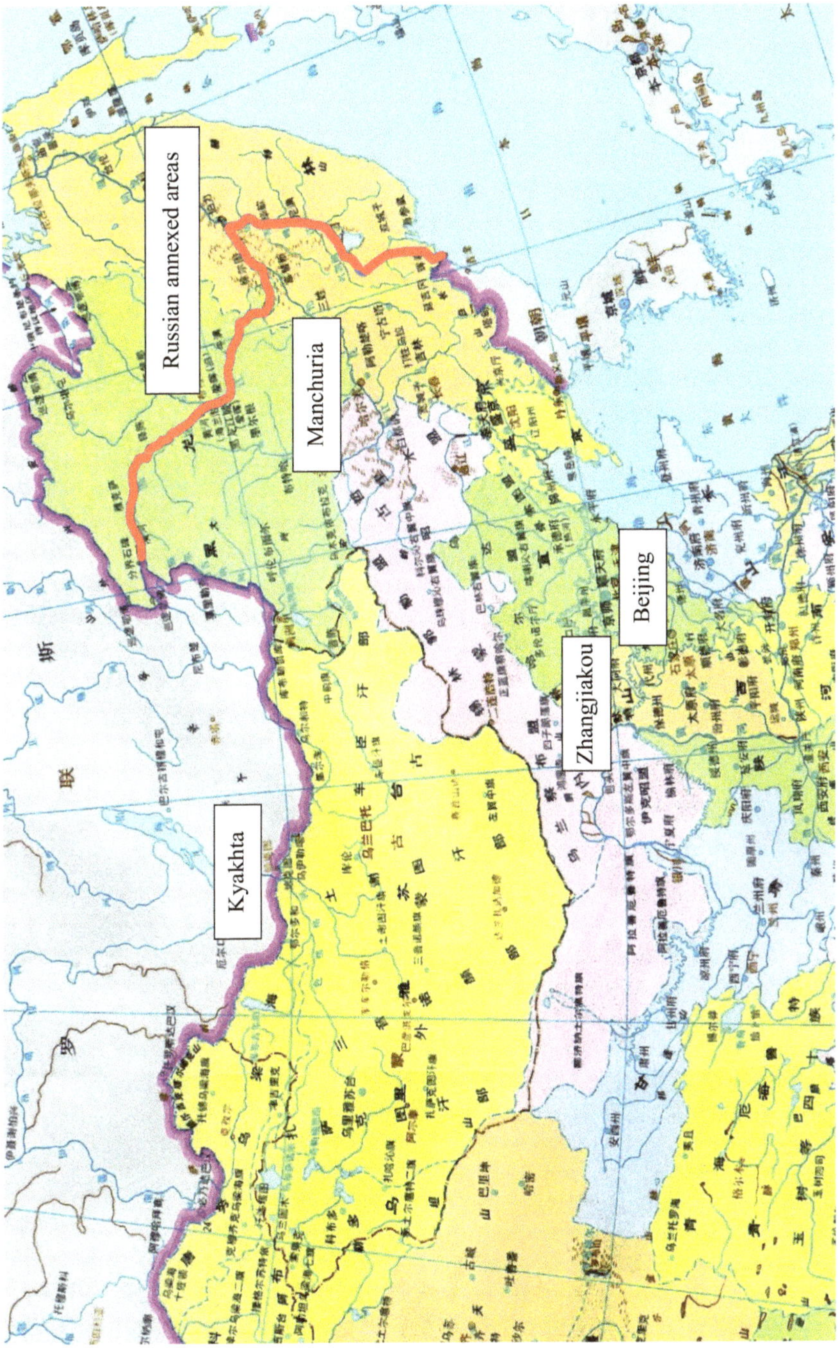

Figure 3.3 Map of Manchuria with Russian annexed area (Baidu public domain)

Beyond strategic and military considerations, Manchuria was also the ancestral land of the Qing imperial family. The Manchu court and elite took its security seriously. A region of 787,300 km^2, Manchuria consists of three large regions: Heilongjiang at the north, Jilin in the centre and Fengtian/Liaoning in the south. Home to nomadic groups like the Manchus, Daur, Nanai, Sibe and Oroqen, Manchuria had also been home to many Chinese fugitives and migrants. Before the Russian annexation in 1858, the Qing administrated a much larger area, including the Outer Khingan Range. This area reached as far north as today's Nikolayevsk-on-Amur, which was called Temple Street by the Chinese/Manchus, and Vladivostok, which is still referred to as Sea Cucumber Bay. After their conquest of the Ming dynasty in 1644, the Manchu elite moved to China, while their homeland became their summer holiday and hunting grounds.[44] Initially, Chinese migration to this region was low and tolerated. But Chinese competition threatened the livelihood of the Manchus, and the Kangxi Emperor discontinued this policy in 1668.[45] As disasters devastated north China from the 1800s, more Chinese escaped to Manchuria.[46] To combat the danger of further Russian annexation, Qing officials asked the court to lift the ban on Chinese migration in order to populate the region. Known as 'moving people to fill the frontier', this policy had been used by earlier dynasties to buttress frontier regions. The ban was relaxed, and thousands crushed into Guandong or East of the Barrier, turning Manchuria from 'space' to 'place' as Mark Elliot has fittingly termed this process.[47]

Li Hongzhang achieved at least two aims with the Inside and Outside the Barrier railway (the Barrier hereafter). He had checked Russian ambition for Manchuria by taking the defence to the border region and had used the region's importance to the Qing court to ensure success of his railway ambition. He could not have chosen a better occasion to propose the Barrier line. In 1889, the young Guangxu Emperor came of age, and it was now time for him to rule on his own. Perhaps this timing explains why he was willing to use extravagant gifts to appease Cixi; he may have wished she would gladly retreat into retirement. Such an outcome would allow for a more aggressive modernisation agenda without her interference, as Guangxu was known to be reform-minded.[48] But this turned out to be wishful thinking as the real power still rested in her hands. The Old Buddha never retired. Guangxu would remain a puppet for whom she pulled the strings. Her lady-in-waiting witnessed it first-hand:

> Kwang Hsu was Emperor, seemingly with the power of life and death over his subjects, with the right to make laws or break them, to interview ministers and foreign diplomats; yet he never decided a single important question without seeking the advice of Old Buddha. He went to the Summer Palace – in those days it was a long, trying journey by sedan chair – sometimes three times a

> week; so that Old Buddha, seemingly taking no part in the affairs of the court, directed the destinies of China just the same. She had once directed the destiny of China through the lips of Tung Chih, from behind the bamboo screen in the rear of her son's throne. She now directed affairs from the Summer Palace, through the lips of Kwang Hsu.[49]

It seemed that Li had prepared for this situation, which explains why, of the six coaches of the Xiyuan railway, five were gifted to Cixi, one to Prince Chun, but none to the Guangxu Emperor. He knew who held the ultimate power and whose favour he had to seek. The Barrier railway was a substantial victory for Li. This once again exposes the problem with late Qing reform and railway development – it was disorganised, sporadic and personal. The Barrier line was known by other names: the Imperial Railway of North China, the Within and Without the Great Wall railway, the Jingfeng [京奉] or Beining [北宁] railroad.[50] From Tangshan, it would journey north through the Mountain–Sea Pass and snake to the outskirts of Fengtian, the old capital of the Qing empire in the heart of southern Manchuria.[51] However, this was not its final destination. The line would stretch further north to Kirin, an important city in middle Manchuria and terminate at Hunchun at the Chinese–Russian–Korean border where the Tumen River flows into the Sea of Japan, a most strategic and sensitive place.[52] A shorter line would branch off from the Mountain–Sea Barrier to Yingkou, then called Newchwang, the port of southern Manchuria.

Li might have secured the commission of the Barrier railway, but the regime he had pinned his hopes on had neither the money nor the expertise to see his projects through. Li was forced to rely on the success of his other reform projects and on profits from the Jintang railway. The chief engineer of the Kaiping Railway Company, Claude Kinder, led the route survey in the spring of 1890.[53] He was joined by a Chinese assistant, Zhan Tianyou, one of the first group of young boys sent to study in the United States in 1872. After graduating from Yale University in 1881, he returned to join the Kaiping coal mine where he laid the tracks of the Jintang railway. Zhan gained his experience from his work with Kinder and would go on to build many more railways on his own. By July 1890, the surveying party had travelled to Yingkou, Kirin, Ningguta and Hunchun. Construction began in May 1891 and was well under way by December 1892.[54] Tracks for the Inside segment, the section between Tangshan and the Barrier, were laid by May 1893, while work on the Outside segment towards Fengtian commenced.

By July 1894, on the eve of the Sino-Japanese War, the Inside segment became operational while the tracks of the Outside segment reached Suizhong county, about 40 miles beyond the Barrier.[55] Though incomplete, the railway was used to quickly move soldiers to Manchuria in August 1894 and later to transport the wounded from the front.[56] Although the rationale Li set out in his memorial on the benefits of railways was justified, the line was, nevertheless, vulnerable to invaders. The Japanese soon captured the track and took possession of the railway's equipment and materials; this amounted to 'six thousand tons of rails' according to Kent.[57] Although the Qing were ardent to resume construction after the war in 1895, they were severely hindered by the indemnities forced upon them by Japan. For the work to continue, the Qing had to seek credit from the British and Chinese Corporation, represented by the Hong Kong Shanghai Banking Corporation (HSBC) and Jardine Matheson – finally providing the reward Keswick had hoped for.[58] But this arrangement would provoke the Russians who considered Manchuria its exclusive sphere of influence.

Kinder's survey, which had penetrated deep into Manchuria, had not escaped the Russians. The rumoured British loan also led to renewed Anglo-Russo animosity, as Russia protested to the British and to the Qing.[59] In his work on the Foreign Office, Nathan Pelcovits argued that Russian 'claims' over Manchuria prompted the British to revise its foreign policy – they would seek a 'Yangzi Protectorate', claiming the expansive Yangzi River valley as its exclusive sphere.[60] This diplomatic row would lead to years of Anglo-Russian wrangling, resulting in an exchange of notes in April 1899. By October 1899, the Barrier line passed Jinzhou, only 20 miles from its destination of Xinmintuen, on the outskirt of Fengtian.[61] But construction was interrupted for a second time by the suppression of the Boxer Rebellion in 1900, during which Russian troops seized Yingkou station and building material, and captured chief engineer Kinder, instigating yet another diplomatic crisis.[62] Construction would be interrupted for a third time by the Russo-Japanese War. The Barrier railway was finally completed at the end of 1911, ten years after Li Hongzhang's death and two months after the fall of the Qing. The Barrier railway exposes not just the Qing's effort at railway development but also the origin of the scramble for Manchuria that will shape history and politics of China for decades to come.

Although he laboured persistently for the Barrier railway, Li had not forgotten the Tongzhou or Beijing segment that had been stifled by his rival, Zhang Zhidong. The Sino-Japanese War had threatened national security, but it had strengthened his hand. To revive the Tongzhou line, Li reached out to Zhang by proposing a line from Tianjin to Lugouqiao in Beijing. This would link up with Zhang's Luhan line, the most salient vertical line that travelled from the capital to south central China. This was a win-win situation. The two rivals and their competing projects were ultimately connected by a railway.[63] The compromise can even be seen in the name of the new line, Tianjin to Lugouqiao, or the Jinlu [津芦], completed in 1897. Again, we see Li's pragmatism, his capacity to compromise in order to find a solution and his efforts at producing results – willing to do anything for his railway projects. But again, we see the characteristics of railway development, being subjected to the endeavours and calculations of different officials who rivalled each other for prestige and power.

The new line was not the extent of Li's ambitions. He began to invest in railway education. Headlines such as 'Railway Harbinger' and 'On the Establishment of Railway Academy by Beiyang' heralded the founding of the Imperial Chinese Railway College, which aimed to educate China's own engineers, technicians and conductors.[64] Headed by Wu Tiaoqing, a comprador who had worked for the HSBC, the school recruited about forty students in 1897.[65] The offer included food, lodging, a monthly stipend of four taels of silver, which increased every year, and a guaranteed job after graduation.[66] As China's first institution devoted to railway education, it has continued to produce experts who now lead cutting-edge research in high-speed rail.[67] Thanks to Li's audacity and perseverance, the humble Tangxu line grew into the Jintang, the Barrier and the Jinlu railways – built and controlled by the Chinese themselves. It seems that railway development was slowly taking off, and Tianjin became the hub of the first rail network. Other developments soon followed. The All-China Railway Company was founded in 1897 with Sheng Xuanhuai as its director. Finally, the Wusong was restored in 1898, a year in which the railway's prospects looked promising as thirty-six locomotives, twelve from Rogers and twenty-four from Baldwin, were sold to China.[68]

The 'Old Buddha'

It seems that Li Hongzhang's ingenious gift to the 'Old Buddha' had changed the fate of railways, but it also marked the beginning of her misappropriation in the final decade of her reign. Her exploitation of the railway is noteworthy because it helps us chart the political biography of railways. Her toy railway was torn up by foreign troops during the suppression of the Boxers. The rebellion broke out in Shandong province in 1899. As the rebels converged upon Beijing in the summer of 1900, the Eight-Country Allied Army gathered on the Bohai Sea outside of Tianjin.[69] They travelled by train first from the port of Dagu to Tianjin, then on the Jinlu line, from Tianjin to Beijing. Clive Bigham experienced this first hand and detailed them in his memoir.[70] The fear of many anti-railway proponents who predicted that foreigners would use railways to move troops during times of war had materialised. The Boxers dismantled railways in order to prevent foreign troops from advancing.[71] This may have slowed their advance, but it did not stop them. Cixi escaped Beijing on 14 August 1900, just before the Allied Army arrived. Disguised as a peasant woman, she fled as fast as her horse-powered carriage could handle. Had she regretted not having a railway at her disposal?

The Boxer Protocol was signed on 7 September 1901 by none other than Li Hongzhang. The humiliating treaty carried with it a staggering indemnity of £450 million taels of silver – £980 million with interest.[72] In October, Cixi began her long journey home with an entourage of 3,000 chariots strong after a sixteen-month exile in northwest China. Cixi left Beijing in haste without any of her belongings; the 3,000 wagons of 'gifts' were extortions from local officials in the provinces she had passed through. Her return journey began in Xi'an, and she traversed Shaanxi and Henan provinces to reach Zhengding in Hebei province three months later. Her arrival in Zhengding coincided with the Luhan railway reaching as far as the city in January 1902. Cixi decided to take the train back to Beijing and accelerate her return to power.[73] Meanwhile, Li Hongzhang had passed away in November 1901 and was succeeded by Yuan Shikai as both governor-general of Zhili and Beiyang commissioner. It was now Yuan's turn to cajole and flatter the Empress Dowager.[74] Yuan prepared a royal train, driven by a Belgian conductor, and accompanied her back to the capital.[75] The last leg of her railway journey from Baoding to the capital (Majiabu station) lasted only a few hours.[76] Her 3,000 chariots of gifts filled twenty-one carriages.

Cixi returned to the palace to find that her toy railway had been destroyed. Her initial support for and later repudiation of the Boxers had brought disaster not just upon herself but also the country at large. The 980 million-tael indemnities were a sum that could not possibly be repaid. In a rare publicity stunt, Cixi issued a decree admitting her mistakes and travelled to repent at the Eastern Tombs, one of two imperial burial sites, in April 1902. She announced that she would visit the Western Tombs in April 1903 by train. This statement acknowledged that the railway was a faster and more comfortable means of travel than a slow horse carriage ride on bumpy roads. Li Hongzhang knew how to please the Old Buddha, Yuan Shikai was even more adept in pleasing the ageing Cixi. The only difference between them was that Li was a Confucian, loyal to the monarchy, who largely looked after the interests of the country, whereas Yuan was an opportunist who would usurp the throne a few years after Cixi's death.

The Western Tombs [西陵] line would travel from Gaobeidian, a station on the Luhan line, to Yixian, hence the name Gaoyi [高易] railway.[77] Yuan Shikai tasked the all-capable Sheng Xuanhuai with the job of building the 22-mile branch line. To do so, he tapped into the profits made by the lines built by Li Hongzhang: the Jintang, the Jinlu and the Barrier railways. Zhan Tianyou, China's first railway engineer, designed and supervised the Gaoyi's construction as thousands toiled to complete it in time for Cixi's excursion to the royal tombs in the spring of 1903.[78] Despite all the money and effort that were poured into the construction of the Western Tombs line, Cixi only used it once every two years, as part of her biennial visits to the Eastern Tombs. This meant that the Gaoyi only operated once in 1903, twice at the most in 1905 before her death in 1908.[79] Preparations for her journey were extensive and often began months in advance. Everything had to halt for her train to pass.[80] There was also talk about a direct line to the Eastern Tombs. Those noisy drilling and engines that disturbed the peace of Manchu ancestors – the argument used against the development of the Kaiping coal mine and Tangxu railway in the early 1880s – was now apparently forgotten.[81] Hardly had the Western Tombs line finished and the Eastern Tombs [东陵] project began, when yet another railway was proposed for Cixi's exclusive convenience and pleasure:

> The Peking-Eho Park railway, recently suggested by Prince Ching and sanctioned by the Empress Dowager, is to be commenced during the middle of next month (March 1903) and is expected to be completed and ready for traffic next November, in time for the Empress Dowager's seventieth birthday anniversary. The money for the construction for this short railway – ten miles – is to be defrayed by 'voluntary' contribution by Princes, Dukes, Nobles, and high ministers at court.[82]

The *Eho* or *Yihe* Gardens were part of the Qing pleasure complex in the northwest of Beijing which Westerners called the 'Summer Palace'. Damaged during the Second Opium War, it had been restored and renovated for Cixi's retirement and delight. Although work had started, the Yihe Park [颐和园] railway was never completed. The 'Old Buddha' was soon distracted by other exotic foreign things; she was converted by the novelty of Western inventions, and her appetite was growing. Her decision to travel only by railway was one indication of this, as was her acceptance of two new ladies-in-waiting in 1903. Yu Der Ling and Yu Rong Ling were sisters who had lived in Japan and France with their father who was a Qing diplomat. Cixi's penchant for luxury and spectacle provided a 'ladder of success' for upstarts like Yuan Shikai and for the two well-travelled Yu sisters.[83] They exposed her to various 'curiosities of European civilisation', including expensive and status-denoting hobbies like piano, ballet and cigarette smoking.[84] Cixi developed a keen fascination with photography, and she had her own cars. She even commissioned a portrait from American painter Katherine A. Carl for display at the St Louis Exposition. Carl spent a few months with Cixi in 1903 and remembered her toy railway:

> Two other modern and novel methods of locomotion had been installed in the grounds of the Sea Palace. There was a small railway, which ran from the

> outer gates to the dwelling Palaces, which had its engine and complete running outfit. This had been constructed by some progressive Mandarins, who wished to get the Empress Dowager's support for some railway scheme, but though she often spoke of how much she had enjoyed her one trip on a real railway, her spirit was too utilitarian to care for toy pleasures. She couldn't stand the puffing of the engine, the tiny cars, and all this trouble for so short and useless a jaunt. There was also in the Sea Palace, as well as at the Summer Palace, a number of automobiles, which had been presented to Their Majesties by Chinese nobles and officials who had been abroad, as examples of the curiosities of European civilization.[85]

Li Hongzhang and M. Thévenet had set the standard; now everyone knew how to pamper the aging Cixi for their own ends. Carl's account confirms that Cixi's toy railway, destroyed during the Boxer suppression, had been quickly restored. Carl also revealed that the Summer Palace was the 'only one in China lighted by electricity'.[86] Once Carl finished the portrait, it was transported to Shanghai for the voyage to the United States:

> A special railway had been laid from the Wai-Wu-Pu to the station outside the Chien-Men, for it was not considered fitting that ordinary bearers transport the picture of Her Majesty. The officials of the Wai-Wu-Pu, as well as many other of the high Officials in Peking, dressed in full dress, accompanied it to the station, and stood to watch the Sacred Picture start off on its long journey to St. Louis. The special train carrying it was met at Tientsin by the Viceroy of the Province, surrounded by all his official staff. It was there placed with great ceremony upon the steamer on which it was to make the journey to Shanghai, and was accompanied from Peking to Shanghai by an official specially appointed for the purpose.[87]

A line built solely for the purpose of transporting her portrait – Cixi now sought to exploit railways in every way possible. Wai-Wu-Pu, or the Ministry of Foreign Affairs, sat in the eastern district of Beijing while the Qianmen station was located in the southern part, and therefore special tracks had to be laid for the portrait train to travel. American missionary Isaac T. Headland furnishes us with an image of officials and a few foreigners, possibly railway and train staff, standing beside the special train before its departure (see Figure 3.4).[88]

Qianmen or 'Front Gate' was not just the entrance to the Forbidden City but also a station on the Barrier line that Li Hongzhang had fought so hard to create. The line had been extended to the front door of the royal palace – all for the convenience of the imperial family. Sarah Conger addressed the construction of this station in her memoir: 'the British are bringing the railroad through the city wall up to the front entrance of the Temple of Heaven'.[89] This also confirms British involvement. Once she discovered the power and prestige that railways could grant, Cixi made sure she enjoyed

Figure 3.4 The Empress Dowager's portrait railway and train (Image courtesy of the Maxwell family, Cadbury research library, Classmark DA26/2/2/3 University of Birmingham and University of Bristol library, HPC, https://hpcbristol.net/visual/Mx04-070, Identifier Mx04-070)

them as much as possible. She travelled to work and to lunch by train, she journeyed to the Western Tombs by train, a new line for the Eastern Tombs and the Summer Palace were planned, and even her portrait had to be carried by a special train. Railways were now at the disposal of a despotic ruler and built at the expense of reform. Needless to say, this kind of extravagance came at a cost to the Qing and ultimately its survival.[90]

Cixi diverted funds to indulge herself, and the more she diverted, the less was available to reform. She tapped into the Sea Defence Fund that had been established for naval development which Li Hongzhang and Shen Baozhen had fought so ferociously over to purchase foreign gunboats.[91] She took charge of the Admiralty's budget set aside to build a national navy supported by railways to renovate the Summer Palace with the excuse that the Kunming Lake could be used as a training ground for the navy.[92] We know from Weng Tonghe's diary that drills were conducted on the lake on 22 April 1889.[93] A large pond within the grounds of the Qing's pleasure palace doubled as a naval base – no wonder the Qing would lose to the Japanese in the First Sino-Japanese War. Cixi squeezed reform projects such as the Barrier railway for the Western Tombs line, taking £100,000 on one occasion alone.[94] She also intimidated Manchu nobles and Chinese officials into contributing to the Summer Palace line. She would continue to starve reform to pay for her luxuries until her last breath in 1908.

Oblivious to the suffering of the ordinary people and to the eminent threats from foreign powers, she never spared herself any pleasure at a time when anti-Manchu revolutionary parties were proliferating and efforts to overthrow the Qing intensified. Despite the lack of academic writings focusing on Cixi, there is no shortage of criticism of her rule in the existing scholarship. Historians are inclined to accentuate her role when discussing the ultimate failure of the late Qing reform.[95] Mary Wright labelled her an 'autocrat'.[96] Sue Fawn Chung attempted to salvage her, arguing that she 'favoured moderate reform'.[97] Isaac T. Headland called her a reactionist, even as he presents her as a reformer.[98] Cixi's reign, when looked at through the history and politics of railways, can be seen in a new light – she was neither a conservative nor a reformer.

Although she was instrumental in launching reform in 1862 when she began her first 'reign from behind the curtain', this was done not because she believed in reform but because she needed allies like the reformist Prince Gong, who commanded respect among the Manchu elite and Chinese officials at a time when she was a novice. While she allowed reform to continue, it would be misleading to label her a reformer. She used reform to divide and rule as she played conservatives against reformers and reformers against each other. Reform gratified her vanity as her thirst for pleasure diverted much-needed funds away from the development of railway and the navy. Staying in power was her ultimate aim and reform was merely a means to that end, not the end itself. She subjected reform to her political desires and pomposity, and she placed her own interest above that of the country. Her egoism wrecked the very empire her reforms were meant to save. The history and politics of railways helps us pinpoint the problems with not just Cixi but also how exactly they led to the collapse of the Qing in 1911, three years after she died.

A railway, like time, has no inherent meaning; we ourselves give meaning to them. Throughout Chapters 1, 2 and 3, we have traced the political genealogy of the railway. From the very beginning, the railway became politicised when its introduction was proposed by the Qing's foreign nemesis, who were pursuing their own interests. For the Qing, it seemed railways stood on the wrong side of history. Many late Qing officials did not seek to reject this technology, but rather to defend the Qing's autonomy, the environment, existing economic life and social stability. Indeed, hindsight would lend support to their arguments. Their fears would materialise with the arrival of railway imperialism. To further complicate matters, the rivalries among officials also politicised railways, as seen in the case of Li

Hongzhang against Shen Baozhen and then Zhang Zhidong. Such rivalries would continue. Furthermore, as this chapter has demonstrated, railway development was subject to the needs and whims of a despotic ruler. The politics of the railways contained dimensions that operated on an international, court, regional and personal levels. The 'personal' mattered a great deal because it meant that politics of railways was subjected to the needs and preferences of the individuals involved. All the lines built so far: the Tangxu, the Jintang, the Barrier, the Jinlu, the Xiyuan and the Western Tombs, all rested in the hands of individuals. This 'personal' aspect exposed the shambolic nature of railway development and late Qing reform, and this would continue in the making of every railway as we shall see in the chapters to come.

This political genealogy of railways opens a unique window to the history of late Qing China. Defeat and humiliation of the two opium wars was not enough to convince the Qing to undertake railway construction as a defence against invaders. The inability to battle the Taiping saw only the development of a few arsenals that were merely improvised to meet the demand of the task at hand. Japanese landing in Taiwan in 1874 led to the creation of the Beiyang and Nanyang Commissioners. In a similar vein, French aggression in the mid-1880s led to the establishment of a Naval Department. The Qing court only knew how to react to emergencies that threatened national security; it had no vision and could not develop the means to protect the country. The few railways they constructed were little more than fragments, and when she was finally persuaded as to their advantage, Cixi was only interested in how they could benefit her personally. Although the Qing court and officials had managed to keep foreigners at bay, they found themselves in even more trouble from the mid-1890s as the scramble for China intensified, ushering in a period of rapid Qing decline and railway imperialism.

Notes

1 Karl A. Wittfogel, *Oriental despotism: a comparative study of total power* (New Haven: Yale University Press, 1957).

2 Gao Zhizhong, *Ming Qing gongci yu gongting wenhua yanjiu* (Beijing: Fangzhi chubanshe, 2014), 172; Wei Kaizhao *et al.*, *Beijing mingyuan qvtan* (Beijing: Jianzhu gongye chubanshe, 1994), 177; Wu Qian, 'Cixi zhuanyong de 'youwang' tielu', *Wenshi Bolan*, vol 21, 5 (November 2013): 21–22.

3 Yang Naiji, 'Taiyechi biande xiaotielu', *Zijincheng*, vol 3 (1982): 6–8, 15 and 'Taiyechi biande xiaotielu buyi', *Zijincheng*, vol 2 (1983): 47; Xin Hao, 'Buxu taiyechi biande xiaotielu', *Zijincheng*, vol 0 (1985): 20–21.

4 Wright, *Last stand.*

5 Wei *et al.*, *Mingyuan*, 28–30; Emily Mokros, 'Reconstructing the imperial retreat: politics, communications, and the Yuanming Yuan under the Tongzhi emperor 1873–4', *Late Imperial China*, vol 33, 2 (2012): 76–118.
6 Dennett, *Americans*, 598.
7 'Tientsin', *Herald*, 8 December 1886: 613.
8 Keswick to Gaston Galy, dated May 12, 1887, PLB, Jardine Matheson Archive.
9 'Article 1', *Herald*, 5 October 1887: 372 and 'Outports: French contracts in the north', 13 April 1887: 409.
10 'Correspondence: a code of professional honour', *Herald*, 11 June 1886: 623.
11 'Tientsin', *Herald*, 6 January 1888: 12.
12 Crush, 'Narrow gauge railways part 2'.
13 The First Historical Archive in Beijing holds two route maps: Central Sea railway construction drawing [中海铺修铁路图样] and North Sea to Central Sea railway construction drawing [北海至中海铺修铁路图样].
14 Wei *et al.*, *Mingyuan*, 175.
15 Liu Yan, 'Yuyuan lide Xiaohuoche', *Zijincheng*, vol 133 (1 November 2005): 114–17.
16 They were the two battalions stationed at West Hill and Fengtai [西山锐健营, 丰台大营].
17 Crush, 'Narrow gauge railways part 2'.
18 Weng Tonghe, *Weng Tonghe riji* (8 vols, Shanghai: Zhongxi shuju, 2012), vol 5, 2277.
19 Dimitri Smirnov, 'The sonic occupation of Central Asia', in *Sonic histories of occupation: experiencing sound and empire in a global context*, eds Russell P. Skelchy and Jeremy E. Taylor (London: Bloomsbury Academic, 2022), 173–94.
20 Melissa S. Dale, *Inside the world of the eunuch: a social history of the emperor's servants in Qing China* (Hong Kong: Hong Kong University Press, 2018); Norman Alan Kutcher, *Eunuch and emperor in the great age of Qing rule* (Oakland: University of California Press, 2018).
21 Sarah Pike Conger, *Letter from China with particular reference to the Empress Dowager and the women of China* (Chicago: A. C. McClurg, 1910), 40.
22 Tientsin to J. Keswick, 21 October 1887, UC, Jardine Matheson Archive.
23 'Jingbao quanlu', *Shen Bao*, 5 April 1900: 14.
24 Wittfogel, *Despotism*, 1.
25 Wittfogel, *Despotism*, 3.
26 Wittfogel, *Despotism*, 101–3 and 106–7.
27 Le Fevour, *Western enterprise*, 94.
28 Le Fevour, *Western enterprise*, 137.
29 Rosenbaum, 'Manchuria Bridgehead'.
30 *Dezong Jinghuangdi shilu* (597 vols, Beijing: Zhonghua shuju, 1986), vol 269, 599; 'Chuanyan guzhi', *Shen Bao*, 18 November 1888: 2 and 'Shu zuobao qingli yangzai hou', 16 June 1890: 1.
31 *Dezong shilu*, vol 282, 757; Qinhuangdao, *Fazhanshi*, 17–18.
32 Rosenbaum, 'Railway enterprise'.

33 'Tielu binggao zhaodeng', *Shen Bao*, 29 October 1888: 1.
34 Wang Ying, *Li Hongzhang yu wanqing zhongwai tiaoyuede yanjiu* (Changsha: Hunan renmin chubanshe, 2011), 207–9.
35 'Jilin biannianshi: Shunzhi shinian (1653)', Zhishuo, Jilin difangzhi, at http://dfz.jl.gov.cn/zsjl/201709/t20170908_5216398.html, accessed 7 October 2021.
36 Jean-Baptiste Du Halde, *Description géographique, historique, chronologique, politique and physique de l'empire de la Chine et Tartarie chinoise* (4 tomes, La Haye: H. Scheurleer, 1736), tome 2, 242–44; Ernst G. Ravenstein, *The Russians on the Amur: its discovery, conquest and colonization* (London: Trubner, 1861), 54–64.
37 'Treaty of peace, friendship, commerce and navigation between Russia and China', in Mayers, *Treaties*, 101–4 and 105–12.
38 'Treaty for the reestablishment of Chinese authority in the country of Ili' in *Catalogue of Treaties 1814 to 1918* (Washington: Government Printing Office, 1919), 105.
39 'E'guo jiangkai huochelu', *Shen Bao*, 28 December 1874: 1, 'Lun xichui xingshi', 12 February 1875: 4 and 'Ji E'bing tuenbian shu', 30 April 1875: 2.
40 Wang Ying, *Li Hongzhang*, 207–8.
41 Liu, *Wenji*, 44.
42 Christopher M. Wyatt, *Afghanistan and the defence of empire: diplomacy and strategy during the great game* (London: I. B. Tauris, 2011), 9–11 and map 2.
43 'Zhang Xiangtao zhijun weichen tielu zougao', *Shen Bao*, 11 August 1889: 2, 'Zongli Yamen zoufu zunchou tielu zhegao', 29 September 1889: 1, and 'Lingkai tielu', 1 May 1890: 2, 'Tielu shuwen', 7 September 1890: 2 and 'Xiboliya tielu daoli jingfei', 17 April 1892: 1; 'Strategic railway lines', *Times*, 14 December 1890: 6.
44 'Lishi renkou: Qingdai qianqi renkou', Shuzi fangzhi, Shengzhi, Renkouzhi, Heilongjiang shizhiwang, at www.hljszw.org.cn/bookinfo/201708230056.html, accessed 4 October 2021.
45 'Dashiji: Qianlong wunian (1740)', Liaoningsheng dang'anguan, at www.lndangan.gov.cn/lnsdaj/index.html, accessed 4 October 2021.
46 Thomas Gottschang and Diana Lary, *Swallows and settlers: the great migration from north China to Manchuria* (Ann Arbor: Centre for Chinese studies, University of Michigan, 2000); James Reardon-Anderson, *Reluctant pioneers: China's expansion northward 1644–1937* (Stanford: Stanford University Press, 2005).
47 Mark Elliot, 'The limits of Tartary: Manchuria in imperial and national geographies', *The Journal of Asian Studies*, vol 59, 3 (August 2000): 603–46.
48 Daniel Barish, *Learning to rule: court education and the remaking of the Qing state 1861–1912* (New York: Columbia University Press, 2022), 83–107.
49 Der Ling, *Old Buddha* (New York: Dodd, Mead, 1928), 186–87.
50 Zhang Jingyu, *Qingmo dongbei tielu kaifang wentizhi yanjiu* (Taibei: Zhengzhi daxue waijiao yanjiusuo, 1961); Cui Zhihai, 'Lun Qingmo tielu zhengcede yanbian', *Jindaishi Yanjiu*, 3 (1993): 62–86.
51 'The imperial Chinese railway of north China', *Herald*, 16 October 1899: 756.

52 Nianshen Song, *Making borders in modern East Asia: the Tumen river demarcation 1881–1919* (Cambridge: Cambridge University Press, 2019).
53 'The Imperial Chinese Railway of North China', *Herald*, 16 October 1899: 756.
54 'The Imperial Chinese Railway', *Herald*, 16 December 1892: 900.
55 Kent, *Enterprise*, 60–75.
56 'Diaobing fufang', *Shen Bao*, 22 August 1894: 2, 'Beidi junrong', 29 November 1894: 2, 'Xiangyong ditong', 14 January 1895: 2 and 'Junshi ming'en', 6 September 1895: 2.
57 Kent, *Enterprise*, 42.
58 'Shanhaikuan to Newchwang railway loan agreement' and 'Preliminary loan agreement for Peking-Newchwang railway', in John Van Antwerp MacMurray, *Treaties and agreements with and concerning China 1894–1919* (New York: Oxford University Press, 1921), vol 1 (1894–1911), 173–81.
59 'Russia and England', *Herald*, 4 July 1898: 36.
60 Pelcovits, *China hands*, 220–31.
61 Conger, *Letters*, 64; 'Huoche tuiguang', *Shen Bao*, 4 October 1899: 2; Kent, *Enterprise*, 57.
62 'The Russians at Newchwang', *Herald*, 29 November 1900: 1134 and 'The recent history of the imperial railway of north China', 9 October 1901: 683.
63 Zhang Zhidong, *Quanji*, vol 2, 267–69.
64 'Tianjin zashu', *Shen Bao*, 6 December 1890: 2, 'Tielu xiansheng', 10 October 1896: 1 and 'Lun Beiyang chuangshe tieluxuetang', 27 November 1896: 1.
65 'Xuesheng daotang', *Shen Bao*, 6 December 1897: 2.
66 'Beiyang tieluxuetang zhaokai xuesheng', *Shen Bao*, 17 April 1898: 5.
67 It has transformed into Xinan Jiaotong University.
68 'The Shanghai-Woosung railway', *Herald*, 19 February 1897: 307, 'The Woosung railway', 8 August 1898: 257 and 'Cycling round the world', 10 October 1898: 691; Charles E. Darwent, *Shanghai: a handbook for travellers and residents* (Shanghai: Kelly and Walsh, 1920), 132.
69 'Peking: the Boxers', *Herald*, 9 May 1900: 823, 'The crisis in the north', 11 July 1900: 77, 'Tientsin', 15 August 1900: 337 and 'A Peking diary', 12 September 1900: 571.
70 Clive Bigham, *A year in China, 1899–1900, with some account of Admiral Sir E. Seymour's expedition* (London: Macmillan and Co., 1901), 168–78.
71 Congers, *Letters*, 90.
72 'Austro-Hungry, Belgium, France, Germany, Great Britain, Italy, Japan, The Netherlands, Russia, Spain, the United States and China Final protocol', in MacMurry, *Treaties*, vol 1, 278–84.
73 'Notes from Peking', *Herald*, 6 March 1901: 428 and 'Notes from Baotingfu', 10 April 1901: 695.
74 'Crisis in the north', *Herald*, 14 August 1901: 309 and 'Notes on native affairs', 31 December 1901: 1269.
75 'Special telegrams', *Herald*, 8 January 1902: 4.

76 'Shengjia rudu', *Shen Bao*, 8 January 1902: 1; 'The imperial entrance into Peking', *Herald*, 8 January 1902: 31; Yang Honglin, *Cixi huiluan: 1901 niande yici teshu luxin* (Beijing: Sanlian shudian, 2017), 164–65.
77 Archibald Little, 'The Hsiling or western tombs', *Herald*, 27 November 1903: 1135.
78 Feuerwerker, *Industrialisation*, 77.
79 Yang, *Huiluan*, 162–63.
80 'The imperial visit to the Western Mausolea', *Herald*, 25 February 1903: 283; 'Zhabei lingchai', *Shen Bao*, 5 March 1903: 2, 'Jingbao quanlu', 23 March 1903: 13, 'Yizhang fanjing', 18 April 1903: 2 and 'Lianggong yelingshi', 19 April 1903: 2.
81 'Dongling xuxiu tielu', *Shen Bao*, 19 February 1905: 4.
82 'The Peking-Eho Park railway', *Herald*, 31 December 1902: 1390; Yan Jiesheng, 'Yiheyuan Cixi zhuanyongde tielu', *Wenhua Jiaoyu* 1 (2010): 72–74.
83 Ping-ti Ho, *Ladder of success in imperial China: aspects of social mobility 1368–1911* (New York: Columbia University East Asian Institute, 1962).
84 Der Ling, *Two years in the Forbidden City* (New York: Moffat, Yard and Co., 1911), 114 and 219; Katherine A. Carl, *With the Empress Dowager* (New York: Century Co., 1905), 20, 26, 37, 176 and 215.
85 Carl, *Empress Dowager*, 290; 'Portrait of the Empress: a critical sitter', *Herald*, 18 January 1939: 116.
86 Carl, *Empress Dowager*, 241.
87 Carl, *Empress Dowager*, 297–98.
88 Isaac T. Headland, *Court life in China: the capital, its officials and people* (New York: F. H. Revell, 1909), 104.
89 Conger, *Letters*, 187–88.
90 'Niangu jiwen', *Shen Bao*, 24 May 1886: 1, 'Jingshi jinshi', 4 August 1887: 1, 'Dayuan qiuhong', 6 September 1888: 1, 'Kangu dagong', 6 April 1889: 2 and 'Dijing jingse', 8 June 1891: 1.
91 Rawlinson, *Naval*, 73.
92 Lei Yi, *Li Hongzhang*, 408–14.
93 Weng, *Riji*, vol 5, 2314.
94 Feuerwerker, *Industrialisation*, 173; Spector, *Li Hung-chang*, 228.
95 Kwang-ching Liu, 'Politics, intellectual outlook and reform: the Tung-wen kuan controversy of 1867', in *Reform in nineteenth century*, 87–100; Rhoads Murphy, *The treaty ports and China's modernisation: what went wrong* (Ann Arbor: Michigan University Papers in Chinese Studies 7, 1970), 68; Wellington K. K. Chan, *Politics and industrialization in late imperial China* (Singapore: ISEAS, 1975).
96 Wright, *Last stand*, 70.
97 Sue Fawn Chung, 'The much-maligned empress dowager: a revisionist study of the Empress Dowager Tz'u-his (1835–1908)', *Modern Asian Studies*, vol 13, 2 (1979): 177–96.
98 Headland, *Court life*, 53 and 69.

Part II

'Tools of empire'

4

'Engines for empire': Russia sets the standard[1]

'My grandfather will remain here forever, so this place is of great significance to me.' These are the words of Ehud Olmert, then Israeli deputy prime minister, when he paid a visit to the Jewish cemetery in Harbin, China on 25 June 2004.[2] His grandfather was one of the thousands of Russian and European Jews who lived and died in the city in the late nineteenth and early twentieth centuries. A Russian colony that grew out of a railway, Harbin tells a very different tale of railway politics as Russia heralded the scramble for railway concessions in China. This would further politicise railway development as the grave concerns of Qing officials would materialise. There is no better example of this than the Russian-built Chinese Eastern [中东] Railway or CER. This chapter examines the rationale for conquest by railways. It advances a model for the study of railway imperialism. It demonstrates the interconnected geopolitics that underpinned the Age of Empire in China. It also turns the spotlight on Chinese officials. They attempted to pit the rapacious foreigners against each other so that China might reap the rewards, epitomised by the Chinese maxim: 'when a snipe and a clam fight, it is the fisherman who benefits'.

'Lamb to the slaughter'[3]

'Li Hung-Chang's Mission to St. Petersburg.' With these words, the *Herald* began to track Li's journey to Russia in March 1896.[4] But why was Li heading to Moscow considering his suspicions of Russian ambition for Manchuria? We cannot understand railway development in China, nor the making of modern China, without acknowledging the significance of the CER. Li's journey signalled an unprecedented change and not just towards Russia. Li would also visit Germany, Holland, Belgium, France, Britain, the United States and Canada. The strategy was clear to those versed in geopolitics: to befriend Russia, Europe and America in order to contain Japan – to 'use foreigners to constrain foreigners', as the Chinese saying goes. To understand this change of policy, we must go back to the First Sino-Japanese War (1894–95). Japan had had their sights set on Korea for a long time. The Donghak rebellion that had broken out in January 1894 was an opportune moment. Although the Korean king had sought the help of the Qing, the Japanese interceded and installed a pro-Japan regime who then requested the Japanese to expel the Chinese.[5] At sea, the Japanese annihilated the Beiyang fleet built up by Li Hongzhang and, on land, they drove the Chinese back to Manchuria.[6]

Their mission was accomplished, but the Japanese Army pressed on. They crossed the Yalu River into Manchuria and down to Liaodong Peninsula, at the end of which lay Lüshun or Port Arthur, named after Lieutenant William C. Arthur who had stationed his fleet there during the Second Opium War.[7] Japan's real intentions were exposed; they desired not just Korea but also Manchuria. The Qing were not prepared for a protracted war. The Empress Dowager had been looking forward to her 60th birthday in 1895. It appears that the celebrations of her milestone birthdays were marred by military conflicts: in 1875, 1885 and again in 1905. As the most senior statesman and diplomat, Li was sent to Japan to negotiate. He signed the humiliating Treaty of Shimonoseki in April 1895, in which Japan procured Taiwan and the Liaodong Peninsula as colonies, and secured trade privileges and 200 million taels of indemnity, the largest amount thus far.[8] Japanese gains alarmed Russia, Russian troops had marched into Manchuria during the conflict, poised to gain from the Qing's misfortune.[9] It also unsettled the Germans, whose interest in this region was growing, and agitated the French (see Figure 4.1). The trio were anxious about Japanese expansion into the strategic Liaodong Peninsula.

Their anxieties led to the 'Triple Invention', which forced Japan to give up Liaodong – a defeat the Japanese would avenge a decade later in the Russo-Japanese War.[10] However, the burden of an extra 30 million taels of indemnity was borne by the Chinese, not the trio.[11] Financially exhausted by the end of the war, the Qing were, nonetheless, required to begin indemnity payments to Japan immediately. The Russians and the French took advantage of the occasion.[12] The *Contract for the Chinese Four Per Cent Gold Loan* was signed in July 1895, and the Russo-Chinese Bank (no Chinese capital) was set up to facilitate transactions.[13] The disastrous Treaty of Shimonoseki earned Li Hongzhang the label of 'country-selling traitor', and he was dismissed as Beiyang commissioner and Zhili governor-general, a post that he had held for nearly 25 years. While this should have signalled the end of his career, Li proved tenacious. As the Japanese threat loomed, the Qing decided to work with the Russians, who invited Li to attend the coronation of Tsar Nicholas II in May 1896. Russia had annexed the area north of the Amur and east of the Ussuri rivers through the Treaty of Tientsin in

Figure 4.1 Map of Liaodong and Shandong Peninsulas (Baidu public domain)

1858 and the Convention of Peking in 1860.[14] Ernst Ravenstein assumed the area to be vast, 'far exceeding the British Islands in extent', whereas Gregory Afinogenov believed it amounted to 'approximately one million km^2 of land'.[15] This acquisition cut the Qing's access to the Sea of Japan and gave Russia a border with Korea, with access to the Pacific Ocean.

Russian expansion was accomplished with ambitious projects such as the Trans-Caspian and Trans-Siberia railways. Ronald E. Robinson discussed the fascination of European geo-strategists with the railway's potential 'for absorbing small states into empires'.[16] One of them was Sergei Witte, Russian Minister of Finance; a 'railway man in the age of empire', as Francis W. Wcislo characterised him from his early career in the Department of Railroad Affair.[17] Chris Miller concurred recently: 'Witte's personality and worldview were shaped by the railroad business'.[18] Witte understood the importance of a railroad linking the capital with the newly conquered and sparsely populated Far East. Both J. N. Westwood and Steven Marks have argued that railways were integral to populating the newly annexed territories, which would help control a 'weak and distant domain'.[19] The details of how to route the line to its destination of Vladivostok generated much debate. Cutting through Manchuria would be the most propitious and the invitation to attend the imperial coronation presented the perfect opportunity to seal the deal. The Qing found itself caught between Russia and Japan in what Ian Nish called a 'Sino-Russo-Japanese triangle'.[20] For the Qing, co-operating with the Russians seemed better than working alone.[21] However, officials like Li Binheng suspected that the war had also exhausted Japan who was unlikely to challenge China again anytime soon. An unnecessary alliance with Russia would only 'lead the wolf into the house', as the Chinese saying goes.[22]

Prince Ukhtomskii, head of the Russo-Chinese bank, whom Kent called the 'most insistent of Russian empire builders', ushered Li onto a Russian steamer in Alexandria, Egypt.[23] Li was brought to St Petersburg where he was 'met with a sumptuous reception by the Czar and Czarina'.[24] After the coronation of Nicholas II in Moscow, the business of negotiation began. But Li had nothing to negotiate for as the treaty had already been drafted by Count Arthur Cassini, the Russian Minister in Beijing.[25] Like Jardine's agent Alexander Michie, Cassini knew how to play the game and ingratiated himself with ranking officials.[26] *The National Observer* called the Cassini Convention 'little less than partition' and declared that it 'amounts to annexation of north China beyond the Great Wall'.[27] Li's job was simply to formalise the treaty. Witte wrote in his memoir:

> As we discussed Sino-Russian relations I stressed how we had helped China preserve her integrity, how committed we were to the principle of the integrity of the Chinese Empire. But I went on to say, it would be difficult for us to assist China as long as we didn't have the railroads to transport our troops from European Russian or from Vladivostok. I pointed out that during the recent Sino-Japanese War the troops we sent from Vladivostok to Kirin did not arrive there until the war had ended, because of the lack of a railroad. For us to increase our troop strength in that region it would be necessary to have a railroad.[28]

Far from the noble aim of preserving Chinese territorial integrity, the main goal of Witte, labelled 'Russia's evil genius', was the expansion of the Tsar's empire into Manchuria.[29] Indeed, the Russo-Chinese Bank was Witte's brainchild, as was the railroad cutting through Manchuria. In his memoir, Witte claimed that Li had been 'promised a substantial sum for using his good office'.[30] Sung-Hwan Chang divulged in passing the figure of $1.5 million, without providing evidence.[31] David Wolff's extensive research in the Russian archive has led him to suggest that 3,000,000 *rubles* were set aside for Li: the first million was endowed by Prince Ukhtomskii in May 1897, whereas the second instalment was scheduled to be paid when the route was approved by the Qing.[32] Witte further claimed that six months after Li's death, in May 1902, the 'Li Hongzhang bribe account' still had funds, and that he had had to obtain permission from the Czar to put this money toward a 'new purpose'.[33] There are many unanswered questions surrounding this Russian slush fund, not least due to a lack of conclusive evidence on the Chinese side. Had the Russians fabricated its existence to discredit Li and hide their complicity or had Li been cunning enough to cover his tracks?

The Li–Lobanov Treaty was signed on 3 June 1896.[34] Witte called it 'an instrument of the highest political and commercial significance'.[35] Mainland scholar Li Cunkuan has described it as a deal between 'wolf' and 'sheep'.[36] The reasons are apparent. While the Treaty tied the fortunes of the Chinese and Russian empires together against Japanese aggression, it also contained provisions that left China in an exposed position. Article III, for example, stated that 'Russian warships could enter any port in times of war', a clause that would be exploited a year later in 1897.[37] The Treaty continued with Article IV:

> In order to facilitate the access of the Russian land troops to the menaced points, and to ensure their means of subsistence, the Chinese government consents to the construction of a railway line across the Chinese provinces of the Amour [i.e., Heilung-kiang] and of Guirin (Kirin) in the direction of Vladivostok.[38]

Heilongjiang and Jilin referred to north and central Manchuria; whereas the 'port' mentioned in Article III indicated the Russians had their eyes on southern Manchuria as well. By laying railway tracks in Manchuria, the Russians were conquering the Qing without conquest, and their gains were manyfold. It would facilitate further expansion down into south Manchuria, north China and the Korean Peninsula. It would also help deter the Japanese and constrain the British, all at the cost of the Qing's sovereignty. The disagreements surrounding the Trans-Siberian route were thus resolved, but the question of its route within Manchuria remained an issue. Russian exploration of Manchuria had intensified in the eighteenth century.[39] In June 1895, two months after the Treaty of Shimonoseki was signed, Russia sent four groups of explorers to Xing'an (Khinga) range, Qiqiha'er, Ningguta and Liaodong bay, to conduct surveys – imposing a treaty was only a matter of time.[40] The Li–Lobanov Treaty signed in June 1896 only legitimised such behaviour on paper as they trampled through Manchuria without restriction. By October 1897, the 'southern route' was proposed. This line passed through the Jilin province, cutting the region in half through the middle and ensuring its centrality. It shortened any future southward expansion into the Liaodong Peninsula, north China and Korea.

St Petersburg favoured this route, but chief engineer Alexander Iugovich expressed grave reservations, as the southern route would 'triple the distance' and the mountainous terrains would present challenges that made little engineering sense.[41] Despite these obstacles, St Petersburg pushed ahead with this route until late 1897, when the Russian Pacific fleet suddenly sailed into and occupied Port Arthur on 14 December.[42] This manoeuvre surprised the Qing court. The Chinese ambassador to Russia, Xu Jingcheng, demanded Russian withdrawal, but his angry protest only resulted in yet another treaty, the *Convention for the Lease of the Liaotung Peninsula* or the Li–Pavlov Agreement, signed in March 1898.[43] The *Herald* called the convention a 'partition of China', as the Qing were forced to lease the Liaodong Peninsula and Port Arthur.[44] With this new foothold in southern Manchuria, Russia had obtained the entire region with two pieces of paper. Witte wrote that he 'resorted to bribery to influence a Chinese decision', and 'I promised a gift of 500,000 rubles to the former [Li Hongzhang] and one of 250,000 to the latter'.[45]

Once again, we see a trail of money; the Rothschild Archive offers some clues as to where they might have ended up. A year after the Russian treaties were signed, Li remitted £4,000 to Lord Rothschild in 1899 to buy shares in the British banking consortium, Peking Syndicate, after it was awarded mining and railway concessions (see Figure 4.2). In his letter to Lord Rothschild dated 6 April 1899, Li wrote that the Syndicate 'may count upon my support in whatever may be necessary for its success here'.[46] Was this money from the Russians? Regardless, he certainly knew who to approach and where to invest. Perhaps Li deserves to be called the real 'evil genius', having managed to shoot four birds with one stone: he had salvaged his career after the disaster of the Sino-Japanese War, made money for himself, found a partner to discourage Japanese incursion and secured the construction of a railway at 'no expense'. Li was, at least, consistent in his efforts to build railways, even when it came at the cost of the Qing's territorial integrity. The *Agreement Concerning the Southern Branch of the Chinese Eastern Railway* was signed in July 1898.[47] While the Li–Lobanov Treaty had sanctioned the creation of a horizontal line cutting through Manchuria to Vladivostok, the Li–Pavlov Treaty now added a vertical trunk line stretching down to Port Arthur. The original hyphen-like railway now resembled the letter 'T' (see Figure 4.3).

Why had the Russian Pacific Fleet suddenly sailed into Port Arthur in December 1897? The rationale behind Article III of the Li–Lobanov Treaty, granting Russian ships entry to 'any port in times of war', was now laid bare – the German landing in Jiaozhou on 14 November 1897. But why should a German landing in Shandong Peninsula prompt the Russian occupation of Liaodong Peninsula and Port Arthur across the Bohai Sea? The

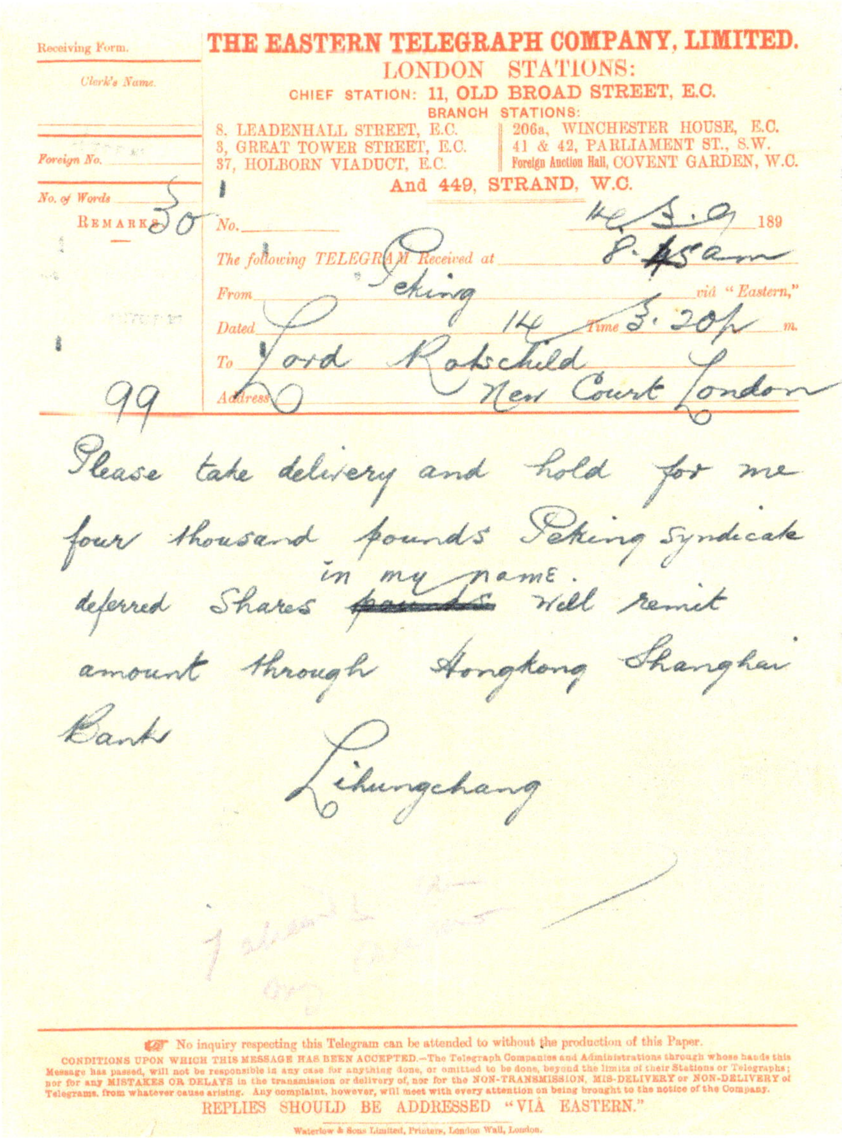

Receiving Form.

Clerk's Name.

Foreign No.

No. of Words 30

REMARKS

THE EASTERN TELEGRAPH COMPANY, LIMITED.

LONDON STATIONS:

CHIEF STATION: 11, OLD BROAD STREET, E.C.

BRANCH STATIONS:

8, LEADENHALL STREET, E.C.
3, GREAT TOWER STREET, E.C.
37, HOLBORN VIADUCT, E.C.
206a, WINCHESTER HOUSE, E.C.
41 & 42, PARLIAMENT ST., S.W.
Foreign Auction Hall, COVENT GARDEN, W.C.

And 449, STRAND, W.C.

No. 14.3.99 189

The following TELEGRAM Received at 8.45 am

From Peking via "Eastern,"

Dated 14 Time 3.20 p m.

To Lord Rothschild

99 Address New Court London

Please take delivery and hold for me four thousand pounds Peking Syndicate deferred Shares ~~pounds~~ in my name. Will remit amount through Hongkong Shanghai Bank

Lihungchang

No inquiry respecting this Telegram can be attended to without the production of this Paper.

CONDITIONS UPON WHICH THIS MESSAGE HAS BEEN ACCEPTED.—The Telegraph Companies and Administrations through whose hands this Message has passed, will not be responsible in any case for anything done, or omitted to be done, beyond the limits of their Stations or Telegraphs; nor for any MISTAKES OR DELAYS in the transmission or delivery of, nor for the NON-TRANSMISSION, MIS-DELIVERY or NON-DELIVERY of Telegrams, from whatever cause arising. Any complaint, however, will meet with every attention on being brought to the notice of the Company.

REPLIES SHOULD BE ADDRESSED "VIÂ EASTERN."

Waterlow & Sons Limited, Printers, London Wall, London.

Figure 4.2 Li Hongzhang's telegraph to Lord Rothschild (XI/109/134, Private correspondence (telegraph), Rothschild Archive)

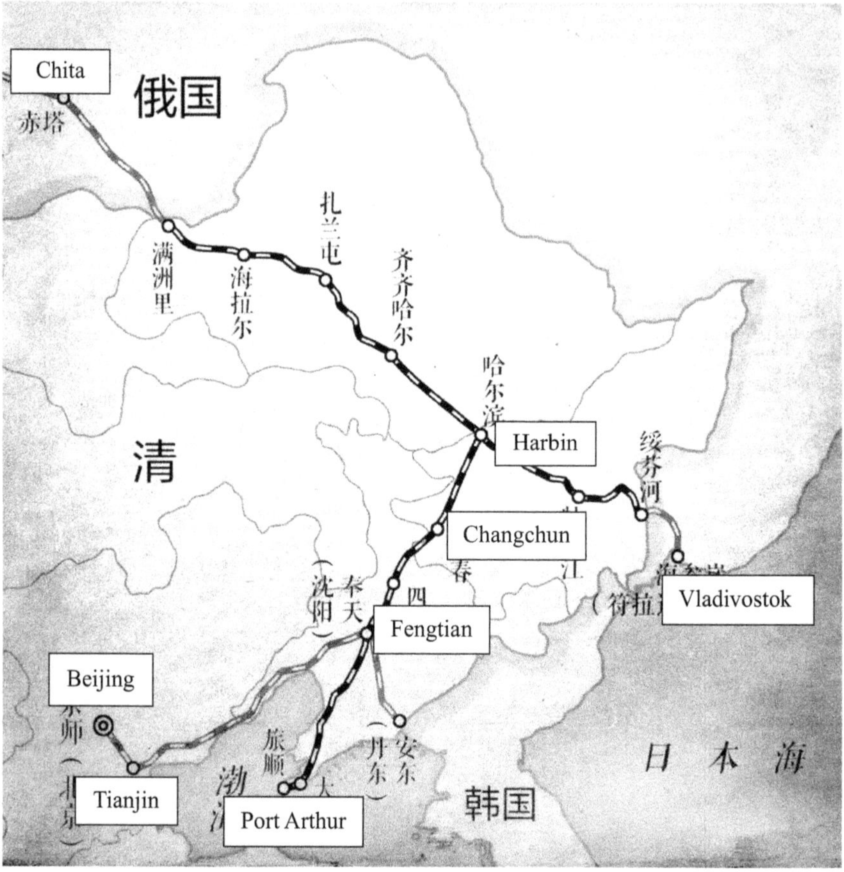

Figure 4.3 Route of the Chinese Eastern Railway (Baidu public domain)

best evidence comes from the Cassini Convention, which made no effort to hide the Russian appetite for Shandong:

> Russia has never possessed a seaport in Asia which is free from ice and open all the year round. If therefore, there should suddenly arise military operations in this continent it will be naturally difficult for the Russian Eastern Seas and Pacific fleets to move about freely and at pleasure. As China is well aware of this, she is willing to lease temporarily to Russia the port of Kiauchou in the province of Shantung, the period of such lease being limited to 15 years.[48]

The Russians wanted more than just Manchuria – they also had their eyes on the Shandong Peninsula, which the Germans also coveted. Witte considered the taking of southern Manchuria as inevitable. He wrote, '[the] transfer of the Kwantung region was unavoidable since we had ships carrying troops in full battle readiness' and 'Had China not yielded, Admiral Dubasov could easily have occupied the territory, for the Chinese had no troops there and the fortress at Port Arthur was a mere toy.'[49] 'Kwantung', or East of the (Mountain–Sea) Barrier, refers to Manchuria in general but southern Manchuria in particular. German landing in Jiaozhou and Russian occupation of Port Arthur aggravated the British. With their dominance in China now seriously challenged, this episode would take Anglo-Russo and Anglo-German rivalries to new heights. Having secured the Liaodong Peninsula and Port Arthur, St Petersburg dropped the original 'southern route' and settled on a northernly horizontal path. The connecting point in the letter 'T' fell on Harbin. The CER entered Manchuria on the Sino-Mongol frontier and exited on the Sino-Korean border. It extended down to the tip of the Liaodong Peninsula at Port Arthur, the pivotal ice-free harbour that overlooks Korea and faces the Shandong Peninsula. The first railway to be financed, built and controlled by a foreign power, the CER inaugurated the age of railway imperialism in China, setting the standard for what was to come.

The 'road to power'[50]

'It may be safely averred that railways, steamships, and telegraphs are combinedly our most powerful weapon in the cause of Inter-Imperial Commerce,' engineer-turned politician Charles Tilston Bright declared.[51] He was right; these technologies, railway in particular, defined the Age of Empire at its zenith. Ronald Robinson described railway imperialism as the use of 'railway loans and concessions to stake out spheres of influence' and to 'cajole their rulers into subordinate alliance'.[52] He concluded that railways 'served as an instrument of informal empire-building' and a 'function of European expansion'.[53] Historians have examined the connection between railways, empire and geopolitics.[54] Daniel Headrick called them 'tools of empire'.[55] Edward Spiers also labelled them 'engines for empire', as they were used by the British army for expansion in India, Central Asia and Africa.[56] Sean McMeekin believes that the Berlin–Baghdad Express marked Germany's bid for world power.[57] Steven Marks has argued that the Trans-Siberian railway enabled Russia to colonise Asia, whereas Tim Zajontz postulated that 'railways have been the single most important infrastructure for the systematic exploitation of Africans by European state and capital'.[58] Maria P. Diogo and Bruno J. Navarro detailed how the Portuguese used railways to conquer Angola and Mozambique.[59] Kees Van Dijk has maintained that the laying of tracks 'pre-empted another power from doing the same, excluding the competitors'.[60] 'If non-Western peoples and rulers resisted the idea, as often happened', as Headrick explained, 'railway promoters knew how to apply guile or force, just as they knew how to overcome natural obstacles with bridges and tunnels'.[61]

This and the following three chapters bring the case of the late Qing to bear on the larger debate by advancing a model in its study. The age of railway imperialism had arrived in China with the Russians. Railway tracks replaced war as an instrument for conquest. Sarah Paine has called the CER the 'apogee of Tsarist imperialism'.[62] It gave birth to a new kind of imperialism, one that was different from the Treaty-port 'settlement' archetype which constrained expansion, one that was less costly than warfare and more profitable to maintain. Building railways allowed foreign powers to grab land without restriction, to station troops in the name of protection, to exploit a region's rich natural and human resources, and to colonise territory through station towns/cities. A proxy for the Russian government in Manchuria, the CER management company had the power to command the military and administrate civilian as well as foreign affairs by issuing visas.[63] No wonder General F. A. Sutton called the CER 'a state within a state'.[64] For Victor Zatsepine, it was 'similar to colonial government'.[65]

Railway construction required the demarcation of significant amounts of land. This was guaranteed by the *Contract for the Construction and Operation of the Chinese Eastern Railway*:[66]

> The lands actually necessary for the construction, operation, and protection of the line, as also the lands in the vicinity of the line necessary for procuring sand, stone, lime, etc., will be turned over to the Company freely, if these lands are the property of the State; if they belong to individuals, they will be turned over to the Company either upon a single payment or upon an annual rental to the proprietors, at current prices. The lands belonging to the Company will be exempt from all land taxes. The Company will have the absolute and exclusive right of administration of these lands. The Company will have the right to construct on these lands buildings of all sorts, and likewise to construct and operate the telegraph necessary for the needs of the line.[67]

Not only was land freed up for acquisition, it was also free from taxation. The CER established a Department of Land Acquisition and Exploitation to lead this. This situation was drastically different from earlier forms of concession in which foreign powers were allotted 'settlements' on which they had to pay taxes and they were confined to a designated area. Venturing outside of these areas could invite the intervention of the Chinese authorities, as in the case of the Wusong. Railways completely changed the rules of the game. The development of railways entitled foreign powers to a 'belt of alienation', what mainland scholars called 'affiliated territory'.[68] This referred to the right-of-way land on both sides of the tracks, normally extending 9.3 or 10 miles out, to be used for stations, switching tracks, repair shops, offices, living quarters and barracks. David Wolff's research has shown that 'at Harbin, the swelling took on proportions more appropriate for city planning than station building'.[69] Entire neighbourhoods were turned into a belt of alienation managed by the CER, to which the Chinese had to apply in order to settle.[70] Olga Bakich reveals that the CER held 'regular auctions for the long-term private lease of lots to Russian and Chinese subjects'.[71] Wolff and Bakich's assessment is borne out by provincial gazetteers, providing us with previously unknown details.

The *Heilongjiang Gazetteer* calculated that Russians took two to three times as much land than was stipulated in the treaties.[72] The *Jilin Gazetteer* details how the Russians turned Kuanchengzi, a small district, into their 'special zone', a country within a country where the Russians exercised power akin to that of a state.[73] Although the treaties did not sanction this kind of land enclosure, the existence of the railway legitimised it as its operation saw the emergence of Russian facilities and colonies along the route. In Dalian, Russians forced out the entire village of Qingniwan to build facilities and residences in 1899. More than 400 angry locals smashed the CER's office in protest but were brutally suppressed.[74] The *Herald* reported this:

> When the Russians took a lease of Port Arthur and Dalienwan, they declared that they still recognise the sovereignty of the Emperor of China. Now we hear from native sources that disturbances have risen through the Russian authorities to exact land taxes from the Chinese, and that 'some tens' of Chinese have been killed. But surely the collection of land tax is the prerogative of the Sovereign, not the lease.[75]

The Chinese had to pay tax to the intruders for their own land. The British jealously watched and complained that the Russians were 'all over Manchuria, parading through the country as if it were their own'; whereas in their own case, 'the only British subjects who were permitted by treaty to own land outside treaty limits are missionaries, [yet] Russians in Manchuria are acquiring land as they please and are already working mines'.[76] John Foster Frazer travelled across Manchuria in early 1902 and observed that 'all the towns along the route are new and Russian. Where there are Chinese towns, they are contiguous to the Russian towns, which are also military centres. For twenty miles each side of the line, the Chinese and Manchus have been driven back.'[77] Further land-grabbing was consolidated in 1907 by a whole series of agreements forced upon the Qing.[78] Russian historian E. Kh. Nilus highlighted its significance:

> by becoming a Manchurian landowner, the CER was but a step from becoming a Manchurian landlord. With the issuance of lease, the CER automatically took upon itself many new obligations to the tenants. This process converted the CER from a railroad into a multifaceted colonial enterprise.[79]

Laying claim to land constitutes territorial expansion; a hallmark of imperialism, it extends one's political authority and economic control. Such claims had taken on different forms in the past: James Cook had claimed the east coast of Australia for Britain in 1770 simply by planting a flag in Botany Bay, whereas the British had secured Hong Kong after defeating the Qing in the First Opium War. By the middle of the nineteenth century, expansionists had a new instrument at their disposal. Railways replaced flag planting and armed conflicts. They empowered Russia to conquer Manchuria without waging a costly war. The Qing had no power or resources to battle a rising Japan. Having the Russian 'wolf' inside the house, as it were, at least served the purpose of deterring the Japanese advance.

The CER was the 'hardware' of the Russian empire, it was reinforced by another. Military aid sat at the heart of the Russo-Chinese alliance, clarified by both the Cassini Convention and the Li-Lobanov Treaty:

> owing to the fact that the said railways will pass for the greater part through barren and sparsely inhabited territory in which it will be difficult for the Chinese authorities to be always able to grant the necessary protection and aid, Russia shall be allowed to place special battalions of horse and foot soldiers at the various important stations for the better protection of the railway property.[80]

The CER legitimised the protection of Russian property and subjects. For Russia, their interests in far-flung Manchuria could only be held by military power. This meant that the CER enjoyed some degree of autonomy that was unavailable in European Russia. To this end, 720 Protective Guards, recruited from the military, arrived in 1897.[81] The number increased to 2,000 in August 1898 when construction began, and quickly to 5,000.[82] The *Herald* reported on the increase of Russian troops in March 1899:

> 30,000 in Port Arthur
> 3,000 at Talienwan
> 2,000 at Kinehon …
> 200 at Newchwang
> 200 at Haicheng …
> 200 at Liaoyang
> 200 at Kirin
> 20,000 at Haipien north of Kirin[83]

Thus, six months after the construction of the CER began, there were at least 55,800 Russian troops in southern Manchuria. We can only imagine as the CER progressed. The Boxer Rebellion provided the pretext for reinforcements.[84] The *Jilin Gazetteer* put the number of troops at more than 100,000.[85] They captured towns/cities, taking over the Barrier railway then under construction. This led to local resistance, which ostensibly gave the

Russians more cause.[86] Mainland scholars have suggested that 170,000 troops marched into Manchuria from different directions.[87] Captain Golitsyn led some of these troops and detailed his campaign in a memoir – they marched to places where there were no Boxer rebels at all.[88] Sarah Paine was absolutely right to call it 'occupation'.[89] By January 1901, the original Protective Guards were succeeded by the Separate Corps of Frontier Guards composed of 25,000 men, a force 'catching up with regular army'.[90] This did not include the naval force in Port Arthur. All foreign troops were to leave China after the Boxer Protocol was signed in September 1901, but the Russians had no intention to depart. The rebellion had strengthened the hand of Witte's rivals in the War Ministry. Admiral Yevgeni I. Alekseev was appointed Viceroyalty of the Far East in 1903, granted supreme authority for affairs east of Lake Baikal. His appointment ended years of internal strife and Witte's control, signalling the rise of the military in Russian political affairs, a development which would have grave consequences. The garrison in Liaoyang in southern Manchuria saw a vast increase of troops, 10,000 men by July 1903.[91]

The presence of the CER bestowed Manchuria with a strategic value, and so made it a prey for other powers. Where there were railways, there were military build-ups and, consequently, conflicts. The Russian troops stationed along the CER militarised Manchuria and led to the outbreak of armed conflicts every decade, beginning with the Russo-Japanese War. The *Herald* publicised the distribution of Russian troops on the eve of this war in February 1904. There were at least 45,400 soldiers, with the majority of them in southern Manchuria, especially in Port Arthur.[92] In addition, there were 'fifty-five companies' Railway Guards consisting [of] 24,000 men' and another 'Railway Corps consisting [of] some four battalions'.[93] The heavy Russian military presence in Manchuria disturbed the British and the Americans. They encouraged Japan to challenge the Russians, which led to the outbreak of the Russo-Japanese War.[94] It is with little wonder that Francis W. Wcislo referred to the CER as a 'road to war'.[95] By the end of the conflict, there were 950,000 troops and 250,000 horses in Harbin.[96] The CER was turning Manchuria into what Owen Lattimore called the 'cradle of conflict'.[97] This kind of military build-up disrupted the CER's normal operation and led to financial losses as Tan Guiluan has detailed.[98]

The 950,000 troops and 250,000 horses could not have survived without the rich natural resources of Manchuria, a 'first-rate granary' blessed with Mongolia's 'abundance of livestock' and resulting in the 'best-fed Russian army ever seen in the field'; it sustained 'not only Russia's Manchurian army, but also the Maritime province'.[99] Francis Younghusband travelled through the region and highlighted the 'richness of Manchuria, the salubrity of its climate, the fertility of its soil, the magnificence of its forests, and the many indications of its mineral wealth'.[100] Masafumi Adasa's research reveals that Nikolai L. Gondatti, governor-general of Priamur, understood this situation very well:

> The name Vladivostok (Ruler of the East), as a fortress of Russian power in the Far East, is ironic. This is not because a [sic] lack of fortification or because of the size of the army, but because it is dependent on Manchuria and China for all its everyday goods and necessities. We do not have our own goods here. With the outbreak of war in China, we will not get Manchurian wheat, we will lose our workforce in the fortress, and we will have to live here without bread, vegetables, and Chinese workers.[101]

Manchuria was the breadbasket of the Russian Far East without which it could not survive. Gondatti could have also mentioned other resources such as the wood often seen piled high alongside the railway stations, which was used in building the tracks, stations, offices and residences, as well as powering the trains. The *Heilongjiang Gazetteer* detailed the plunder of timber prior to the Russo-Japanese War.[102] After the conflict, when the Russian sphere was reduced to mid and north Manchuria, the CER imposed more agreements on Heilongjiang and Jilin provinces on timber-felling.[103] The next twenty years saw the establishment of thirty lumbering sites; the CER felled nearly all the trees within the railway zone.[104] Russian logging thus marked the first systematic destruction of prehistoric forests and the biodiversity these forests hosted.[105] This kind of deforestation was common in the Age of Empire: Pallavi V. Das details how railway construction saw deforestation in British India and the 'exhausted state of Punjab Forest by the early 1860s'.[106] This marks an important shift when compared to early treaties. The Treaty of Nanking had not entitled the British to log and exploit natural resources wherever they saw fit. Railways changed the nature and dynamics of imperialism because more demands could be added, justified by their construction and operation.

Manchuria was not just blessed with forests but also with coal deposits.[107] The importance of this resource can be seen in the *Statutes of the Chinese Eastern Railway*, which entitled the CER to 'exploit, in connection with the railway or independently of it, coal mines' and also 'other enterprises – mining, industrial and commercial'.[108] Although mining in the region dates back to at least the Ming dynasty, its production was small in scale with dated technology that could not meet the demands of the railway and its growing population.[109] As construction progressed, the CER imposed not just logging but also mining agreements. It took over existing mines and opened new ones within and without the 10-mile beltway on both sides of the rail.[110] Opened in 1901, the Jalainur (Zhalainuo'er) coal mine produced 16,457 tons of coal in 1903; this jumped to 447,183 tons in 1906, with an annual average of 250,000 by 1931.[111] Jalainur was only one of many mines that supplied the CER, the Russian communities in Manchuria, as well as customers in Siberia.[112] Manchuria fed the Russians along the CER, in Siberia and kept them warm. In Jilin province, the CER aimed at nothing short of a coal monopoly, and only Russians could mine and sell coal. Like railways, coal had become a strategic weapon, fuelling the Age of Empire.

Coal and railways were what I call the 'twin engines' of empire. The coal could not have been mined and the CER built without tens of thousands of Chinese labourers; this could also be said for the Trans-Siberian railway, with a 150,000–strong Chinese workforce.[113] Russian railways exploited not just natural but also human resources in the region and beyond. Most of them came from the Shandong and Hebei provinces in north China; they constituted 70 to 90 per cent of the coolies. Chinese workers were preferred because they were extremely cheap. Historians have offered different statistics on the numbers recruited. Xie Xiantian's research into Chinese archives suggests that in 1900 alone, more than 100,000 were recruited.[114] David Wolff put the number at 75,000.[115] His research in the Russian archive led him to believe that a total of 270,298 Chinese labourers worked on the construction of the CER.[116]

John Fraser noted their horrific working and living conditions while travelling through Manchuria in 1902, as did Hungarian contractors who worked on the site.[117] Even the slightest fluctuation in worker numbers perturbed the Russians.[118] The CER pressured the Russian Eastern Asiatic Steamship Company to open a special route between Yantai in Shandong and Vladivostok to facilitate the coolie trade.[119] Chinese coolies were also a source of concern because they would 're-establish China on Russian soil' constituting 'another Yellow Peril'.[120] Despite these fears, the CER finished before schedule; the key factor in this, as Wolff concluded, was 'Chinese labor'.[121] As Lewis Siegelbaum put it: 'The Chinese – construction workers, diggers, porters, merchants and peasants – were quite simply indispensable to the continuation of the rapid pace of economic activity occasioned by the construction of the Trans-Siberian Railway'.[122]

If the railway and the military presence can be considered the 'hardware' of empire, then cultural penetration can be considered its 'software.' The CER powered the colonisation – Russification to be more precise – of Manchuria. The grandfather of Ehud Olmert was only one of the thousands who came to work.[123] The CER attracted not just Russians but also Ukrainians, Georgians, Poles, Hungarians and Armenians, among others, many of them Jewish as it offered them, as Joshua Fogel puts it, an 'opportunity unavailable elsewhere in the lands of the Tsars: a heaven relatively free from the virulent strains of antisemitism so prominent in eastern Europe at the time'.[124] Some worked on the railway, some in its affiliates and still others came to make a living, occasionally a fortune. They settled in the towns/cities that grew along the railroad. Many congregated in Harbin, the largest metropolis and heart of the Russian community. It is no exaggeration to say that the history of CER is the history of Harbin.

Designed as a Russian city, the primary and most prominent architectural structure was the Orthodox Christian Church from where the main avenues fanned out. Soon Russian shops, restaurants, pubs, banks, residences, schools, hospitals and post offices emerged. There were also parks, theatres, a circus, cinemas, opera houses, taverns, casinos, brothels and police stations. Churin (Tschurin), a Russian department store, opened in Harbin, with more than forty branches spread out across the region.[125] Rivalling Sincere in Shanghai, it drove fashion trends in Manchuria and beyond for many years. While known as a 'company town', Harbin was also referred to an 'amazing' and 'international city', as well as a 'Manchurian Petersburg' or 'Manchurian Paris'.[126] Harbin was a Russian metropolis as Russian architecture, culture and language dominated its landscape and soundscape for decades. The *Herald* reported:

> The city is at its gayest from 1 to 4 a.m. It is only during these hours that the cabarets are full. One goes to the movie or theatre at 7 p.m., is invited to tea at 11, starts for the cabaret for the midnight, goes home at dawn, breakfasts at eight or nine, and then to business until one. It is astonishing how often this can be done without perceptible decline in vigour. This 'Paris of the Plains' is the centre of imports from all the countries of Europe, and the wares of Berlin, London, Paris and New York are cheaper there than home, while any make of motor-car is at your service at what seems to be a song. The Harbin International Club (thanks, Watson) is one of the high spots of Asia ...[127]

While life in Harbin could not have been better for Russians and Europeans, the Chinese were excluded, pushed into a separate area which the Russians called 'China Street'. The Chinese of Russian Harbin were treated like foreigners in their own country. Despite this, Russian and Chinese lives and histories became entangled.[128] Historians of China generally agree that communism travelled to China with students returning from Japan and France in the late 1910s. But Manchuria may have been the first place it made its appearance through the First Russian Revolution. David Wolff gave a precise date 'the far left made its political debut in Harbin on November 27, 1905'.[129] The *Heilongjiang Gazetteer* revealed that May Day was celebrated in Harbin as early as 1907 and that CER workers organised strikes during the Russian Revolution in 1917.[130] The CER would help spread communism and enable the USSR to exert its influence on China.

The 'Manchuria Bridgehead'[131]

The CER helped the Russians to conquer and colonise in ways that a 'settlement' never could without waging a costly war. Li Cunkuan rightly compared Russia to a 'wolf', but it was not the only power circling China as its prey.[132] There was a pack of wolves converging on the late Qing, the 'lamb' to continue the metaphor.[133] Just as Japanese gains at the end of the First Sino-Japanese War had alarmed the 'Triple' allies, so, too, would Russian acquisitions agitate the British. While the emergence of an Anglo-Russo rivalry can be seen from the Crimean War, it would continue over their scramble in Central Asia. After annexing Merv (Turkmenistan), the Russians pushed down to the Afghan border, where the British rushed to buttress Afghanistan and northwest India with strategic lines.[134] Ronald E. Robinson summarises the role of railways in their entanglement:

> The imperial propensities of locomotion were demonstrated on a vast scale in the era of transcontinental lines and continental partitions that followed. By rail, for example, the British Raj rivetted military control on the Indian subcontinent and defended its northern frontiers against Russian lines approaching from the Caspian Sea and Tashkent, a conflict which led the two Powers to divide much of central Asia and Persia into exclusive railway zones by 1907.[135]

By the 1890s, this kind of Anglo-Russo railway rivalry had moved to China. Francis Younghusband travelled to Manchuria in 1886, calling it a 'land of promise', declaring that 'it is our business to see that Manchuria remains as open to our trade as it now is; that we continue to in the future enjoy all the rights we at present possess'.[136] Britain had been engaged in a thriving trade in Newchwang (Yingkou) since the close of the Second Opium War.[137] But the CER, terminating at Dalian–Port Arthur, would divert the traffic and threaten British dominance. Nathan Pelcovits commented on the 'struggle for domination of China by means of railway concessions'.[138] Summing up the situation at the time, he stated that 'to the three sorespots of railway politics', that is Anglo-German, Anglo-French and Anglo-Belgium, a fourth was added – Russian objection to a railway from the Mountain–Sea Barrier to Newchwang. This is the branch line of the Barrier railway for which the Qing planned to borrow from the British to build.

This development upset the Russians who considered Manchuria its exclusive sphere. Percy Kent believed that it was British involvement in the Barrier line that prompted Russia's move to lend Qing China a 'helping hand' at the end of the First Sino-Japanese War.[139] The *Herald* referred to the CER as the means of the 'Russian acquisition of Manchuria'.[140] Whereas the *Guardian* reported Russian activity in Newchwang:

> First, there is no Russian trade here, but great Russian railway activity. There is no Russian Consul, therefore appeal in the case of complaints would have to be made to Russian military authority. The second thing is the disregard of the maritime Customs by steamers carrying Russian railway material. Cargoes were landed without Customs supervision and payment of duty, impunity being secured by the fiction that the railway is partly Chinese, whereas not a single share is held by the Chinese. The railway is a purely Russian military and strategic line. The third thing which struck Lord Charles Beresford was the Russian action in securing positions of advantage, all making for military domination. Thus, there are armed Cossacks in the treaty ports, Cossacks patrolling the railway, and Russian military posts along the railway throughout Manchuria.[141]

The CER justified every Russian undertaking on Chinese soil. Unsettling the British further, Russia decided to build a branch line from the CER terminus in Port Arthur to Newchwang in 1899. They also forced the Qing to refuse the loan from the British to build the Barrier to Newchwang extension.[142] The *Herald* concluded that 'Hostility to England is the alphabet of Russia policy'.[143] Although they failed to secure this line, Britain nevertheless prevailed in the end, as the '"compensation" that China had to grant to the British were of far greater importance to Britain's Far East interest than the loan itself'.[144] Anglo-Russian rivalry explains why the Russians refused to withdraw from Manchuria after the conclusion of the Boxer Protocol despite having signed the *Treaty of Russian Withdrawal* in 1902.[145] To defend what it considered its interest in Manchuria, Britain decided to work with Japan, forming the Anglo-Japanese Alliance in 1902.[146] Britain and the United States encouraged Japan to take on the Russians, leading to the outbreak of the Russo-Japanese War in 1904.[147] The British were pleased that the Japanese drove the Russians away from southern Manchuria.[148] But they did not anticipate that the Japanese would drive them out of business in Newchwang.

The war resulted in the Portsmouth Treaty, brokered by President Theodore Roosevelt. The CER enabled Russia to thwart the British, to impede the Germans, as well as whetting the appetite of Americans.[149] The Treaty heralded American entry into the scramble for railway concessions in Manchuria with the region now in the sights of several powers, even though American railway interest in China goes back to Oliver Bradford and Ulysses Grant.[150] Owen Lattimore saw Manchuria as the 'cradle of conflict' in 1932, we see its origin in the battle over a railway since the 1890s. Britain quickly revised its foreign policy, which now sought to prevent Russian expansion to south of the Great Wall, just as Russia was determined to forestall Britain's northward move into Manchuria. After much diplomatic wrangling, Count Mouravieff, the Russian Minister for Foreign Affairs, and Sir Charles Scott, the British Ambassador in St Petersburg, signed a note of exchange in April 1899:

1. Great Britain engages not to seek for her own account, or on behalf of British subjects or of others, any railway concessions to the north of the Great Wall of China, and not to obstruct, directly or indirectly, applications for railway concessions in that region supported by the Russian Government.
2. Russia, on her part, engages not to seek for her own account, or for Russian subjects or of others, any railway concessions in the basin of the Yangtze, and not to obstruct, directly or indirectly, applications for railway concessions in that region supported by the British Government.[151]

Philip Joseph was accurate in outlining that railways were the main point of contest for the two empires.[152] Russia claimed Manchuria as its exclusive sphere; as did Britain with the entire Yangzi River valley, from Sichuan through south–central China to Shanghai. This would put Britain in conflict with Belgium, Germany and France as they secured rights to finance and build various lines in north and south–central China. This exchange was merely a settlement on paper, for it did not prevent Britain from expanding beyond the Great Wall. Neither did it constrain the Russians from expanding to the south of the Great Wall as their quarrels spilled over to other railways. According to the *Herald*: 'A much greater scheme is within the Russian plot. The Central Asian Railway, from the Caspian district, is to pass through Kansu, Shensi, and Shansi, and join the Lu-Han railway about Chengting. The Lu-Han railway is under the thumb of the Russo-Chinese Bank, and is already made from Chengting to Peking.'[153]

The British feared that Russia would encircle China, along with existing British interests and claims, not just with the CER but also with the Central Asian railway by joining the Zhang Zhidong proposed Luhan railway. Their scramble quickly moved to the Luhan when others seized the opportunity. We know this from Sheng Xuanhuai's letters and telegraphs that Britain, Russia, the United States, France and Germany all fought for the right to finance the Luhan. Sheng used such words as 'ganging up', 'conspirative' and 'intimidating' to describe them and asked the Qing regime to intervene.[154] To avoid further trouble, Zhang Zhidong awarded the loan and contract to the small and seemingly benign Belgium.[155] Much to British frustration, Belgium brought in the French to undertake the Luhan. Moreover, the Russo-Chinese bank, with a large amount of French capital, secured the right to finance a major branch line from the Luhan, the Zhengtai [正太] railway.[156] This would further Anglo-Russo animosity and galvanise

Anglo-French rivalry. Branching out from the Luhan in Zhengding, Hebei province, this line travels west to Taiyuan, China's coal capital and gateway to the vast northwest. Russian intention, according to Kent, was to extend the line onto Xi'an 'where it would in course of time effect a junction with a continuation of the Russian Central Asiatic railway from Tashkent by Vernoe and Kuldja'.[157] But this, as Henry Picot described, 'forms part of train of Arabian Nights' fancy, having no chance of realisation in the present or the near future'.[158]

Britain exaggerated Russian ambition; so did Russia in its response. The Russians protested and hassled the Chinese when they heard of rumours that the Qing had signed a secret agreement with the British for the Jingzhang [京张] railway, a line from Beijing to Zhangjiakou or Kalgan, a city to the northwest of the capital that had grown out of Sino-Mongol-Russian trade. This led to yet another round of diplomatic wrangling and the 'Exchange of Notes between China and Russia regarding Pledging of Peking – Kalgan and other Railways north of Peking,' where Prince Qing, then head of the *Zongli Yamen*, took an extraordinary step to reassure the Russians in June 1902:

> We dignitaries desire to stipulate that the branches to the north of Peking, as well as the line from Peking to Chang-chia-k'ou (Kalgan), must be constructed by the Chinese Government without interference by foreigners and with Chinese capital, without the application of foreign capital for this purpose; and these lines, or the revenues derived therefrom, must never serve as security for loans placed in foreign nations. In accordance with what is set forth in the report, we request you to refer the matter to the English Minister for examination and reply.[159]

Caught between Britain and Russia, the late Qing had to appease both powers when planning its own railway. Such was the extent of Anglo-Russo railway rivalry – they fought line by line from Manchuria to north and northwest China. Britain would continue to battle not just Russia, but also Germany and France, to retain what it considered its dominant place in China.

A railway helped Russia to acquire territory, station troops, exploit resources and colonise Manchuria. It also enabled them to race ahead. Manchuria had no need for a railway at that time; the CER was built to serve Russian expansion in the Far East. The fears of anti-railway sceptics had now been realised. The CER robbed the Qing of its sovereignty and undermined its territorial integrity. China fell into the 'dependency' trap predicted by Yan Mao. Xu Zhixiang's long memorial was particularly concerned with two main issues: that railways destroy existing livelihoods and damage the environment. The truth of this could not have been more obvious with the CER. It ruined and forever changed the lives of the indigenous people who had lived there for thousands of years through logging, mining, industrialisation, urbanisation and destruction of the ancient landscape.[160] Xu and Yan did not live to see this. But Li Hongzhang might have interpreted it differently. The CER had helped him to regain his political prominence, secure another railway at 'no expense', use the Russians to deter the Japanese and check British ambition.

Intense rivalry among competing powers drove railway imperialism. It opens a new window onto the history and politics of China, and sheds new light on the Age of Empire. But the railway is itself neutral. As a tool, it carried all sides towards their destinations equally. The intermittent debate on imperialist-built railways has focused on whether they benefited China. While earlier scholarship emphasised the detrimental effects, recent scholarship has argued the opposite. At the cost of the Qing's sovereignty and the environment, the CER did build up the region's infrastructure. Some believed that it was a 'railway that has achieved an agricultural and industrial revolution in Manchuria'.[161] But these benefits were not what Russia set out to accomplish. Chao Kang referred to Manchuria as a 'frontier economy'.[162] Rather than a frontier, the CER had turned Manchuria into a 'bridgehead' as it led to heightened rivalry among imperialist powers. The CER would be tied to all major events over the next half a century, as Manchuria became the 'cradle of conflict'. Never before had a railway dictated the history of China to this extent. If opium defined China's problems and complexities in the late Qing; the railway did so to an even greater extent. Russia refashioned the practice of imperialism, other powers would extend its scope through the construction of more railways, the focuses of the following three chapters.

Notes

1 Spiers, *Engines*.
2 Li Fangchao, 'Israel Deputy PM visits grandpa's Harbin grave', *China Daily*, 26 June 2004, at www.chinadaily.com.cn/english/doc/2004-06/26/content_342861.htm, accessed 12 December 2021.
3 Lee Welling Squier, *A lamb to the slaughter: an American girl's experience in the Orient from the China-Japan war to the relief of Peking* (Greenburg: Patriot Publishing, 1901).
4 'Li Hung-Chang's mission to St. Petersburg', *Herald*, 6 March 1896: 349; 'Fuxiang xingcheng', *Shen Bao*, 5 March 1896: 1; 'Tientsin: the viceroy Li Hung-Chang', *Herald*, 20 March 1896: 447; Cai Er'kang, *Li Hongzhang lipin oumeiiji* (Changsha: Yuelu Chubanshe, 1986), 41–57.
5 'Japan', *Herald*, 20 July 1894: 94 and 'The declaration of war by Japan', 17 August 1894: 275.

6 'Japan', *Herald*, 10 August 1894: 222 and 'Japan', 31 August 1894: 349.
7 'The Japanese demand Port Arthur', *Herald*, 7 December 1894: 94.
8 'The treaty of peace', *Herald*, 19 April 1895: 585 and 'The Shimonoseki negotiations', 31 May 1895: 813. For the treaty, see MacMurray, *Treaties*, vol 1, 18–24.
9 'Lun xingjun yijiandie weixian', *Shen Bao*, 11 January 1895: 1, 'Er'jun nanxialun', 2 June 1895: 1 and 'Lun Er'guo diaobingshi', 28 September 1896: 1.
10 'Japanese imperial proclamation regarding retrocession of Liaotung peninsula', in MacMurray, *Treaties*, vol 1, 52–53.
11 'Japanese imperial proclamation regarding retrocession of Liaotung peninsula', in MacMurray, *Treaties*, vol 1, 50–52.
12 Weng, *Riji*, vol 6, 2846, 2848, 2888–89, 2899, 2903, 2906, 2923–24 and 2939.
13 MacMurray, *Treaties*, vol 1, 35–42.
14 Gregory Afinogenov, *Spies and scholars: Chinese secrets and imperial Russia's quest for world power* (Cambridge: Belknap Press of Harvard University Press, 2020), 234–55.
15 Ravenstein, *Russians*, preface and Afinogenov, *Spies*, 236.
16 Robinson, 'Introduction', in *Railway imperialism*, 1.
17 Francis W. Wcislo, *Tales of imperial Russia: the life and times of Sergei Witte 1849–1915* (Oxford: Oxford University Press, 2011), 69.
18 Chris Miller, *We shall be masters: Russian pivots to East Asia from Peter the Great to Putin* (Cambridge: Harvard University Press, 2021), 118.
19 J. N. Westwood, *A history of Russian railways* (London: George Allen and Unwin, 1964), 109; Steven G. Marks, *Road to power: the Trans-Siberian railroad and the colonization of Asian Russia 1850–1917* (Ithaca: Cornell University Press, 1991), 13–27.
20 Ian Nish, *The history of Manchuria, 1840–1948 vol I & II: a Sino-Russo-Japanese triangle* (Folkstone: Renaissance Press, 2016); Ernest M. Satow, *Korea and Manchuria between Russia and Japan 1895–1904* (Tallahassee: Diplomatic Press, 1966).
21 'Lun zhongguo yifu'er zhishiji', *Shen Bao*, 31 March 1901: 1.
22 Li Bingheng, *Li Bingheng ji* (Beijing: Zhonghua shuju, 2013), 588–93.
23 Kent, *Enterprise*, 41; 'Li Hung Chang in Russia', *Times*, 2 May 1896: 5.
24 'Latest intelligence: the Chinese envoy', *Herald*, 15 May 1896: 742.
25 MacMurray, *Treaties*, vol 1, 79–81; 'Er'lu shuwen', *Shen Bao*, 12 August 1891: 1.
26 Weng, *Riji*, vol 5, 2779 and vol 6, 2939.
27 A. C. Hide, 'The alleged Cassini Convention', *The National Observer*, 12 December 1896: 84–85; 'The Cassini convention', *Herald*, 30 October 1896: 739 and 742.
28 Sidney Harcave (translator), *The memoir of Count Witte* (Armonk: M. E. Sharpe, 1990), 232.
29 Harcave, *Memoir*, introduction (xiii).
30 Harcave, *Memoir*, 237.

31 Sung-Hwan Chang, 'Russian designs on the Far East', in *Russian imperialism from Ivan the great to the revolution*, ed. Taras Hunczak (New Brunswick: Rutgers University Press, 1974), 299–321 (318).
32 Wolff, *Harbin*, 23 and 28; 'Li Hung Chang accused', *Times*, 4 April 1898: 7.
33 Harcave, *Memoir*, 83.
34 'The secret treaty', *Herald*, 1 November 1895: 718 and 'The secret treaty', 6 March 1896: 347.
35 Harcave, *Memoir*, 237.
36 Li Cunkuan, 'Guoji waijiaoshi shangde dapianju: lun Guangxu zhong'e miyue', *Shehui Kexue Zhanxian*, 2 (1987): 182–91.
37 MacMurray, *Treaties*, vol 1, 81.
38 MacMurray, *Treaties*, vol 1, 81.
39 Dukes, *Manchuria*, 8 and 12–13; Afinogenov, *Spies*, 139–58.
40 'Lun xingjun yijiandie weixian', *Shen Bao*, 11 January 1895: 1 and 'Dashiji: Guangxu er'shiyinian (1895)', Liaoningsheng dang'anguan, at www.lndangan.gov.cn/lnsdaj/index.html, accessed 5 October 2021.
41 Wolff, *Harbin*, 21.
42 'The occupation of Port Arthur', *Herald*, 24 December 1897: 1113; 'Russia to keep Port Arthur', *Times*, 23 February 1898: 7.
43 MacMurray, *Treaties*, vol 1, 119–21; 'Russian demands', *Herald*, 28 March 1898: 505.
44 'The partition of China', *Herald*, 28 March 1898: 515.
45 Harcave, *Memoir*, 277.
46 Harcave, *Memoir* (letter), King did discuss this letter, at www.rothschildarchive.org/materials/ar2007pekin.pdf, accessed 21 July 2023.
47 MacMurray, *Treaties*, vol 1, 154–57.
48 MacMurray, *Treaties*, vol 1, 79–81 (80).
49 Harcave, *Memoir*, 273–77 (277).
50 Marks, *Road to power*.
51 Charles Bright, *Imperial telegraphic communication* (London: P. S. King & Son, 1911), XIII.
52 Robinson, 'Introduction', in *Railway imperialism*, 1.
53 Robinson, 'Introduction', in *Railway imperialism*, 2.
54 Otte and Neilson, *Paths of empire*.
55 Daniel Headrick, *The tools of empire: technology and European imperialism in the nineteenth century* (New York: Oxford University Press, 1981).
56 Spiers, *Engines*, 76–95, 96–114 and 115–48.
57 Sean McMeekin, *The Berlin-Baghdad express: the Ottoman empire and Germany's bid for world power* (Cambridge: Harvard University Press, 2021).
58 Tim Zajontz, 'Railway imperialism in East Africa: laying the track for exploitation', in *The Oxford handbook of economic imperialism*, eds Zak Cope and Immanuel Ness (New York: Oxford University Press, 2022), 515–34.
59 M. P. Diogo and B. J. Navarro, 'Re-Designing Africa: railways and globalization in the era of the new imperialism', in *Technology and globalization: networks*

of experts in world history, eds D. Pretel and L. Camprubi (Cham: Palgrave Macmillan, 2018), 105–28.

60 Kees Van Dijk, *Pacific strife: the great powers and their economic and political rivalries in Asia and the Western Pacific 1870–1914* (Amsterdam: Amsterdam University Press, 2015), 23.

61 Daniel R. Headrick, *The tentacles of progress: technology transfer in the age of imperialism 1850–1940* (New York: Oxford University Press,1988), 49.

62 Paine, *Imperial rivals*, 178.

63 The Management Bureau was headed by Dmitry L. Horvat from 1903 to 1920.

64 F. A. Sutton, 'How to kill communism', *Herald*, 18 June 1927: 533.

65 Zatsepine, *Beyond the Amur*, 105.

66 MacMurray, *Treaties*, vol 1, 74–77.

67 MacMurray, *Treaties*, vol 1, 74–77 (76).

68 Wolff, *Harbin*, 27; Cheng Weirong, *Jindai dongbei tielu fushudi* (Shanghai: Shehui kexue chubanshe, 2008), 19–20.

69 Wolff, *Harbin,* 28.

70 'Tudi chanquan zhidu, Qingchao shiqi: Sha'er zhongdong tielu zhandi', Shuzi fangzhi, Shengzhi, Tudizhi, Heilongjiang shizhiwang, at www.hljszw.org.cn/bookinfo/201708230068.html, accessed 21 June 2021 and 'Statutes of the Chinese Eastern Railway', in MacMurray, *Treaties*, vol 1, 74–77 (77).

71 Bakich, 'Russian city'.

72 'Tielu yange: Zhongdong tielude tequan', Shuzi fangzhi, Shengzhi, Tieluzhi, Heilongjiang shizhiwang, at www.hljszw.org.cn/bookinfo/201708230070.html, accessed 21 June 2021.

73 'Kuanchengzi banjia', Zhishuo at Jilin difangzhi, at http://dfz.jl.gov.cn/zsjl/201712/t20171201_5216560.html, accessed 7 October 2021.

74 'Dashiji: Guangxu er'shiwunian (1899)', at Liaoningsheng dang'anguan, at www.lndangan.gov.cn/lnsdaj/index.html, accessed 2 December 2021.

75 'Chinese sovereignty in the Liaotung Peninsula', *Herald*, 27 February 1899: 330.

76 'Russian Manchuria', *Herald*, 30 May 1898: 913 and 'Russian progress', 13 March 1899: 452.

77 John Foster Frazer, 'Across Manchuria', *Herald*, 19 February 1902: 343.

78 'Statutes of the Chinese Eastern Railway Company Note 5', 'Contract for the appropriation of lands in Heilungkiang province' and 'Contract for the appropriation of lands in Kirin province', in MacMurray, *Treaties,* vol 1, 84–90 and 663–71.

79 Wolff, *Harbin*, 28.

80 MacMurray, *Treaties*, vol 1, 79–81 (80).

81 'Tielu yange: Zhongdong tielude tequan', Shuzi fangzhi, Shengzhi, Tieluzhi, Heilongjiang shizhiwang, at www.hljszw.org.cn/bookinfo/201708230070.html, accessed 9 June 2021.

82 Wolff, *Harbin*, 67.

83 'Russian progress in Manchuria', *Herald*, 13 March 1899: 452.

84 'Junshi, Zhujun: Sha'er zhanlingjun', Shuzhi fangzhi, Shixianzhi, Mudanjiangshi xiajun, Heilongjiang shizhiwang, at www.hljszw.org.cn/bookinfo/201710200378.html, accessed 6 October 2021.

85 'Mayiling kang'er zhanjidi', Tushuo at Jilin difangzhi, at http://dfz.jl.gov.cn/ybjl/201107/t20110720_5217114.html, accessed 7 October 2021.
86 'Sanxing shixian', *Shen Bao*, 18 August 1900: 1.
87 Cheng, *Fushudi*, 130.
88 B. B. Golitsyn (translated by Li Shuxiao and Tian Yigeng), *Zhongdong tielu huludui canjia 1900nian manzhoushijian jilue* (Beijing: Shangwu yinshuguan, 1984), 112–13 and 310–15.
89 Paine, *Imperial rival*, 211–19.
90 Wolff, *Harbin*, 67.
91 'Russians in Manchuria', *Herald*, 17 July 1903: 161.
92 'Distribution of Russian troops', *Herald*, 12 February 1904: 310.
93 'Distribution of Russian troops', *Herald*, 12 February 1904: 310.
94 'Russians in Manchuria', *Herald*, 30 July 1902: 226 and 'The Russian position', 30 December 1903: 1390.
95 Wcislo, *Tales*, 180.
96 Wolff, *Harbin*, 121.
97 Lattimore, *Cradle of conflict*.
98 Tan, *Zhongdong*, 304–52.
99 Wolff, *Harbin*, 121.
100 Francis Younghusband, *Among the celestials: a narrative of travels in Manchuria, across the Gobi Desert, through the Himalayas to India* (London: John Murray, 1898), 65.
101 Masafumi Adasa, 'Making a Vancouver in the Far East: "the trinity transportation system" of the Chinese Easter Railway 1896–1917', in *Russia and its northeast Asian neighbors: China, Japan and Korea 1858 to 1945*, ed Kimitaka Matsuzato (Lanham: Lexington Books, 2017), 65–88 (73).
102 'Tielu yange, Zhongdong tielu he minzu tielu shiqi: Zhongdong tielude fushu qiye', Shuzi fangzhi, Shengzhi, Tieluzhi, Heilongjiang shizhiwang, at www.hljszw.org.cn/bookinfo/201708230070.html, accessed 7 October 2021.
103 'Contract for the felling of timber in Kirin province' and 'Contract for the felling of timber in Heilungkiang province', in MacMurray, *Treaties*, vol 1, 671–74 and 721–24; Tan, *Zhongdong*, 473–515.
104 'Kaifa jianshe, Kaifa yange: Qingchao qianqi', Shuzi fangzhi, Shengzhi, Linyezhi, Heilongjiang shizhiwang, at www.hljszw.org.cn/bookinfo/201708230044.html, accessed 1 December 2021; Cheng, *Fushudi*, 183–89.
105 Wang Xiliang, 'Jindai zhongguo dongbei cenlinde zhiminkaifa yushengtai kongjian bianqian', *Heilongjiang Shenhuikexueyuan Lishiyanjiu*, 1 (2017), at www.hlass.org.cn/skyj/jphc/cglb/2017/05/19370.htm, accessed 1 December 2021.
106 Pallavi V. Das, *Colonialism, development, and the environment: railways and deforestation in British India 1860–1884* (New York: Palgrave Macmillan, 2015), 55.
107 'Dashiji: Qianlong sanshiqinian (1772)' and 'Jiaqing shiqinian (1812)', Liaoningsheng dang'anguan, at www.lndangan.gov.cn/lnsdaj/index.html, accessed 4 October 2021.
108 MacMurray, *Treaties*, vol 1, 84–91 (84 and 90).

109 'Dashiji: Daoguang shibanian (1838)', Liaoningsheng dang'anguan, at www.lndangan.gov.cn/lnsdaj/index.html, accessed 4 October 2021.
110 'Agreement for coal mining in Heilungkiang' and 'Agreements regarding the operation of coal mines in Heilungkiang province', in MacMurray, *Treaties*, vol 1, 661–62 and 658–61.
111 'Tielu yange, Zhongdong tielu he minzu tielu shiqi: Zhongdong tielude fushu qiye', Shuzi fangzhi, Shengzhi, Tieluzhi, Heilongjiang shizhiwang, at www.hljszw.org.cn/bookinfo/201708230070.html, accessed 1 December 2021; Tan, *Zhongdong*, 269–85.
112 Cheng, *Fushudi*, 204–8.
113 'Eastern Siberia', *Herald*, 16 October 1896: 661.
114 Xue Xiantian, *Zhongdong tielu hulujun yu dongbei bianjiang zhengju* (Beijing: Shehui kexue wenxian chubanshe, 1993), 38.
115 Wolff, *Harbin*, 32–33.
116 Wolff, *Harbin*, 43.
117 'Across Manchuria', *Herald*, 12 March 1902: 496 and 'Yungpingfu Chihli', 8 Apr 1904: 736; Istvan Santha, 'Hungarian witnesses of infrastructure construction in Manchuria (1877–1931): the case of the Chinese Eastern Railway', *Inner Asia*, vol 16, 1 (January 2014): 152–77.
118 'The Variag will be saved', *Times*, 25 February 1904: 2.
119 'Manchurian topics', *Herald*, 17 January 1908: 110.
120 'Russia's Chinese worry', *Times*, 5 February 1892: 2 and Lewis Siegelbaum, 'Another "Yellow Peril": Chinese migrants in the Russian Far East and the Russian reaction before 1917', *Modern Asian Studies*, vol 12, 2 (April 1978): 307–30.
121 Wolff, *Harbin*, 43.
122 Siegelbaum, 'Yellow Peril'.
123 'Russian Manchuria', *Herald*, 30 May 1898: 913; Shao Da, 'Finding family roots in Harbin's Jewish cemetery', Zhongguowang, 14 September 2004, at www.china.org.cn/english/2004/Sep/106964.htm, accessed 1 July 2018.
124 Joshua A. Fogel, 'The Japanese and the Jews: a comparative analysis of their communities in Harbin, 1898–1930', in *New frontiers: imperialism's new communities in East Asia, 1842 to 1953*, eds Robert Bickers and Christian Henriot (Manchester: Manchester University Press, 2017), 88–108.
125 Mark Gamsa, *Harbin: a cross-cultural biography* (Toronto: University of Toronto Press, 2020), 116–40; 'Zhongda shilue 44: Qiulin gongsi fenhao', Shuzi fangzhi, Shixianzhi, Daxing'anling gaikuang shangjuan, Heilongjiang shizhiwang, at www.hljszw.org.cn/bookinfo/201710200418.html, accessed 6 October 2021.
126 Bakich, 'Russian city'; James Hugh Carter, *Creating a Chinese Harbin: nationalism in an international city 1916–32* (Ithaca: Cornell University Press, 2002); 'Harbin and its inhabitants', *Herald*, 12 January 1929: 82.
127 'Seeing Harbin', *Herald*, 10 April 1926: 61.
128 Dan Ben-Canaan *et al.* eds, *Entangled histories: the transcultural past of northeast Asia* (Cham: Springer, 2013), 35–58.

129 Wolff, *Harbin*, 130; Henry Reichman, 'The 1905 revolution on the Siberian railroad', *Russian Review*, vol 47, 1 (1988): 25–48.
130 'Xin minzhu zhuyi gemin shiqi zhongda huodong: fandi fanfengjian junfa douzheng', Shuzhi fangzhi, Shengzhi, Gongcandangzhi, Heilongjiang shizhiwang, at www.hljszw.org.cn/bookinfo/201708230024.html, accessed 10 October 2021.
131 Rosenbaum, 'Manchuria Bridgehead'.
132 Li Cunkuan, 'Dapianju'.
133 Squier, *Lamb*.
134 Spiers, *Engines*, 79–92; Christopher M. Wyatt, *Afghanistan and the defence of empire: diplomacy and strategy during the great game* (London: I. B. Tauris, 2011), 62–64, 80, 105, 109, 116 and 162; John Fisher, 'Lord Curzon and British strategic railways in Central Asia before, during and after the First World War', in *Paths of empire*, 137–55.
135 Robinson, 'Introduction', in *Railway imperialism*, 2.
136 Younghusband, *Travels in Manchuria*, 74–75.
137 'Newchwang', *Herald*, 24 July 1891: 112 and 'The French Mail', 13 June 1898: 1016.
138 Pelcovits, *China hands*, 233.
139 Kent, *Enterprise*, 41–43.
140 'The Russian acquisition of Manchuria', *Herald*, 9 June 1893: 830; 'Fang'E ce', *Shen Bao*, 28 November 1892: 1 and 'Lun shiju zhibian', 26 April 1895: 1.
141 'Russian Activity at Newchwang', *The Guardian and the Observer*, 10 November 1898: 9.
142 'Niuzhuang tielu', *Shen Bao*, 25 November 1898: 1 and 'Lun Ying'Er buqia', 14 August 1898: 1.
143 'Russia, England, and the East', *Herald*, 14 August 1896: 287.
144 Zen Sun, *British interests*, 31.
145 MacMurray, *Treaties*, vol 1, 326–29.
146 John Cooper, *The unexpected story of the Nathanial Rothschild* (London: Bloomsbury, 2015), 161–62.
147 Moore, *Defining and defending*, 53–83.
148 'The end of Russian rule at Newchwang', *Herald*, 5 August 1904: 313.
149 'Mr. Hitchcock to Mr. Day, Embassy of the United States, St. Petersburg, May 25, 1898', No. 459, Office of the Historian, Department of State, USA, at https://history.state.gov/historicaldocuments/frus1898/d459, accessed 11 August 2021.
150 Austin, *Ulysses S. Grant*, 89–90.
151 'Identic notes exchanged with regards to railway interests in China', in MacMurray, *Treaties*, vol 1, 204.
152 Philip Joseph, *Foreign diplomacy in China 1894–1900* (London: G. Allen and Unwin, 1928), 387–98.
153 'Opening the door in north China', *Herald*, 6 March 1901: 422.
154 Sheng Xuanhuai telegraphs to Zhang Zhidong on 19 March 1897, sxh12-0137 and cuhk:1726962-01; to Li Hongzhang and Weng Tonghe on 3 March 1898, sxh12-0212 and cuhk:1727147–01 and 02; to Zhang Zhidong on 1

May 1898, sxh13-0015 / cuhk:1727193-01 and 02; to Wang Wenshao and Zhang Zhidong on 7 June 1898, sxh13-0019 and cuhk:1727203–01 and 02, Sheng Xuanhuai Archive, The Chinese University of Hong Kong. For more, see https://repository.lib.cuhk.edu.hk/.

155 'The Lu-Han railway', *Herald*, 22 May 1899: 909.

156 'Zhengtai tielu xingche xiangxi hetong', *Shen Bao*, 28 December 1902: 2; 'Russia and China loan contract and operating contract for the Chengtingfu-Taiyuanfu railway', in MacMurray, *Treaties*, vol 1, 356–69.

157 Kent, *Enterprise*, 170–71.

158 Henry Philip Picot, *Railways in western Asia* (London: Central Asia Society, 1904), 29.

159 MacMurray, *Treaties*, vol 1, 335–36 (335).

160 Ruth Rogaski, *Knowing Manchuria: environments, the senses and natural knowledge on an Asian borderland* (Chicago: University of Chicago Press, 2022).

161 'Past and present of the C. E. railway', *Herald*, 13 Sep 1924: 417.

162 Kang Chao, *The economic development of Manchuria: the rise of a frontier economy* (Ann Arbor: Center for Chinese Studies, University of Michigan, 1982).

5

'By order of the Kaiser': Germany outmanoeuvres[1]

> Summarizing my observations, I find here a coalfield of considerable extent, which has a very considerable future ahead of it. So far, there has been a complete lack of drilling attempts. I found only one attempt made at drilling at all in the region. The storage is not disturbed much, the angles of incidence are small, there is no interference other than by very recent structures. The region needs nothing but a firm intervention of foreign enterprise and the construction of a railway line in order to achieve a significant upswing. The presence of rich iron ore will be particularly favourable to it.[2]

Ferdinand von Richthofen drafted this entry in his diary on 30 March 1869. The 'place' was Yizhou (see Figure 5.2), Shandong province, where his prediction would materialise in the 1890s. Richthofen was a geographer who spent more than three years researching in China, having served under the first Prussian mission to the Qing led by Count Friedrich zu Eulenburg.[3] Upon his return to Germany, he authored many books, coining the phrase the 'Silk Road'.[4] He became the Chairman of the Berlin Geographical Society, the Rector of Berlin University and also advised the German *Kolonialrat* (Colonial Council). His geographical work earned him a place as an Honorary Member of the North China Branch of the Royal Asiatic Society.[5] But his research served to fuel and facilitate German expansion in China. This chapter focuses on German-built railways, charting how the 'Pacific strife' enabled Germany to lay claim to, militarise, exploit and colonise Qingdao, and Shandong more generally.[6] The geopolitical rivalry that defined and accompanied the introduction of the CER intensified with the arrival of Germany.

Cousins racing for 'a place in the sun'[7]

Having concluded his Russian trip, Li Hongzhang arrived in Germany on 13 June 1896, accompanied by his German advisers, Gustav von Detring and Constantin von Hanneken.[8] Pampered by his hosts, Li was presented with his favourite cigar and songbird (the thrush) upon his arrival at the hotel.[9] After an audience with Kaiser Wilhelm II, Li 'took five o'clock tea with the Empress of Germany at the tennis lawn in Potsdam', before attending the parade of imperial guards.[10] The *Herald* referred to him as the 'uncrowned King of China'.[11] He visited military sites, including the naval base in Kiel, the weapons maker Friedrich Alfred Krupp and Aktien-Gesellschaft Vulcan Stettin, where nine of his Beiyang Navy ships were made. He cultivated a friendship with Otto von Bismarck who referred to Li as the 'Bismarck of the East' and entertained the Chinese guest at his home.[12] Did Li know that Bismarck was the architect of German expansion? Li admired German military prowess and weapons. Li's justification for his amiable attitude towards Berlin rested on the fact that the Germans had not waged wars on China and had even lent the Qing a helping hand during the 'Triple Intervention'. But Germany had not been idle. Without having participated in the Second Opium War, Eulenberg's mission managed to procure their share of all privileges ceded to Britain, France, Russia and the United States. But the Germans were hungry for more.

Calls for a 'German Macao or Hong Kong' had become more persistent by this time, having initially emanated after the unification of Germany in 1871. George Steinmetz has written expertly on German relations with China at this juncture.[13] The German Navy spearheaded the race as they began to survey ports along the Chinese coast. The *Wolff*, for example, sailed into Newchwang and conducted a survey on the Liao River as early as 1881.[14] Newchwang was a British stronghold contested by the Russians, but with German interest now also piqued, the scramble for Manchuria had begun in earnest. One of the catalysts for the call of a German colony in China was the Sino-French War where Germany was the 'main exporter of weapons to China'.[15] The call became so unrelenting that it alarmed the Chinese Ambassador to Germany, Xu Jingcheng, who expressed his concern to the Qing court by underlining the importance of naval defence in Jiaozhou on the south side of the Shandong Peninsula in 1886.[16]

The ambition to catch up with his cousins on the British and Russian thrones, as well as his neighbour to the west, became an obsession for Wilhelm II.[17] British dominance, and French and Russian gains strengthened German resolve to expand their territories and influence. This determination was reinforced after the First Sino-Japanese War, a conflict that, according to Julius Becker, transformed 'Wilhelmine foreign and world policy' due to Berlin's fears of the 'prospect of being left empty handed in an upcoming scramble for China'.[18] Kees Van Dijk believes that it was German ambition that 'set in motion a development that culminated in the dividing up of the Western Pacific'.[19] Edmund Friedrich Gustav von Heyking, then German Ambassador to the Qing, outrightly asked the *Zongli Yamen* for a harbour in 1895.[20] When this was refused, Heyking suggested using missionary incidents as a pretext. The Chinese refusal proved inconsequential, as the German East Asian squadron visited a number of ports on the Yellow Sea coast, including Jiaozhou and Haizhou (Lianyungang) in January 1896.[21] This squadron was under the leadership of Rear Admiral von Tirpitz, a man Klaus Mühlhahn described as the 'main architect' of German colonial enterprise in China.[22] Wilhelm II instructed Admiral Eduard von Knorr 'to prepare a plan for the occupation of Jiaozhou Bay' in November 1896, four months after Li left Germany.[23] The choice of Jiaozhou had its roots in Richthofen's influential works and the resources Shandong had to offer. Germany was ready to strike.

Was Li Hongzhang aware of this rapid development in German policy while he was in Europe? He had raised the issue of Jiaozhou's defence during and after the Sino-French War in July 1886, buttressing its fortifications in 1892 and reiterating its importance in 1894, just before the outbreak of the First Sino-Japanese War.[24] He clearly understood the importance of Jiaozhou, but he also relished the visit to Germany and enjoyed the deference that had been afforded him there.[25] Li believed that Sino-German relations were affable, failing to see that they had very different aims: while the Qing looked for allies, Germany was searching for opportunities to expand. The *Herald* discussed the complaint by a German newspaper (*Vossische Zeitung*) immediately after Li left:

> too much has been done, perhaps more than is compatible with the dignity of a Power so civilised as Germany, certainly more than was wise in the case of the extremely shrewd representative of the most arrogant nation of the Yellow race. In the German capital, Li Hung-chang was overwhelmed with honours as though he himself was the almighty ruler ... In the provinces every effort has been made to outbid Berlin ... Prince Bismarck could not have been more splendidly received in Stettin, Essen, Cologne, and other centres of German industry than was Li Hung-chang, for whom was invented in Germany, not in China, as may have been imagined, the flattering appellation of the 'Chinese Bismarck'.[26]

Li's visit produced no immediate benefit for Germany. However, the opportunity for expansion into China soon transpired into the kind of missionary incident Heyking had hoped for. On 1 November 1897, two German missionaries, Richard Henle and Francis Xavier Nies from the Society of the Divine Word, were murdered in Juye, southwest Shandong.[27] The Juye Incident could have been resolved diplomatically were it not for the opportunity it presented to the Germans. Wilhelm II claimed in his memoir that during his visit to Russia in August 1897, he had expressed his desire for Jiaozhou Bay to Nicholas II. He recorded that 'The Tsar stated that he had no interest in the territory south of the Tientsin-Peking line, which meant that there was no reason why he should place obstacles in our path in Shantung.'[28] Chris Miller believes that the Kaiser encouraged Russian expansion into China because 'the more resources Russia devoted to the Far East, the few [sic] it would deploy in Europe'.[29] But it also means that he could follow his cousin to East Asia. When news of the killings arrived, Tirpitz, now State Secretary for the Imperial Navy, found the perfect excuse for action.[30] The Kaiser took personal charge, instructing Admiral Otto von Diederichs, then stationed in Wusong, Shanghai, to set sail for Jiaozhou.[31] On 14 November, 'By order of the Kaiser', Diederichs seized Jiaozhou on the southeast coast, far from Juye in the southwest interior.[32] The two unfortunate missionaries were quickly forgotten and became mere footnotes in history.

Russia had also set its sights on Jiaozhou, as can be seen from the Cassini Convention. The German landing in Jiaozhou in November 1897 prompted the Russian Navy to occupy Port Arthur on 14 December, forcing the Li–Pavlov treaty on the Qing which led to the vertical segment of the CER. Is it possible that the Tsar had encouraged his cousin to occupy Jiaozhou in order to create the justification needed to extend the CER and, more importantly, to gain Port Arthur? Wilhelm II had done his cousin a favour by giving Russia the pretext to secure their interests in southern Manchuria. Bernhard von Bülow, then Foreign Minister, was pleased with the result, calling it 'one of the most significant actions in modern German history, and which secured for us "a place in the sun" in the Far East, on the shores of the Pacific Ocean'.[33] But he seemed to have forgotten all about this in 1906, during his tenure as Chancellor, when he stated 'Germany had never been guilty of any adventure or policy of territorial aggression in the Far East' in a speech.[34]

Jiaozhou sat in a most strategic place (see Figure 4.1). The naval thoroughfare to Tianjin, throat of Beijing and home to the Beiyang Navy, was now controlled by Russia on the Liaodong Peninsula, Britain on the north side of the Shandong Peninsula (Yantai, procured with the *Chefoo Convention* in 1876) and now Germany on the south side, which also overlooks Korea and Japan in the distance. The golden triangle of northeast Asia had been claimed by three European powers, with Japan in control of Korea. Richthofen called Jiaozhou the *Eingangspforte* or 'entrance gate' of Shandong.[35] The geographer was right. Sitting between the Bohai and Yellow Seas, the Shandong Peninsula looks at an arm stretching out to afford the shortest route to north China. He highlighted the region's resources, 'the position of the beds near the sea coast will result in the mines of Shandong being the first in China to be made available for foreign use'.[36] Richthofen's geographical surveys furnished Germany with the most reliable information for its expansion.[37] Maria M. Ogilvie Gordon concluded that Richthofen's field work was the 'basis for the action of Germany'.[38] Steinmetz believes that Richthofen made 'one of the most significant German contributions to an explicitly colonial frame of China in the decades leading up to the annexation of Qingdao'.[39] Shellen Wu has called researchers like Richthofen 'agents of science and empire'.[40] They were not just scholars but also 'spies' as Gregory Afinogenov fittingly labelled them.[41]

'Infrastructures make empires'[42]

An undeniable game changer, German railway imperialism would further develop the model advanced in Chapter 4. German land acquisition commenced on the day they landed in Jiaozhou on 14 November 1897, long before any diplomatic negotiation began and the *Convention Respecting the Lease of Kiaochow* was signed in March 1898.[43] The Qingdao Municipal Archives holds public notices issued by the German Navy announcing the freeze of land sale.[44] Using the slain missionaries as a pretext, Admiral Diederichs issued a proclamation in early December demarcating the boundary of German-held territory, stating that 'they are to be held by the German forces, until the case of the murder of our German missionaries in Shantung be settled'.[45] But the Germans would remain for seventeen years until November 1914, when they were defeated by Anglo-Japanese forces. Between December 1897 and January 1898, the German Navy classified land beyond Jiaozhou. They reached as far as Jimo, a town to the north. They based themselves in the local Confucius Temple, the most sacred place to the Chinese people, damaging the statue of the ancient philosopher: Chinese sources indicate that, in an act of extreme disrespect, they gouged out his eyes.[46] Locals retaliated by attacking the Germans with rudimentary weapons, and one German soldier was killed.[47]

This was only the beginning of Shandong resistance to German aggression. This kind of German brutality was also experienced during the construction of railways in Namibia and Tanzania.[48] In Shandong, it would ultimately culminate in the Boxer Rebellion. The Germans went as far west as Yizhou and as far north as Qingzhou.[49] The *Lease* signed in March 1898 gave the Germans an 'area of 553 square kilometres encompassing the village of Qingdao, several larger towns, and 275 tiny villages'.[50] This was hardly what they would secure. Section II, 'Railway and Mining Affairs,' stipulates that:

> The Chinese Government sanctions the construction by Germany of two lines of railway in Shantung. The first will run from Kiao-chau via Weihsien,

> Tsingchofu, Poshan, Tzechwan, and Tsowping to Tsinan and the boundary of Shantung. The second line will run from Kiao-chau to Ichowfu, and from there to Tsinan via Laiwuhsien.[51]

This was the Jiaoji [胶济] railways, from Jiaozhou to Jinan (see Figure 5.2), the capital of Shandong province, with the second line making a detour to Laiwu and Yizhou, the place rich with coal which Richthofen mentioned in his diary.[52] Prussian victory in the Franco-Prussian War had proved the power of railways in armed conflicts.[53] Just as the route of the CER revealed Russian designs for not just Manchuria, but also north China and Korea, the route of the Jiaoji railway illuminated German plans for further penetration into not just the Shandong interior but also north and east-central China. The Jiaoji railway commenced at the southeast end of the Peninsula, marching west to the interior, crossing all major cities to arrive at provincial capital. Richthofen travelled to Shandong via the Grand Canal. Accessing this economic artery was reinforced by the second line that dipped down to Yizhou, next to Zaozhuang which sits on the Grand Canal with access to Tianjin, Beijing, Nanjing, Suzhou, hence Shanghai.

Like a dagger, the Jiaoji railway stabs into the heart of north China, with its tip pointing at the Grand Canal and its handle resting firmly on Jiaozhou. As with the CER, the Jiaoji provided a licence to requisition more land, as stipulated in Article IV:

> The Chinese Government shall allow German subjects to hold and develop mining property for a distance of 30 *li* from each side of those railways, and along the whole extent of the lines. The following places where mining operations may be carried on are particularly specified: Weihsien and Poshan along the line of the northern railway from Kiau-chau to Tsinan, and from Ichow, Laiwuhsien, etc., along the southern or Kiao-chau-Ichow-Tsinan line.[54]

Thirty Chinese *li* is the equivalent of 15 km or 9.32 miles. With the *Lease* specifying access on 'each side of those railways', this meant that the Germans could legally claim nearly 20 miles of land fanning out from the lines as its belt of alienation where they also secured mining rights. Just as with the CER, all kinds of facilities arose to support the construction and operation of the Jiaoji, such as the Hauptreparatur-Werkstätte Syfang der Schantung-Eisenbahn or the Sifang Main Repair Factory of the Shandong Railway. The *Regulation on the Purchase of Land* was published in September 1898, followed by others regarding taxation.[55] They stipulated that the German authority had the first right to purchase land and subjected the Chinese to strict controls. Germany became a land owner in Shandong; they appropriated land and made the Chinese pay taxes for their land.

As the route survey gathered pace and construction began, the Germans, as Steinmetz detailed, 'destroyed farmers' irrigation systems, divided their fields, violated ancestral burial sites and generally infuriated villagers, who responded by sabotaging the railway tracks and destroying offices'.[56] The *Herald* reported German atrocities, including the burning of two villages in May 1899.[57] While the Qing court may have been easy to bully, the people of Shandong, robbed of their dignity and livelihood, found ways to retaliate. When they reached Gaomi in June 1899, the Germans were met with a rebellion as they based themselves in the local school. Villagers destroyed railway works, uprooted telegraph poles and surrounded the foreigners with weapons ranging from kitchen knives to sickles. Mühlhahn believes that the Germans killed at least 450 people, many women and children.[58] When they finally left Gaomi, they destroyed the school and burnt its library. Both the *Herald* and *Shen Bao* detailed the resistance at Gaomi.[59] Such was the legacy of the Gaomi uprising that the event was fictionalised in *Sandalwood Punishment*. A descendant of the Gaomi rebels, its author, Mo Yan (pen name), became China's first winner of the Noble Prize in Literature in 2012.[60]

This kind of resistance against railway construction at Gaomi constituted a notable precursor to the Boxer movement.[61] Historians of China have frequently attributed the outbreak of the rebellion to missionaries who had been operating in Shandong.[62] Although there had certainly been friction with the locals, it was the heavy-handed military-style aggression that accompanied railway construction which contributed to increasing resistance and the uprising. We can also see evidence in *Shen Bao*, which reported that Germans retaliated in the name of suppression, killing as many as 500 villagers in one campaign alone in November 1900.[63] Such aggression explains why the Boxers attacked railways, telegraph lines in addition to churches, missionaries and converts.[64] German railway imperialism sheds light on the origin of the Boxer Rebellion.

The German Navy spearheaded and executed the capture of Jiaozhou, and asserted its power through the control of both the railways and the colony.[65] Klaus Mühlhahn has argued that 'neither economic nor diplomatic but military considerations were at the centre of Tirpitz's blueprint for China'.[66] Indeed, Tirpitz called this expansion the 'Empire of the Navy'.[67] Steinmetz is to the point: 'Kiaochow was administrated directly by the German navy rather than the Foreign Office, an anomaly within the German colonial empire.'[68] It was no accident that the governors were all ranking naval commanders. Like Witte, Tirpitz took personal control, 'reading and commenting on almost every report from China' even after he became naval minister.[69] He fired any governor who disagreed with him, including Carl Rosendahl and Paul Jaeschke.[70] The Navy-steered 'empire' placed enormous importance on the bay area, which can be seen from the clause 'Rights ceded to German troops' in the *Lease*:

> His Majesty the Emperor of China, guided by the intention to strengthen the friendly relations between China and Germany, and at the same time to increase the military readiness of the Chinese Empire, engages, while reserving to himself all rights of sovereignty in a zone of 50 kilometres (100 *li*) surrounding the Bay of Kiao-chau at high-water, to permit the free passage of German troops within this zone at any time, as also to abstain from taking measures, or issuing any Ordinances therein, without the previous consent of the German Government, and especially to place no obstacle in the way of any regulation of the water-courses which may prove to be necessary.[71]

As with the Russians, the Germans' intentions 'to increase the military readiness of the Chinese Empire' sounded noble. The German Navy would militarise the region, guarding Jiaozhou with the latest battleships, further buttressed by fortresses on land and along the coast. The most important of these was the Bismarck Fortress on Jiaozhou hill, standing high above the bay overlooking the Yellow Sea. The Germans fortified it with Krupp cannons – an ironic twist considering Li Hongzhang's appreciation of Krupp weapons. They also dug deep into the hill and constructed an underground fortress that was divided into three stories. The top floor was a watchtower, a 360-degree revolving observatory that took in the surrounding view from land to sea. The middle level was the command centre, furnished with everything needed for war, from communication hub and hospital to emergency exit and sewage system. The lower level was a depot for ammunition and food supplies, including cells for prisoner interrogation. This was an independent citadel purposely built to function autonomously during war, and indeed as it would during both world wars. The entrance was protected by three separate iron gates that were bullet-proof, gas-proof and water-proof. There was also a reservoir, providing enough water for hundreds of soldiers. To protect this stronghold, elaborate and secret trenches and shelters were constructed around the hill hidden among rocks and trees.[72]

There were over forty forts named after high-ranking officials in German politics and military, with more planned.[73] Jiaozhou and southeast Shandong had become a gigantic military fortress. The Krupp cannons of the Bismarck Fortress pointed to the sea. Who, then, were the Germans most afraid of? Certainly not the Chinese. This kind of military build-up was a trademark of railway imperialism. Jiaozhou and the Jiaoji railways became the prey of Germany's rivals: the British and the Japanese. Unlike those two powers and the Russians, who could call for help nearby, the Germans were limited in number: a few thousand scattered in Shandong and on the sea. This would have adverse consequences as we move into the 1910s, when the Germans found themselves increasingly isolated in world politics. Bülow realised that in 1914: 'We are now vulnerable at sea. We have entrusted millions to the ocean, and with these millions the weal and woe of our countrymen.'[74] The Jiaoji railways turned Shandong into another 'bridgehead'.

The German Shandong Railway Mining Company was established immediately after the *Lease* was signed in March 1898.[75] As the route survey stretched westward, it signalled to a more important matter without which the railway could not run: coal. The importance of this resource was underscored by Richthofen, who mentioned it many times in his diary.[76] The two lines travelled closely alongside coal deposits, the two largest being Weifang and Zibo. Thanks to Richthofen, the Germans had made the most economically informed choice for their railways. The Chinese had been mining coal in Shandong for nearly 2,000 years and there were at least 200 pits in 1840. Shandong merchants had managed a thriving business before the Germans arrived. A correspondent of the *Herald* saw an 'out-crop of coal' being mined in Qingzhou in 1896.[77] Alexander Armstrong, the principal of the Chefoo School founded by China Inland Mission, knew that:

> Coal is in great abundance throughout the province. The principal fields are in the vicinities of Poshan, Wei-hsien, I-chou, I-hsien, and Teng-chou. The first named is of the most importance. The Poshan coal and coke are excellent in quality. The mines are well managed, their product being sent over a large area of Shantung, and some of it even exported.[78]

Shandong's coal was exported before the German landing in November 1897 with Jinan alone having sixteen working mines.[79] Richthofen was correct that Shandong was thus the perfect coaling station, and the Germans were already drilling in Weixian by February 1899.[80]

> Of great importance for the railway is the progress made in the exploitation of neighbouring coal districts. From the nearer coal mines at Weihsien the first load of 150 tons was carried to Tsingtao on October 30th, 1902. The quality of the coal is, on account of its greater heating power and more sparing development of soot, rather superior to the Japanese coal now prevalent in the ports of Eastern Asia. A second district of coal fields is situated near the branch line (in course of construction) to the Poshan valley ...[81]

Shandong's coal was superior to its Japanese counterpart. The German company opened new mines and they also took over existing mines that had been under Chinese control, including Zibuo. The Zibuo and Fangzi coal mines supplied the Germans with more than 3,414,632 tons of coal between 1905 and 1913.[82] They also took over mines beyond the 30 *li* or 9.32 miles remit prescribed in the *Lease*, by extending existing railways.

Zaozhuang on the far southwest corner of Shandong is an example. Situated roughly 256 *li* or 79 miles from Yizhou where the second German railway passed, it sat outside the defined railway zone and was managed by the Chinese until it was taken over. It would become the most profitable mine for the Germans, as *Shen Bao* reported: 'Germany has its hands on the huge profit of coal in Shandong.'[83] They would procure rights to operate more mines and build more railways.[84] This meant further penetration into the interior and more profit. They included the Zhangbo [张博], Zhangdian to Boshan, railway, completed in 1904; and the Linzao [临枣], from Lincheng to Zaozhuang line, finished in 1912. Further plans included the Shide [石德] line, from Shijiazhuang on the Luhan railroad to Dezhou in Shandong. The outbreak of WWI put a hold on this important line; but WWII would revive its construction.

Railways had become an instrument for empire – as had coal. They were what I term the 'twin' engines of empire. Historians have dissected the intimate relationship between coal and empire. On Barak has argued that 'Britain's industrialisation and imperialism were not separate processes, both – not just the former – were predicated on coal.'[85] As the empire expanded, Britain secured 'the most complete chain of coaling stations all round the world,' from Gibraltar and Port Said to Mokha and Bombay.[86] The French and the Russians also secured their own chain of stations, and now the Germans, and soon the Japanese, would follow suit.[87] The race for coal led to a scramble that was equally if not more ferocious than that over railway concessions. Like railways, coal had become a strategic weapon, hence a source of competition and conflict. Barak called it 'coalonialism,' believing that 1820 to 1911 constituted the 'age of coal'.[88] Quan Hansheng and Tim Wright pioneered the study of the history of coal in China.[89] Shellen Wu has authored excellent works on how coal fuelled China's modernisation.[90] Victor Seow has examined the development of the Fushun coal mine.[91] But less has been written about how coal worked in tandem with railways to fuel imperialism.

Indeed, commercial considerations were important to the Germans. As Tirpitz recorded in his memoir, 'If German trade was ever to cease being the go-between for English and Chinese products and to begin putting German wares on the Asiatic market, it needs its own Hong Kong just as our squadron did.'[92] The Jiaoji railways would facilitate German trade and turn his 'empire of the Navy' into a profitable enterprise. In 1902, 251 ships with 368,057 tons of goods sailed into Jiaozhou. This number increased every year, reaching 936 ships with 1,309,294 tons by 1913, just before the outbreak of the WWI.[93] The Germans pushed for more branch lines into the interior and other rights, including the right to produce salt – a most profitable businesses reserved for the Qing government.[94] An Imperial Maritime

Customs was established in Jiaozhou. The Qingdao Municipal Archive compiled its tax income (silver tael):

192,918 in 1902
1,900,000 in 1913
1,197,000 in 1914[95]

The 1914 figure is for August, the eve of WWI in China. The value of import and export from Qingdao rose from 3,950,000 in 1900 to 59,160,000 (tael) in 1913.[96] As for the total profit of the Jiaoji railway from its completion in 1904 to 1914, the passenger share was about 30 per cent with 70 per cent in freight, half of which was coal.[97] The railway's net income was more than 1,912,295 (tael) in 1905, reaching more than 3,730,242 by 1911.[98] Despite the interruption by WWI, Sino-German trade flourished into the outbreak of WWII, due to a 'conjunction of interests between a heavily industrialised Germany in search of new export market and sources of new raw materials, and a China committed to rapid industrial development', as William Kirby concluded.[99]

Like the CER, the Jiaoji railways could not have been built, nor the coal mined or trade carried out, without workers, and Shandong was famous for its coolies. Alexander Armstrong noticed that:

> In this part of China the natives are generally of larger and stronger build than those further south. Their power of endurance is much greater, though they lack the manner of the educated southerners. Amongst the men, many work in the fields. All the different handicrafts are represented, and foreigners are filled with wonder to see what work joiners, carpenters, masons, blacksmiths, etc., produce with even the most primitive tool. Others of the men become merchants, hucksters, boatmen, mule drivers, barrowmen, soldiers, priests and beggars.[100]

Possibly due to his encounter with Chinese coolies during his stay in California, Richthofen paid special attention to Shandong men and their 'diligence and frugality', stating they had 'well-built bodies and tough muscles'.[101] The coolie trade may have begun in Guangdong as thousands were recruited to build railways in north America, but Shandong soon became synonymous with the 'coolie'. Thousands of Shandong men were put to work constructing the Jiaoji railways, to toil in the German coal mines and to serve the Germans in the colony.[102] Mainland researchers have recently claimed that those who built the Bismarck Fortress were put to death at its completion in order to keep the project a secret.[103] Moreover, thousands more were sent to the Pacific German colonies of New Guinea and Samoa. Andreas Steen and John A. Moses highlight the role of coolies in the German imperial project; Steen believes that 1.6 million were taken out of China.[104]

Shandong coolies were also exported to Manchuria and Siberia to build the CER and the Trans-Siberian railways. The Russian Eastern Asiatic Steamship Company opened a special service between Yantai and Vladivostok to facilitate the coolie trade in 1908.[105] The Allies also came to recruit in Yantai and Jiaozhou during WWI.[106] As a result, more than 140,000 labourers, mostly from Shandong, were despatched to Europe.[107] Some 94,500 of these laboured for the British Chinese Labour Corp, while 30,000 worked for the French.[108] They built roads and railways, dug trenches, carried weapons and served as nurses, blacksmiths, bricklayers, carpenters, fitters and masons. Hundreds died; some were buried in France and Britain. However, this sacrifice did not earn the respect of the Allies, who, at the end of WWI, were ready to hand German railways, mines and colony in Shandong to Japan.[109]

The Jiaoji railways ushered in what the *Herald* called the 'Germanisation of Shandong'.[110] The German Navy gave Jiaozhou a new name: Qingdao (Tsintau/Tsingtao), which translates as 'green isle'. It would become their proud 'musterkolonie' or 'model colony'.[111] To build a German city upon the old Jiaozhou, the Chinese were 'driven out' and their villages razed to the ground.[112] The Germans soon established an exclusive 'villa district', which boasted a modern water and sewage system.[113] The Chinese were forbidden from residing in this district by the 'General Regulations for the Maintenance of Public Health', which cited 'sanitary reasons' for the proscription.[114] Much like Harbin where the Chinese were pushed into a ghetto called 'China street', the Germans created a 'Chinesen-Stadt' or Chinese city. Insults such as 'dirty' and 'pestilential filth' characterised German descriptions of the Chinese, leading to racial segregation.[115] German residents enjoyed all kind of privileges, whereas the Chinese were subjected to both Chinese and German law and punishment. As Steinmetz revealed, even German children could whip the Chinese.[116] For minor offences, 'a policeman can impose a punishment of ten strokes on the spot without any trial'.[117] German attitudes towards the Chinese was epitomised by Elisabeth von Heyking's famous words: 'Whatever the Chinese might have been in the past, today they are nothing but dirty barbarians who need a European master and not a European ambassador – the sooner the better.'[118]

Germanisation transformed the look of Qingdao, the terminus of the railways and heart of Deutsch–China. It became the perfect summer destination for Europeans.[119] A correspondent of the *Herald* paid a visit to Qingdao in April 1903 just before the Jiaoji railway was completed: 'The impression made on the writer, upon approaching Tsingtao from the sea for the first time, was that a collection of toy houses set down on a shore of glittering yellow sand: so clean and new of red brick and white plaster.'[120] By the time the first railway was completed in June 1904, Qingdao was a fully fledged German city with public institutions, churches, schools, hospitals, factories, shops and restaurants. The German bank issued its own currency for use in the colony and railway zone. Qingdao looked like a European city, a 'Switzerland of the East' (see Figure 5.1).

Labelled 'German Hong Kong', Qingdao was, and still is, famous for its beer.[121] The population had grown from a 'couple of thousand' in 1898 to 55,700 by 1913.[122] For Germans, 'Far away at home in Qingdao' was an apt description of the city.[123] Sabina Groeneveld has studied German memoirs and recorded their impressions:

> The city is half a holiday colony, half a big exhibition, where everything is complete, all can be seen in action, everything is new and perfect to the last detail. The only things missing is the vibrant, active life of at least ten thousand Europeans. ... The traveller, who is used to the half-civilized and adventurous cities of the East Asian coast, ... gets from this Tsingtau a strange and nearly unnatural impression, it seems that Tsingtau is like a piece of Germany that fell from the sky.[124]

As a matter of fact, 'A piece of Germany that fell from the sky' still characterises Qingdao today.[125] Over time, the German imperialist attitudes began to change. Both Steinmetz and Mühlhahn agree that after 1905, the Germans began to see the Chinese more favourably.[126] This was due to three factors: first, economic, as they needed the collaboration of the locals; second, Chinese resistance; and third, geopolitics, as Germany found itself increasingly isolated in Europe.[127] It seems ironic that isolation

Figure 5.1 'Switzerland of the East' today (Baidu public domain)

at home directed German attention toward the Chinese. The founding of the Deutsch-Chinesische Hochschule or German-Chinese College in 1909 is a compelling example of this. The Germans finally realised that the Chinese were not their adversaries as they faced stiff competition from other powers, the British in particular.

The Shandong bridgehead: genesis

> Lines extending from Berlin, Vienna, and Constantinople into the Balkans did much to bring about partition there, while the projection of the trans-Siberian line to the Pacific, together with that of the Berlin-Constantinople line to Baghdad, drove the Powers to partition the Chinese and Ottoman empires with exclusive railroad rights.[128]

The partition of Manchuria and Shandong certainly followed the tracks laid for railways. German gains in Shandong served to whet the appetite of others. If the CER turned Manchuria into one 'bridgehead', the Jiaoji railways would turn Shandong into another. By the 1890s, Germany had risen to be a major competitor in China. This agitated the British who were trying to maintain its position of dominance. Already challenged by the Russians in north China and the French in the south, the British now had to battle the Germans in north–central and east–central China. Upon hearing of the German *Lease*, the British demanded a new naval station in Zhoushan in March 1898 in an effort to shore up its strategic possessions.[129] They called Jiaozhou a 'Chinese Pondicherry', in reference to the much sought-after Bengal region.[130] China was not the only place the Germans and the British battled it out. Africa was another arena where the ardent British imperialist, Cecil Rhodes, used the railway to expand north from the Cape Colony to the Boer Republics and ultimately Cairo, but his efforts were frustrated by the Germans, the Belgium King and the Portuguese.[131] They also battled each other in the Pacific; Kees Van Dijk has written about their fight for Fiji and Samoa.[132]

The *Boxer Protocol* awarded Germany the second largest indemnity of more than 90 million taels of silver.[133] Despite this, the Germans were not content. Richthofen had already outlined what Germany's next steps should be:

> The importance of Kiautschou as a seaport is not limited to the province of Shantung. Efforts must be made to transform it into the maritime gateway to the north-eastern provinces of China and beyond. This, too, is only possible by means of railways, ... as the rivers are frozen in winter and Tientsin is closed to shipping. For the same reason, Beijing is also cut off from lighter traffic in winter and can only be reached by arduous land routes. The nearest port from

which these areas can be accessed by railways during the cold season is that of Kiautschou.[134]

Here the geographer sounds more like a shrewd businessman or a visionary politician. Tianjin was the hub of an emerging railway network in north China, with access to Manchuria and Beijing. As Richthofen had foreseen, the Jiaoji railway provided the best means to expand further into north China. But expanding the network would put them in conflict with the British. The British had been talking about a line centred in Nanjing that would stretch northward to Beijing via Tianjin and southward to Shanghai.[135] Lying between these cities, Shandong was considered by the German Reich as its exclusive sphere.[136] Just as the Russian push into Manchuria led to increased Anglo-Russo rivalry, the German drive into north China would fuel Anglo-German competition. While the Germans began to build the Jiaoji, construction of the Luhan line finally commenced in 1898. Connecting the Jiaoji railway with the Luhan would lead not just to Beijing, but also to Hankou, the 'great emporium of central China'.[137] This would open new horizons, but covering the long distance from the Jiaoji to any point on the Luhan would be costly and would take skilful negotiation with the Belgians who had been awarded the rights to finance and build it. This meant the Germans needed to find a shorter and German-controlled route.

Officials in Jiangsu province secured the permission to build the Jinpu [津浦] railway from Tianjin to Pukou, a district in Nanjing on the north side of the Yangzi River, in 1898. The Jinpu would be a significant development due to its link with Tianjin, which meant access to the Barrier railway, the Jinlu and the Luhan railways. Connecting with Nanjing was equally significant, as the city sat in the heart of the most developed region, with easy access to Shanghai. If they could procure the rights to build the Jinpu line, the Germans would have obtained access to not just Manchuria and the capital but also to south–central and eastern China – an exponential gain. The planned Jinpu line would have to pass through Shandong, and this provided the best opportunity. But the Jiangsu authority was leaning towards an Anglo-American syndicate with Rong Hong (Yung Wing) as their agent.[138] The Germans and the British jostled hard for as long as a decade for the rights to finance and construct the Jinpu line. Sean McMeekin has argued that the Baghdad railway outlined Germany's 'bid for world power'.[139] The Jinpu was now compared to the 'Baghdad railway of the Far East'.[140] They battled each other until the *Agreement for the Imperial Chinese Government five-percent Tientsin-Pukou Railway Loan* was signed in January 1908.[141] Their solution was to cut the Jinpu in half:

This loan is designed to provide capital for the construction of a government railway line from a point connecting the Imperial Railways of North China

> at or near Tientsin through Techou and Tsinanfu to Ihsien near the southern frontier of Shantung, hereafter known as the Northern Section of the Tientsin-Pukou Railway Line, and from Ihsien to a point at or near Pukou (opposite Nanking on the Yangtze-kiang), hereafter known as the Southern Section of the Tientsin-Pukou Railway Line.[142]

The 'Northern Section', from Tianjin to Yixian (Yizhou in Richthofen's diary) in Shandong province, was to be funded by Germans, whereas the 'Southern', from Yixian to Pukou in Nanjing, would be financed by the British, who also received permission to build the Nanjing-Shanghai line in 1898 – a double win.

Further benefit would soon present itself along the line. The Jinpu would pass Xuzhou on its way south. Xuzhou was significant because the Bianluo [汴洛], Kaifeng to Luoyang, railway was near completion, and it would be extended to Xuzhou. Though very short, this horizontal line was paramount because it crossed the Luhan railroad in Zhengzhou. Connecting the Jiaoji with the Luhan had been Richthofen's vision. There were plans for further extension: from Zhengzhou, the Bianluo would travel east, passing Xuzhou, and terminate at Haizhou on the Yellow Sea coast, which the German cruiser *Prinzess Wilhelm* had explored in 1896.[143] On the west side, the line would criss-cross north central China into the northwest and terminate at Lanzhou, Gansu province, gateway to Central Asia, turning it into the Longhai [陇海] railway. Tapping into the Longhai would give Germany access to the vast northwest interior. When the construction of the Jinpu began in 1908, the British had finished the Shanghai to Nanjing line. Once completed, the Jinpu would have access to the financial capital of China, with the caveat that although the two lines would meet in Nanjing, they were separated by the mighty Yangzi River. Modern bridge building, just like railway construction, was another venture that needed funds and expertise, which the late Qing lacked. Thus, the trains had to be ferried across the Yangzi by gigantic ships until 1968, when a bridge was finally built over the river in Nanjing, turning the Jinpu into the Beijing–Shanghai or Jinghu [京沪] railway.

E-tu Zen Sun was right to conclude in 1954 that the 'foreign acquisition of railway rights in China reached a climax in 1898'.[144] Indeed, the German landing in Jiaozhou and the resultant *Lease* of March 1898 were extremely significant for two major reasons. First, it created a domino effect, prompting the Russians to take Port Arthur, and impose another treaty and a north–south trunk line on the Qing, also in March 1898. The French quickly acquired the *Concession for Railway from Tongking to Yunnan* in April.[145] Not to be outdone, the United States, in the guise of the American China Development Company, secured the *Contract for Canton-Hankow Railway* in April.[146] Likewise, the Japanese obtained the

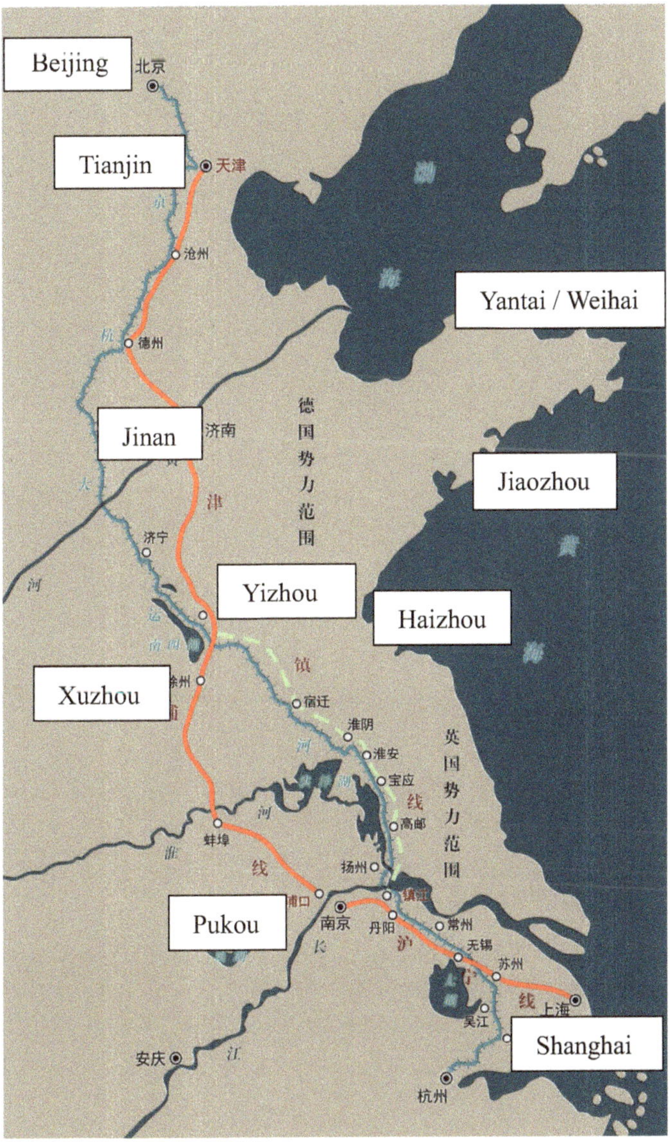

Figure 5.2 Route of the Jinpu railway (Baidu public domain)

Declaration Concerning Non-Alienation of Fukien.[147] Japan was already planning the conquest of southern China in 1898. Belgium procured the contract to finance-build the Luhan railway in June and soon the Bianluo line.[148] Kees Van Dijk differentiates between three types of colonial powers operating in China: the three old empires of Britain, France and Russia; the three 'relative newcomers' of the United States, Germany and Japan; and the three 'not very powerful' ones – the Netherlands, Spain and Portugal.[149] Belgium was a most important power when it comes to railways as they built the cross-like Luhan and Bianluo lines. Did Li Hongzhang receive rebates for all these concessions awarded in 1898? Perhaps his telegraph and letter to Lord Rothschild in April 1899 for investment in Pekin Syndicate offers an answer to that.

Second, the German entry into the scramble for China only helped Britain to become the biggest beneficiary, procuring New Territory as part of Hong Kong (86 per cent of the colony), mining privileges and the *Lease of Wei-hai Wei* on the eastern tip of the Shandong Peninsula next to Yantai (see Figure 5.2).[150] Shandong was now home to two arch rivals, which would turn the region into another 'cradle for conflict'. Aside from the Jinpu and the Shanghai to Nanjing lines, the British secured more railway concessions: the Daokou (Henan province, rich with coal) to Zezhou (Shanxi province, rich with coal and iron) [道泽] line, the aforementioned Bianluo eastward extension to Xuzhou [开徐], the Pukou to Xinyang [浦信] line with the hope of further extension to Sichuan and the Suzhou–Hangzhou–Ningbo [苏杭甬] line, not to mention the Canton–Kowloon [广九] railway and the Yunnan–Burma railroad, which was then in negotiation.[151] The Pekin Syndicate backed by the Rothschild family and its competitor the British and Chinese Corporation led by Jardine and HSBC were engaged in good railway business.[152] Their gains encompassed north, east, south and southwest China. Zen Sun puts it perfectly: 'Britain's railway interests in China were rounded out and strategically placed over practically the entire country'.[153] Although the British had less mileage, their dominance took the form of capital. Between 1898 and 1912, British shares in railway loans totalled £18,660,000, compared to those of Germany at £9,500,000, Belgium at £6,000,000, France at £5,000,000 and Japan at £2 million.[154] British dominance manifested in the form of railway loans. They had learnt their lesson and avoided the troubles that come with construction and management.

Wherever there were railways, there was rivalry and ultimately conflict. German entry into China did not just galvanise great power competition, but it further politicised railway development and reform in the late Qing. So many foreign powers imposed so many unwanted treaties on the Qing in the spring and summer of 1898; they enraged many officials and scholars. The Guangxu Emperor had surrounded himself with reformers like Kang Youwei and Liang Qichao who persuaded him to address the rapidly deteriorating situation as a matter of urgency by issuing many edicts, starting on 11 June 1898, to promote radical reform. This was a comprehensive program that constituted an overhaul of the late Qing polity, economy, education and military. For example, it encouraged private enterprise, allowing ordinary people to open mines and build railways, a change of the old 'government supervised commercial management' system.[155] But this attempt was cut short 103 days later by the Empress Dowager, hence the name 'Hundred Days Reform'. She beheaded six reformers, put Guangxu under house arrest and ostentatiously returned to rule. German aggression allows us a unique window into not just the origin of the Boxer Rebellion, the doomed 'Hundred Days Reform' but also the scramble for China at the turn of the century that delivered the final blow to the Qing.

The German colonial enterprise in Shandong would have been very different without the Jiaoji and Jinpu railways. They enabled the Germans to occupy many parts of the province, to militarise Jiaozhou, to exploit Shandong's rich natural and human resources, to colonise the extended Jiaozhou region and to expand far beyond the Peninsula province to north, east and south–central China. The prescient fears of late Qing officials materialised with a vengeance. The unintended consequence of imperialist-built railways is not the main focus of this book, but it is certainly a subject that requires more studies. Elisabeth Köll remains the only scholar to have written about the Jinpu from the perspective of a business institution, arguing that it 'created new infrastructure necessary for economic growth and social mobility in Shandong', which resulted in 'increased commercialised agriculture within the province and economic integration with neighbouring provinces'.[156] The Jinpu was destined to become one of the most important north-south economic arteries of modern China.

This was also the case with the Jiaoji as it modernised Shandong's infrastructure and enabled China to acquire the hardware needed for industrialisation, even though the opening was 'passive'.[157] By 1911, local merchants had come to see the benefits of railways, leading them to campaign for a line to be extended to their locality.[158] Local demand, stressed by Li Hongzhang in the 1860s, had finally emerged in the early twentieth century. The Jiaoji's legacy continues to this day. Many of China's bullet trains make their debut in Qingdao, manufactured in the old Hauptreparatur-Werkstätte Syfang der Schantung-Eisenbahn transformed and now state-owned CRRC Qingdao Sifang Company. Qingdao remains an economic powerhouse, leading China's high speed rail production. The CER and Jiaoji railways remind us that the 1890s were far removed from the 1870s, when the Qing resisted the encroachment of foreign railways. The political biography of railways helps us to pinpoint the Qing's loss of control and its inability to uphold territorial integrity, a fact that could not have been more obvious in the case of Japanese railway imperialism.

Notes

1 Terrell D. Gottschall, *By Order of the Kaiser: Otto von Diederichs and the rise of the imperial German Navy, 1865–1902* (Annapolis: Naval Institute Press, 2003).

2 Ferdinand von Richthofen, *Ferdinand von Richthofen's tagebücher aus China band 1* (Berlin: Dietrich Reimer, 1907), 166–67.

3 Bernd Martin, 'The Prussian expedition to the Far East (1860–1862)', *Newsletter for Modern Chinese History*, 6 (1988): 38–52.

4 Daniel Waugh, 'Richthofen's 'Silk Roads': toward the archaeology of a concept.' *The Silk Road*, vol 5, 1 (Summer 2007): 1–10.

5 'Meeting of the North China Branch of the Royal Asiatic Society', *Herald*, 29 April 1881: 418.

6 Van Dijk, *Pacific strife.*

7 Bernhard von Bulow, *Imperial Germany* (New York: Dodd and Mead, 1914), 115.

8 'Li Hung-Chang's tour', *Herald*, 24 December 1896: 1096.

9 Cai, *Li Fuxiang lipin oumeiji* (Taibei: Wenhai chubanshe, 1972, 14.

10 Li Hung-Chang and jumboism', *Herald*, 3 July 1896: 27.

11 Henry Norman, 'Li Hung-Chang', *Herald*, 15 February 1889: 185.

12 'Li Hung-Chang and Bismarck', *Herald*, 7 August 1896: 243.

13 Steinmetz, *Devil's handwriting*, 426.

14 'Newchwang', *Herald*, 18 October 1881: 403.

15 Steinmetz, *Devil's handwriting*, 426.

16 'Chushi deguo dachen Xu Jingcheng zhe, 13 March 1886', Qingdao jiyi, Qingdao Dang'an xinxiwang, at www.qdda.gov.cn/qddaxxw/qddaxxw/csjy/ztjyk/jzwsj/2020/10/22/402811867546a63d01754ee406d1291a.html, accessed 13 August 2021 and 'A secret treaty between Germany and China', *Herald*, 16 February 1887: 162.

17 John C. G. Rohl, *Wilhelm II: the Kaiser's personal monarchy 1888–1900* (Cambridge: Cambridge University Press, 2004).

18 Julius L. Becker, '"To Grab, When the Grabbing begins" German foreign and colonial policy during the Sino-Japanese war of 1894/95 and the Triple Intervention of 1895', *The International History Review*, vol 44, 1 (2022): 1–20.

19 Van Dijk, *Pacific strife*, 21.

20 Steinmetz, *Devil's handwriting*, 427.

21 'Ichoufu-Shantung: German naval visit', *Herald*, 24 January 1896: 123.

22 Mühlhahn, 'Negotiating the nation', in *Twentieth century colonialism*, 37–56 (39).

23 Steinmetz, *Devil's handwriting*, 434.

24 'Zhili zongdu Li Hongzhang chouyi Jiao'ao gonghan,16 July 1886' and 'Zhili zongdu Li Hongzhang deng pian, 29 May 1894', Qingdao jiyi at Qingdao dang'an, at www.qdda.gov.cn/qddaxxw/qddaxxw/csjy/ztjyk/jzwsj/2020/10/22/402811867546a63d01754efabf0f29b6.html and www.qdda.gov.cn/qddaxxw/qddaxxw/csjy/ztjyk/jzwsj/2020/10/23/402811867553a0b601755427372100dd.html, accessed 4 December 2021.

25 'Li Hung-Chang and jumboism', *Herald*, 3 July 1896: 27.
26 'The progress of H. E. Li Hung-Chang', *Herald*, 28 August 1896: 349.
27 'The murder of the two German missionaries', *Herald*, 17 December 1897: 1096.
28 Wilhelm II, *The Kaiser's memoir* (New York: Harper Brothers, 1922), 67 and 'Germany visits Russia', *Herald*, 13 August 1897: 294.
29 Miller, *Masters*, 129.
30 Alfred von Tirpitz, *My memoir* (New York: Dodd and Mead, 1919), vol 1, 91–117.
31 A Harding Ganz, 'The German navy in the Far East and the Pacific: the seizure of Kiautschou and after', in *Germany in the Pacific and Far East 1870–1914*, eds John A. Moses and Paul M. Kennedy (St Lucia: Queensland University Press, 1977), 115–36; Gottschall, *By order of the Kaiser*, 131–80 and Patrick J. Kelly, *Tirpitz and the Imperial German Navy* (Bloomington: Indiana University Press, 2011), 103–65.
32 'News from Kiaochou Bay', *Herald*, 17 December 1897: 1097 and 'German at Kiauchou', *Herald*, 10 December 1897: 1022.
33 Bulow, *Imperial Germany*, 115.
34 'Prince Bulow's speech', *Herald*, 30 November 1906: 516.
35 Ferdinand von Richthofen, *Schantung und seine eingangspforte Kiautschou* (Berlin: Dietrichs Reimer, 1898).
36 'Baron Richthofen on the coal beds of China', *Herald*, 29 January 1874: 85 and 'Baron Richthofen on China', 8 January 1874: 25.
37 'Letter by Baron von Richthofen', *Herald*, 15 September 1871: 1.
38 Maria M. Ogilvie Gordon, 'Ferdinand von Richthofen's Diaries from China', *Scottish Geographical Magazine*, 24, 6 (1908): 310–17 (315).
39 Steinmetz, *Devil's handwriting*, 405; Sebastian Conrad, *German imperialism: a short history* (Cambridge: Cambridge University Press, 2014), 128.
40 Shellen Wu, *Empires of coal: fuelling China's entry into the modern world, 1860–1920* (Stanford: Stanford University Press, 2019), 96–128.
41 Afinogenov, *Spies*.
42 Peter H. Christensen, *Germany and the Ottoman railways: art, empire, and infrastructure* (New Haven: Yale University Press, 2017), 1.
43 MacMurray, *Treaties*, vol 1, 112–17.
44 'Tudi zhidu', Qingdao jiyi at Qingdao Dang'an, at www.qdda.gov.cn/qddaxxw/qddaxxw/csjy/ztjyk/qdafs/2021/02/19/40281a6277b472120177b96e095f0424.html, accessed 25 April 2021.
45 'The Germans at Kiauchou bay', *Herald*, 3 December 1897: 1005.
46 Steinmetz, *Devil's handwriting*, 455.
47 'The murder of a sentry at Kiaochou bay', *Herald*, 2 February 1898: 193.
48 Bradley D. Naranch, '"Colonized body", "Oriental machine": debating race, railroads, and the politics of reconstruction in Germany and East African, 1906–1910', *Central European History*, vol 33, 3 (2000–01): 299–338.
49 'News from Kiaochou', *Herald*, 7 January 1898: 22, 'Ichowfu, Shantung', 16 May 1898: 841 and 'Chingchoufu, Shantung', 24 October 1898: 768.

50 Steinmetz, *Devil's handwriting*, 435; 'Chengshi guanli jigou', Qingdao jiyi at Qingdao dang'an, at www.qdda.gov.cn/qddaxxw/qddaxxw/csjy/ztjyk/qdafs/2021/02/01/40281a62775b242001775b7036a40080.html, accessed 29 April 2021.
51 MacMurray, *Treaties*, vol 1, 115; Wang Tieya, *Zhongwai jiuyuezhang huibian* (3 vols, Beijing: Sanlian shudian, 1957), vol 1, 738–40.
52 MacMurray, *Treaties*, vol 1, 116.
53 Dennis E. Showalter, 'Railroads, the Prussian army, and the German way of war in the nineteenth century', in *Paths of empire*, 21–44.
54 MacMurray, *Treaties*, vol 1, 115–16.
55 'Tudi zhidu', Qingdao Dang'an, at www.qdda.gov.cn/qddaxxw/qddaxxw/csjy/ztjyk/qdafs/2021/02/19/40281a6277b472120177b96e095f0424.html, accessed 25 April 2021.
56 Steinmetz, *Devil's handwriting*, 456; Yuan Rongsu, *Jiao'ao zhi* (12 vols, Taibei: Wenhai chubanshe, 1969), vol 2, 17.
57 'The Germans in Shantung, inhumane barbarism', *Herald*, 1 May 1899: 764 and 'German movement in Shantung', 15 May 1898: 841.
58 Mühlhahn, 'Negotiating the nation', 37–56 (46) and 'Weihsien, Shantung: attack on German engineers', *Herald*, 21 February 1900: 306.
59 'The Kiauchou trouble', *Herald*, 10 July 1899: 81; 'Gaomin zhaoshi', *Shen Bao*, 28 June 1899: 1, 'Gaomin bujing', 9 March 1900: 1 and 'Deren canbao', 15 December 1900: 1.
60 Yixu Lu, 'Germans and the death-throes of the Qing: Mo Yan's The sandalwood torture', in *German colonialism revisited: African, Asian and Oceanic experiences*, eds Nina Berman, Klaus Mühlhahn and Patrice Nganang (Ann Arbor: University of Michigan Press, 2014), 271–83.
61 'Daofei shisheng', *Shen Bao*, 8 January 1900: 2, 'Quanfei choujiao', 8 May 1900: 2 and 'Yihequanfei fenhui chezhanshi', 8 June 1900: 2.
62 Karl J. Rivinius, 'The Boxer movement and Christian missions in China', *Mission Studies*, vol 7, 1 (1990): 189–217; Joseph W. Esherick, *The origins of the Boxer uprising* (Berkeley: University of California Press, 1988), 38–95; Paul Cohen, *History in three keys: the boxers as event, experience and myth* (New York: Columbia University Press, 1997), 38.
63 'Bingjian yunji', *Shen Bao*, 16 June 1900: 2.
64 Congers, *Letters*, 94.
65 Dirk Bonker, 'Global politics and Germany's destiny from an "East Asian perspective": Alfred von Tirpitz and the making of Wilhelmine navalism', *Central European History*, vol 46, 1 (March 2013): 61–96.
66 Mühlhahn, 'Negotiating the nation', 37–56 (39).
67 Tirpitz, *My memoir*, vol I, 91.
68 Steinmetz, *Devil's handwriting*, 436.
69 Mühlhahn, 'Negotiating the nation', 37–56 (39).
70 Beiträge zur Geschichte Tsingtaus–1897 bis 1953, at https://tsingtau.org/rosendahl-carl-1852-1917-gouverneur/ and https://tsingtau.org/jaeschke-paul-1851-1901-gouverneur/, accessed 2 March 2021.

71 MacMurray, *Treaties*, vol 1, 113.
72 'Diary of a German officer in Tsingtao', *Herald*, 22 May 1915: 550.
73 Jork Artelt, *Tsingtau: Deutsche stadt und festung in China 1897–1914* (Dusseldorf: Droste Verlag, 1984, translated/published by Qingdao dang'anguan 2011), 166–67; Hu Wenben, *Diguozhuyi yu Qingdaogang* (Jinan: Shandong renmin chubanshen, 1983), 17–18.
74 Bulow, *Imperial Germany*, 22.
75 'Chinanfu: opening of the Shantung railway', *Herald*, 10 June 1904: 1243.
76 Richthofen, *Tagebucher aus China band I* (Berlin: Dietrich Reimer, 1907).
77 'Chingchoufu, Shantung: coalmine', *Herald*, 21 August 1896: 310.
78 Alex Armstrong, *Shantung: a general outline of the geography and history of the province* (Shanghai: Mercury, 1891), 5.
79 'Notes from native papers', *Herald*, 14 February 1890: 182 and 'Chinanfu mineral resources', 26 July 1895: 137.
80 'Weihsien, Shantung: drilling', *Herald*, 20 February 1899: 290.
81 'The Shantung railway company', *Herald*, 14 August 1903: 346.
82 Hu, Qingdaogang, 32; 'Kuangwu gongsi baogao chumei xiangshu', *Shen Bao*, 2 Feb 1910: 12.
83 'Dewo Shandong meijinzhi dali', *Shen Bao*, 21 October 1906: 3.
84 'German railway agreements: China and a Quid Pro Quo', *Herald*, 10 January 1914: 115.
85 On Barak, *Powering empire: how coal made the Middle East and sparked global carbonisation* (Oakland: University of California Press, 2020), 3.
86 'Our position in the Mediterranean,' *Herald*, 27 October 1887: 462 and 'Coaling stations', 27 June 1890: 793.
87 'Pondicherry', *Herald*, 8 May 1885: 532 and 'A prospective Russian coaling station', 2 October 1901: 626.
88 Barak, *Powering empire*, 3 and 'Outsourcing: energy and empire in the age of coal 1820–1911', *International Journal of Middle East Studies*, vol 47, 3 (2015): 425–45.
89 Quan Hansheng, *Zhongguo jinjishi luncong* (Hong Kong: CUHK, 1972); Tim Wright, *Coal mining in China's economy and society 1895–1937* (Cambridge: Cambridge University Press, 1984); Wang Shouqian, *Mei yu zhengzhi: wanqing minguo fugongsi kuang'an yanjiu* (Beijing: Shehuikexue wenxian chubanshe, 2009); Jeff Hornibrook, 'Riding the tiger: merchant-state alliance in a coalmine modernisation', *Business History*, vol 45, 2 (April 2003): 35–51.
90 Shellen Wu, 'The search for coals in the age of empires: Ferdinand von Richthofen's odyssey in China, 1860–1920', *The American Historical Review*, 119, 2 (April 2014): 339–63.
91 Victor Seow, *Carbon technocracy: energy regimes in modern East Asia* (Chicago: University of Chicago Press, 2021).
92 Tirpitz, *My memoir*, vol I, 91.
93 'Qingdaogang rugang chuanboshu tongjibiao (1902–13)', Qingdao jiyi at Qingdao Dang'an, at www.qdda.gov.cn/qddaxxw/qddaxxw/csjy/lsjcxxjyk/qdjcsj/jgq/jtys/2004/10/11/402811867574f174017582f5805b0b85.html, accessed 16 April 2021.

94 'Deshang yaoqiu Qingdao zhiyan quanli', *Shen Bao*, 18 September 1907: 11 and 'Deren qiangduo Yanqing luquan', 7 October 1907: 11.
95 Qingdao dang'anju, *Diguo zhuyi yu jiaohaiguan* (Beijing: Dang'an chubanshe, 1986), 159.
96 Xu Zeng'e, 'Zhongguo shouhui Jiaoji tielu shimuo', Qingdao jiyi at Qingdao dang'an, at www.qdda.gov.cn/qddaxxw/qddaxxw/csjy/ztjyk/qdhghcs/2020/11/05/402811867574f1740175962ab9e94638.html, accessed 16 August 2021.
97 Qingdao dang'anju, *Jiaohaiguan*, 132.
98 Qingdao dang'anju, *Jiaohaiguan*, 132.
99 William C. Kirby, *Germany and Republican China* (Stanford: Stanford University Press, 1984), jacket.
100 Armstrong, *Shantung*, 20–21.
101 Steinmetz, *Devil's handwriting*, 410.
102 Steinmetz, *Devil's handwriting*, 439.
103 'Zuobian Zhongguo: Qingdao xiuzhu paotaide laogong xiansuo', at China Central Television, https://tv.cctv.com/2009/12/04/VIDE1355520589238453.shtml, accessed 29 April 2021.
104 Andreas Steen, 'Germany and the Chinese coolie: labor, resistance and the struggle for equality, 1884–1914', in *German colonialism revisited*, 147–60; John A. Moses, 'The coolies labour question and German colonial policy in Samoa, 1900–14', in *Germany in the Pacific*, 234–62; Van Dijk, *Pacific strife*, 43–62.
105 'Manchurian topics: Chinese coolie immigrants', *Herald*, 17 January 1908: 110.
106 'Recruitment of Chinese Labour Corp from Wei-hai Wei', WO 32/11345, The National Archive, Kew.
107 'Shanghai and the war', *Herald*, 24 August 1918: 481.
108 Brian Fawcett, 'The Chinese Labour Corp in France, 1917 to 1921', *Journal of the Hong Kong Branch of the Royal Asiatic Society*, vol 40, 1 (January 2000): 33–111.
109 Chen Hansheng, *Huagong chuguo shiliao* (Beijing: Zhonghua shuju, 1980); Frances Wood and Christopher Arnander, *Betrayed ally: China in the great war* (Barnsley: Pen and Sword Military, 2016).
110 'The Germanisation of Shantung', *Herald*, 31 July 1903: 261.
111 Mechthild Leutner, *Musterkolonie Kiautschou: die expansion des Deutschen Reich in China 1897 bis 1995* (Berlin: Verlag, 1997).
112 Steinmetz, *Devil's handwriting*, 446–47.
113 Agnes Kneitz, 'German water infrastructure in China: colonial Qingdao 1898–1914', *NTM Zeitschrift fur Geschichte der Wissenschaften, Technik und Medizin*, vol 24, 4 (December 2016): 421–50.
114 Mühlhahn, 'Negotiating the nation', 37–56 (42).
115 Sheila F. Weiss, 'The race hygiene movement in Germany, 1904–1945', in *The wellborn science: eugenics in Germany, France, and Russia*, ed Mark B. Adams (New York: Oxford University Press, 1990), 8–68.
116 Steinmetz, *Devil's handwriting*, 457–58.
117 Mühlhahn, 'Negotiating the nation', 37–56 (41).
118 Steinmetz, *Devil's handwriting*, 361 and Ulrike Brisson, 'Elizabeth von Heyking: China through the eyes of a female aristocrat', in *Sino-German encounters*

and entanglements: transnational politics and culture 1890 to 1950, ed Joanne Miyang Cho (Cham: Palgrave Macmillan, 2021), 29–53.
119 'Tsingtao', *Herald*, 24 July 1901: 156.
120 'Tsingtao', *Herald*, 23 April 1903: 804.
121 Matthew A. Yokell, 'Investing in German 'Hong Kong': the building of a global economic presence in Qingdao 1898–1914', in *Sino-German encounters*, 77–106.
122 John E. Schrecker, 'Kiautschou and the problem of German colonialism', in *Germany in the Pacific*, 185–208 (191).
123 Sabina Groeneveld, 'Far away at home in Qingdao (1897–1914)', *German Studies Review*, vol 39, 1 (February 2016): 65–79.
124 Groeneveld, 'Far away at home'.
125 Yixu Lu, 'Colonial Qingdao through Chinese eyes', in *The cultural legacy*, 127–43; Orazio Coco, 'German imperialism in China: the leasehold of Kiaochou Bay, 1897–1914', *The Chinese Historical Review*, vol 26 (2019): 156–74.
126 Klaus Mühlhahn, 'A new imperial vision?: the limits of German colonialism in China', in *German colonialism in a global age*, eds Bradley Naranch and Geoff Eley (Durham: Duke University Press, 2020), 129–46.
127 Steinmetz, *Devil's handwriting*, 490–500.
128 Robinson, 'Introduction', *Railway imperialism*, 2.
129 'A naval station at Chusan', *Herald*, 7 March 1898: 378 and 'Tsingtao', 21 November 1900: 1083.
130 'The French Mail', *Herald*, 13 June 1898: 1060.
131 W. Travis Hanes III, 'Railway politics and imperialism in Central Africa, 1889–1953', in *Railway imperialism*, 41–69.
132 Van Dijk, *Pacific strife*, 63–80.
133 MacMurray, *Treaties*, vol 1, 278–320.
134 Richthofen, *Eingangspforte*, 290–91.
135 Stephenson, *Report*, 7 and 33.
136 Mühlhahn, 'Negotiating the nation', 37–56.
137 Kent, *Enterprise*, 5.
138 'No German monopoly in Shantung', *Herald*, 15 August 1898: 278. Rong was the first Chinese to graduate from Yale University.
139 McMeekin, *Berlin-Baghdad express*, 4.
140 T. G. Otte, '"The Baghdad railway of the Far East": the Tientsin-Yangtze railway and Anglo-German relations, 1898–1911', in *Paths of empire*, 112–36.
141 'The Tientsin-Chinkiang railway', *Herald*, 22 August 1898: 333 and 'Railways in China', 6 May 1910: 292; MacMurray, *Treaties*, vol 1, 684–93; Lee, *Exploitation*, 49–53.
142 MacMurray, *Treaties*, vol 1, 685.
143 'Ichoufu-Shantung: German naval visit', *Herald*, 24 January 1896: 123.
144 Zen Sun, *British interests*, 6.
145 MacMurray, *Treaties*, vol 1, 124–25 and 128–30.
146 MacMurray, *Treaties*, vol 1, 519.
147 MacMurray, *Treaties*, vol 1, 126.

148 'Belgium and China: loan contract, and operating contract for the Peking-Hankou Railway', in MacMurray, *Treaties*, vol 1, 135–45; Johan Mattelaer *et al.*, *A Belgium passage to China* translated by Liu Yue *et al.*, *Bilishi Zhongguo xiri zhilu* (Beijing: Shehui kexue wenxian chubanshe, 2021).
149 Van Dijk, *Pacific strife*, 21.
150 MacMurray, *Treaties*, vol 1, 130–31, 31–34, 700–2, 152–53, 181–83 and 183–88.
151 MacMurray, *Treaties*, vol 1, 402–8, 179–81 and McLean, 'Townley Agreement'
152 Quan, *Jinjishi*, 752–64.
153 Zun Sun, *British interests*, 43–44.
154 Zun Sun, *British interests*, 10–11.
155 Jian Bozhan, *Wuxu bianfa ziliao congkan* (2 vols, Shanghai: Renmin shudian, 2000), vol 2, 253–58.
156 Köll, 'Tianjin-Pukou', in *Manchurian railways*, 123–42 (141).
157 Wu-san Dai, 'Modern railway technology transfer to China: the case of Kiaotsi railway, 1898–1914', *East Asian Science, Technology and Society*, vol 10 (February 2016): 211–14; Shellen Wu, 'Engineering empire: German influence on Chinese industrialization, 1880–1925', in *Exploration and entanglement: Germans in the Pacific world*, eds Hartmut Berghoff, *et al.* (Oxford: Berghahn Books, 2018), 153–170; Philipp Mahltig, 'German railways in China: technology as a site of knowledge', *Journal of Transport History*, vol 40, 3 (December 2019): 301–21; Wang Bin, *Beidongde kaifang: Qingmo jiaoji tielu jiansheshi* (Shandong: Keji chubanshe, 2022).
158 'Yixian shenshang lizheng qicunlu zhan', *Shen Bao*, 21 September 1911: 11.

6

'Military preparedness in civil garb': Japan catches up[1]

A large crowd of Japanese gathered at Harbin railway station waving their national flag on 26 October 1909. As the train pulled to a stop, an elderly man stepped off onto the platform. Seconds later, three shots were fired at him and Ito Hirobumi, Japan's four-time prime minister and most senior statesman, was killed instantly.[2] The assassin was An Jung-geun, a Korean nationalist. The reason for Ito's visit, given the animosity of the Russo-Japanese War, brings us to another chapter in the political biography of railways in China. Daniel Headrick was exact in stating that 'among non-Western peoples, only the Japanese showed any real enthusiasm for railway'.[3] Indeed, following its victory in the Russo-Japanese War, Japan would procure the southern section of the Russian-built CER and the German constructed Jiaoji railways in November 1914. This chapter analyses how railways drove Japanese expansion into China and defined Japanese imperialism. It exposes the scramble for railway concessions among the Western powers and a rising Japan. Japan would perfect the game, using railways to partition, militarise, exploit and colonise Manchuria and beyond.

Revenge on the 'triple intervention'

> In recent years, the Great Powers have been industriously expanding their claims in East Asia. Each seizes even the slightest of opportunities to acquire concessions for fear of being outdone by the other. For this reason, we too must avail ourselves of the opportunity and develop our national strength by endeavouring, in this instance, to expand our claims in Manchuria, Korea, and the Maritime Territory.[4]

Sounding much like his German counterpart, Bernhard von Bulow, these were the words of Jutaro Komura, the Japanese Foreign Minister (1901–06). Six years after Qing China launched the Tongzhi Restoration, Japan embarked on the Meiji Restoration in 1868. Japan's reform was top-down as its ruling elite were determined to catch up with Europe and the United States.

While the Chinese settled for 'self-strengthening', Japan aimed to 'enrich the country and strengthen the military'. Japanese efforts were characterised by a 'commitment to join rather than overthrow the very international system that had threatened its existence'.[5] This 'commitment' can best be seen through the prism of railways. Japan's first railway, from Yokohama to Tokyo, opened in the summer of 1872, and it was the emperor himself who led ranking officials to take the first ride.[6] Japan made its first attempt at Taiwan in 1874, and it secured the island after the First Sino-Japanese War. Although it had to relinquish the Liaodong Peninsula as a result of the 'Triple Intervention', the opportunity for revenge presented itself a decade later. The *Boxer Protocol* should have ensured the withdrawal of foreign troops from China, but the Russians did not pull out. Kimitaka Matsuzato called Russia the 'decisive factor' in the making of modern East Asia, as their refusal to leave planted the seeds of a larger conflict.[7] The Russians had marched south in the name of protecting the CER, taking control of the Barrier railway then under construction and occupying stations.[8]

Chinese protest led to the *Treaty of Russian Withdrawal* in April 1902.[9] Although the Russians agreed to retreat, this did not materialise. Japan opened negotiations with St Petersburg.[10] Japan was in a strong position to do this, having formed a coalition with Britain in 1902.[11] Jacob Schiff, a Frankfurt-born Jewish American banker, pledged financial assistance in the conflict with Russia.[12] When the Russians ignored demands to withdraw, the Japanese launched a surprise attack on the Russian fleet at Port Arthur on 8 February 1904 – a tactic that they would employ against Britain and the United States in WWII.[13] This led to the outbreak of the Russo-Japanese War. Japanese forces quickly moved the war to land and northwards, engaging the Russians along the CER, and turning the railway into a battleground (see Figure 6.1). Caught unprepared, the Russians underestimated Japanese ferocity. Reinforcements from European Russia were slow to arrive as the Trans-Siberian railway was yet to be completed. They lost most battles but managed to hold on to territory north of Kuanchengzi in Changchun. Christian Wolmar writes of 'how wars were won and lost on the railways'.[14] It is not an exaggeration to say that the CER was the railway that won the war, whereas the Trans-Siberian was one that lost it.

By the summer of 1905, neither side could annihilate the other and both were exhausted; the war had entered a stalemate. In an attempt to promote his government's Open Door Policy, President Theodore Roosevelt stepped in at this crucial moment to broker a peace accord. The Treaty of Portsmouth was signed in September 1905 by Jutaro Komura and Sergei Witte.[15] The Russians and the Japanese cut the vertical CER in half:

> The Imperial Russian Government engage to transfer and assign the Imperial Government of Japan, without compensation and with the consent of the

Figure 6.1 Japanese military railway during the Russo-Japanese War (Photograph by James A Ricalton. Image courtesy of Jamie Carstairs and University of Bristol library, HPC, https://hpcbristol.net/visual/JC-s190, Identifier Jc-s190)

> Chinese Government, the railway between Chang-chun (Kuan-cheng-tzu) and Port Arthur and all its branches, together with all rights, privileges and properties appertaining thereto in that region, as well as all coal mines in the said region belonging to or worked for the benefit of the railway.[16]

The Russians retained the vertical line north of Changchun (see Figure 4.3) whereas the Japanese secured what lay south – 704.3 km of the total 2,500 km milage.[17] Renaming their section the South Manchurian [南满] Railway, the Japanese changed the gauge from Russian (1.524 mm or 5 ft) to Japanese standard (3 ft 6 in) and then to British standard (4 ft 8½ in) in order to connect with the Barrier and other railways in north China.

Matsusaka articulates that Japan's subjugation of Manchuria 'began in 1905 with the acquisition of a railway concession'.[18] Japanese railway imperialism was distinguished by the dominant role of the Army, with the support of the South Manchurian Railway Company (SMRC).[19] Like the German Navy, the Japanese Army controlled the railways, station towns and the belt of alienation. The shareholders of the SMRC included the Tokyo government, who owned a 50 per cent stake, with an additional 1 per cent reserved for the imperial household. Goto Shinpei, a doctor trained in Germany who then served in Taiwan, brought his training and experience to shape the SMRC as its first director.[20] The company procured: an area of '1,300 square miles' known as the Leased Territory and a railway zone of '100 square miles'.[21] This facilitated the gradual colonisation of a region of 365,300 km^2, almost the size of Japan itself (377,975 km^2), with the SMRC functioning as its government. Emer O'Dwyer believed that this was the 'largest company in the whole of Japanese history'.[22] It would consolidate Japan's hold on Liaodong Peninsula, steer colonisation in Manchuria and support the conquest of China Proper and Southeast Asia.

What the Japanese had not procured through the Treaty of Portsmouth, they obtained by coercing the Qing court. In December 1905, they forced a series of agreements upon China, including the *Treaty and Additional Agreement Relating to Manchuria* and a *Secret Protocol*, where they extracted more privileges that ranged from Treaty-port cities, mining and logging rights, and tax exemptions.[23] The SMRC opened many coal mines and factories, developing the region's agricultural economy to feed the growing Japanese Army. It exploited the rich natural and human resources of Manchuria and eastern Mongolia. It Japanised the region through immigration: 300,000 Japanese out of an original target of 5 million, were sent to settle the region, displacing thousands of Chinese in the process. The SMRC set up branch offices in cities like Beijing, Tianjin, Shanghai and Nanjing, and gathered intelligence for the Army. It played a major role in the establishment of the puppet state of Manchukuo in 1932.[24] It aided the full invasion of north China in 1937, facilitating the Nanjing Massacre and installing the Wang Jingwei collaborator regime in 1940. *Shen Bao* compared the SMRC to the East India Company, but it was really a 'cabinet agency'.[25] The Japanese would be more aggressive in implementing their imperialist policies on China. They would perfect railway imperialism.

'Instrument of territorial control'[26]

'Territory rather than trade lay at the heart of Japanese aspirations in Manchuria', Matsusaka argues in his excellent book on Japanese Manchuria.[27] Acquiring the South Manchurian Railway (SMR hereafter) meant that Japan inherited not just a railway but, more importantly, territory. Soon they pushed for a multitude of new railways for a simple reason: more railways, more land. These railways marked their efforts at 'reterritorializing', to borrow a term from Tony Ballantyne and Antoinette Burton, southern Manchuria, as they 'quickly staked out land sites and built their administration offices and other structure[s]'.[28] They demanded three new railways in the *Additional Agreement*.[29] The first was the Anfeng [安奉], a line from Andong on the Korean border to Fengtian, the heart of southern Manchuria (see Figure 6.2). This 188-mile light rail was neither part of the old CER, nor sanctioned by the Portsmouth Treaty. Assembled during the Russo-Japanese War to move troops and weapons, it should have been dismantled at the end of the war.[30] But the *Additional Agreement* legitimised its existence: 'The Imperial Chinese Government agrees that Japan has the right to work and maintain the military railway line constructed between Antung and Mukden to improve the said line so as to make it fit for the conveyance of commercial and industrial goods of all nations.'[31]

The second line, the Jichang [吉长], extended from Kirin to Changchun where the CER and SMR met. The *Herald* called this line 'the railway with a purpose' because it ran parallel to the horizonal CER on its way to Vladivostok.[32] There was no local demand for such a line, it would only facilitate Japanese expansion into Russian northern Manchuria. However, according to the *Secret Protocol,* the Chinese were expected to build it:

> The railway between Changchun and Kirin will be constructed by China with capital raised by herself. She, however, agrees to borrow from Japan the insufficient amount of capital which amount being about one-half of the total sum required. The contract concerning the loan shall, in due time, be concluded, following *mutatis mutandis*, the loan contract entered between the board of the Imperial Railway of North China and the Anglo-Chinese Syndicate.[33]

Although the Russians had tried to exclude the British from Manchuria, the Japanese ensured its stay in the region, at least for now. The British would benefit from the third line, the Xinfeng [新奉], from Xinmintun, the terminus of the Barrier railway outside of Fengtian to the centre of the city. The 96 km railway, constructed by the Barrier Railway Company, was taken over by the Japanese and changed into narrow gauge during the Russo-Japanese War. But the Qing had to sell it at a price dictated by the Japanese.[34] Moreover, the line was to be a Japanese monopoly as the *Secret Protocol* stipulated:

> The Chinese Government engage, for the purpose of protecting the interest of the South Manchuria Railway, not to construct, prior to the recovery by them of said railway [Mukden-Xinmintuen], any main line in the neighbourhood of and parallel to that railway, or any branch line which might be prejudicial to the interest of the above-mentioned railway.[35]

The short line is significant because it linked the SMR with the Barrier railway and the Luhan, hence access to north and south-central China. The Barrier line passed Tianjin, terminus of the Anglo-German Jinpu line that travelled to Nanjing and Shanghai. These Japanese demands revealed the extent of their ambitions for not just north Manchuria but also north, east and south–central China. Just as Japanese gains in the First Sino-Japanese War had alarmed the 'triple' allies, Japanese appetite after the Russo-Japanese War provoked American reaction.[36] The United States had been expanding to the Asia Pacific with the acquisition of Hawaii and the Philippines in 1898. President Theodore Roosevelt tried to contain Russian expansion by promoting the Open Door Policy.[37] This explains why the US encouraged Japanese to take on Russia, which led to the Russo-Japanese War. The US might be glad that Japan had squeezed Russian out of southern Manchuria; they now faced a new power with growing ambition. The two signed a series of pacts, including the Root–Takahira Agreement where Japan recognised America's annexation of Hawaii and the Philippines, and the US acknowledged Japan's control over Korea and Manchuria.

Japanese gains renewed American railroad interest in China.[38] Railroad mogul Edward Henry Harriman visited Japan and China in 1905, seeking to conquer the world with a Trans-Asiatic line. He believed he could secure trackage over the Trans-Siberian line from the Baltic, purchase the CER and the SMR, and establish a line of steamers to link with the Transcontinental Railroad.[39] To further complicate matters, the British firm of Pauling and Company managed to secure a contract for an extension from Xinmintuen to Fakumen, to facilitate trade at Newchwang.[40] This irked the Japanese as it would divert traffic from the SMR and their preferred destination of Port Arthur. If Harriman was naïve in thinking that money could dictate outcomes in East Asia, even more so was the Secretary of State, Philander C. Knox, who believed that the US could neutralise railways in Manchuria under the banner of 'Dollar Diplomacy'.[41] He proposed an international consortium for a loan to China to buy both CER and SMR in 1909.[42] When the Russians and the Japanese vehemently opposed any American initiative, Knox then endorsed another line from Jinzhou on the Barrier line to Qiqihar on the horizontal CER, which would terminate at Aihui on the Sino-Russian border (see Figure 6.2).[43]

Parallel to the vertical SMR and CER, the Jinzhou to Aihui line undermined both Russia and Japan. This American proposal prompted Ito's visit to Harbin in 1909, where he was due to meet with Russian representatives – a meeting impeded by his assassination. In their tussle for Manchuria, the politics of railways had divided Russia and Japan, but it also had the potential to bring the two arch rivals together. Bruce Elleman has expertly written about the collaboration of the duo in their attempts to 'keep China from

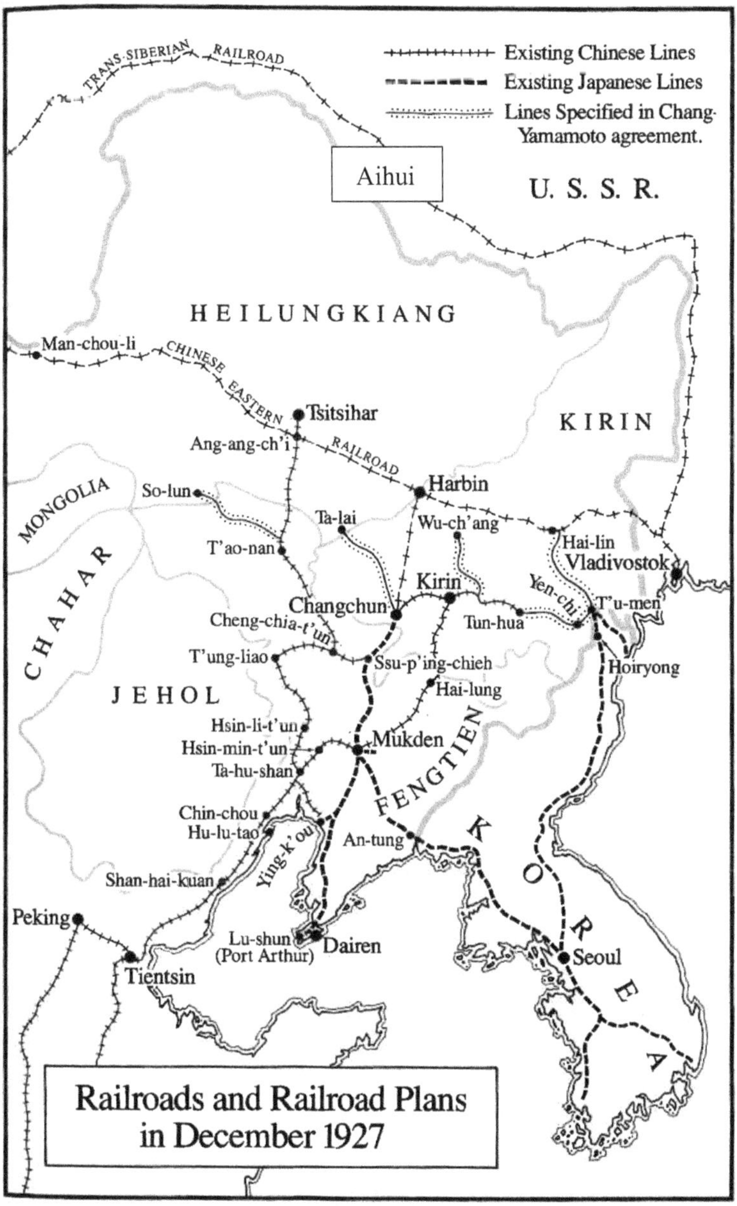

Figure 6.2 Railways in Manchuria by 1927 (Gavan McCormack, *Chang Tso-lin in northeast China, 1911–1928: China, Japan and the Manchurian idea* (Stanford: Stanford University Press, 1977), 218)

being dominated by a third foreign power'.[44] Russia threatened the American initiative with their plan to build a line from Kyakhta on the Qing–Russian border to Zhangjiakou (see Figure 3.3) and then on to Beijing.[45] British investors were keen to join the American-led project, the China Association put pressure on the Foreign Office to intervene.[46] Such was the scope and scale of the jostle for railway concessions by the late 1900s. Fortunately, both the Harriman and the Knox schemes failed to materialise because the Americans had failed to comprehend the complex geopolitics of railways in Manchuria.

It was at this crucial juncture that both the Guangxu Emperor and Empress Dowager Cixi died in November 1908; the Qing dynasty collapsed after the Nationalist Revolution in October 1911. But the revolution did not change the fortune of railways. Instead, the disintegration that followed made it easier for the Japanese to stake their claims. Although Dr Sun Zhongshan was sworn in as president of the Republic of China in Nanjing on 1 January 1912, the six-year-old Puyi Emperor still sat on the throne in Beijing, and real power rested in the hands of Yuan Shikai. Sun agreed to relinquish the presidency if Yuan could force the boy emperor to abdicate. This materialised in February 1912. The Japanese soon presented the *Exchange of Notes in Regard to the Construction of Certain Railways in Manchuria* to the Yuan regime in October 1913.[47] They demanded five new railways into Russian north Manchuria and east Mongolia.[48]

Three of the five lines centred on Taonan in the northwest corner of Jilin province. These lines travelled from Siping to Taonan [四洮], from Changchun to Taonan [长洮] and from Taonan to Rehe/Chengde [洮热], the Qing's summer resort to the north of Beijing. The Japanese were already preparing for the conquest of north China in 1913 – railways outlined their ambition. Taonan, though small, sat in a strategic location, just below north Manchuria, neighbouring east Mongolia, and with access to north China (see Figure 6.2). The remaining two lines centred on Hailong (Meihekou) in east Jilin province: from Kaiyuan in Liaoning province to Hailong [开海], and from Kirin to Hailong [吉海]. Sitting in the most sensitive zone between Russian north and Japanese south, these lines constrained Russian access to east Mongolia, north China and the Korean Peninsula.

Hardly were the five lines built, when the Japanese sought to extend their influence through courting and coercing of Zhang Zuolin, the inspecting general of Manchuria. Zhang, in return, used railways to entice as well as to bargain with the Japanese for support in order to consolidate his rule in Manchuria and expand to north China.[49] The Japanese proposed a multi-phased, multiple-line plan in 1925 consisting of five main lines: Dunhua to Tumen [敦图], Changchun to Dalai [长大], Taonan to Suolun [洮索], Yanji to Hailin [延海] and Kirin to Wuchang [吉五] (see Figure 6.2).[50] Dubbed the 'new Manchuria–Mongolia five line', they were sanctioned by the *Japanese-Zhang Secret Agreement*. They reached as far northwest as Solun in east Mongolia and Manzhouli on the Sino-Russian border, and as far north-east as Hoeryong on the Sino-Korean border and Chongjin, a Korean city that borders Russia and faces the Sea of Japan.[51] They would be interlaced with the SMR through more extensions. The purpose was clear, to 'hard-wire north Manchuria and Inner Mongolia for military operation and rapid occupation'.[52]

Not content with these substantial gains, the Japanese Army tabled, in the same year, a much grander strategy: a four-phased twenty-year scheme called the 'Manchuria–Mongolia railway network' that aimed to build a total of 5,500 miles:[53]

161 miles per year for the first phase 1926–28
230 miles per year for the second phase 1928–33
300 miles per year for the third phase 1933–37
347 miles per year for the fourth phase 1938–44

Railways defined Japanese imperialism: in 1925, the Japanese Army was already planning the conquest of not just Russian north Manchuria but also Mongolia and north China, with a plan hinged on railways. Railways provide a unique window through which we can measure the extent of Japanese imperialism, but more research is needed to understand how these lines were built and facilitated their conquest. Goto Shinpei dreamed big, arguing that Japan was 'an indispensable junction for the transpacific and Europe-Asia traffic', with the SMR serving as an intermediary with Europe and the Japanese mainland facilitating traffic with north America.[54] Goto was thinking as big as Harriman. But Harriman had failed; and so, too, would he.

Tirpitz called Jiaozhou and the Jiaoji railways the 'empire of the Navy'; southern Manchuria and the SMR network were undoubtedly the empire of the Japanese Army. A history of Japanese railway imperialism in China is a history of the Army's drive to dominate domestic politics and overseas expansion, in competition with the Navy. This is a common theme with both the CER and the Jiaoji, as the military apparatus ultimately triumphed over more moderate agencies like the foreign ministry. The Japanese Army exaggerated the threat from Russia and lobbied for the Russo-Japanese War. It often acted before seeking permission, initiating surprise attacks which caught its own government off guard to make war a *fait accompli*. It is no wonder then that Danny Orbach has described the Army as Japan's 'curse'.[55] Military presence had not been sanctioned by the *Additional Articles* of the Portsmouth Treaty:

> The Higher Contracting Parties reserve to themselves the right to maintain guards to protect their respective railway lines in Manchuria. The number of

> such guards shall not exceed fifteen per kilometre and within that maximum number, the Commanders of the Japanese and Russian Armies shall, by common accord, fix the number of such guards to be employed as small as possible having in view the actual requirements.[56]

The word 'guards' was simply another term for military presence as far as the Japanese Army was concerned. The precept to keep that force 'as small as possible' was reduced to a joke as the railway guards would transform into the most formidable of Japanese war machine, the Kwantung/Guandong Army (GA hereafter). It intervened in warlord conflicts, lending a helping hand to Zhang Zuolin in exchange for his endorsement of their railway scheme. But Zhang was not obliging, and so Colonel Komoto Daisaku staged the Huanggutun Incident, blowing up the train Zhang was travelling on as it approached Fengtian, killing him on the SMR in early June 1928.[57] Using railways to fabricate incidents and kill was a common practice, as Alvin Coox points out.[58] When his son, Zhang Xueliang, took over the Manchurian regime, the GA discovered that the son was even more obstinate. The junior Zhang planned his own railways, which the GA saw as an 'iron ring' around the SMR, driving it to the 'jaws of death'.[59] To justify military intervention, the GA exaggerated the situation, declaring that the 'Chinese are building railways to encircle and, in effect, drive the SMR to destruction.'[60]

Military intervention materialised on 18 September 1931, when GA engineers, under the direct leadership of Commander Ishiwara Kanji, staged an explosion on the SMR and claimed it was a Chinese attack.[61] Labelled the Manchurian or Mukden Incident, it ushered in a full-scale Japanese invasion of north Manchuria.[62] Railways were paramount to the invasion. Ito Takeo, senior researcher at the SMRC Research Department, believed that the incident 'could not have arisen without the assistance of the SMR structure'.[63] Stacie E. Goddard calls the conquest of Manchuria 'Japan's folly' as it would lead to Japan's isolation and drag them into decades of armed conflict.[64] Railways had become the most effective instrument to stage *fait accompli*. No wonder that Sadako N. Ogata called this 'defiance in Manchuria'.[65] The SMR and its branch lines helped the GA to move quickly into north Manchuria, where fighting centred on railway lines and towns.[66] Russian Harbin fell into Japanese hands in February 1932 and the GA quickly installed the puppet state Manchukuo in March.[67] Chinese protest and international outrage led to the Lytton Commission.[68] The Lytton Report detailed Japanese aggression and rebuffed its spurious claim of self-defence.[69] It also reminded the world that the Japanese attacked Shanghai between 28 January and 3 March 1932, where hundreds were killed.[70] The League of Nations condemned Japanese aggression and refused to recognise the puppet state. In response, the Japanese delegation simply announced its exit from the League; Japan no longer needed the great powers.

The GA soon demanded that its Commander-in-Chief be 'invested with the authority to direct and supervise the economic programme for Manchuria' and that the supervision of the SMRC be 'transferred to the Commander-in-Chief of the Kwantung Army from the Minister of Colonies'.[71] The GA won the battle for Manchuria, defeating the civil ministries.[72] Now they ruled the railway empire and answered directly to the Prime Minister. After expelling the Russians from north Manchuria, the GA constructed more railways and feeder lines to the Russian border.[73] They established a multitude of underground fortresses along the border, the largest being Dongning, calling it the 'Maginot Line in the East'.[74] This great wall of fortresses were served by military railways. From roughly 10,000 soldiers on the eve of the Mukden Incident in 1931, the GA grew to more than 1 million by 1941. Those captured by Russian troops in August 1945 amounted to nearly 600,000.[75] Supported by the SMRC Research Department, the GA conducted surveys and collected intelligence.[76] They also undertook chemical and biological research, including the infamous 731 Unit that experimented on live humans.[77] Among the six Class A criminals who were sentenced to death by hanging at the International Military Tribunal for the Far East was the GA Chief of Staff, Tojo Hideki, and its intelligence chief, Doihara Kenji.

'Construction is the synonym of Manchuria', David Tucker argues in his study of Japanese labour policy and practice.[78] The multitude of railways and fortresses could not have been constructed without Chinese labourers. Despite the push to hire Japanese workers, the SMRC justified its stance of using Chinese: 'It is hardly necessary to point out that business in Manchuria can make good use of Chinese labor, which is generally acknowledged throughout the world for its low price.'[79] Most of them came from north China. The *Herald* believed that there were 'no fewer than 100,000' Shandong coolies in southern Manchuria by 1924.[80] To drive up

Table 6.1 Chinese labourers sent to Manchuria by the Japanese Dadong Company 1926–37

Year	Figure	Year	Figure
1926	646,617	1932	372,629
1927	1,043,772	1933	568767
1928	967,154	1934	627,322
1929	941,661	1935	440,540
1930	673,392	1936	359,761
1931	416,825	1937	322689

the numbers of coolies, the SMRC established the Dadong Company, an operation that recruited labourers but also spied.[81] Its branch in Qingdao alone sent millions to Manchuria between 1926 and 1937 (see Table 6.1).[82]

Elisabeth Köll highlights how the Jinpu railway 'played an important role in transporting' them from Shandong to Manchuria.[83] Mainland scholars have provided new data on those from north China:

1938: 492,300+
1939: 98,5600+
1940: 1,318,900+
1942: 1,000,000 (locally recruited in Manchuria)
1943: 1,200,000 (locally recruited in Manchuria)
1944: 1,300,000 (locally recruited in Manchuria)
1945: 1,600,000 (estimated based on data from previous years)[84]

Mainland historians believe that 170,000 were put to work building the Dongning fortress.[85] Many of them died, either from exhaustion, disease, malnutrition, accident and abuse.[86] Others were murdered at the end to prevent information about the fortress from leaking out.[87] Tucker believes that of the 5.5 million non-agricultural workers in Manchuria, about 500,000 worked in construction, including railways.[88] Edwin Pauley, US representative to the Allies Reparations Committee, compiled a report in 1946, he believed that up to 300,000 were employed in coal mines.[89] The Archives of Qingdao also reveal that the Japanese sent at least 40,000 Chinese to Japan during WWII.[90] Chinese coolies had become synonymous with railway construction and mining not just in Manchuria and Shandong but also in North America, Australia and South Africa. If opium smoking identified the Chinese by the late nineteenth century, so did the tens of thousands of coolies.

'There is only one way in which to make Japan strong and rich, and that is to use the resources of China. And there is no better way to begin than to develop that unopened treasure house within our sphere of influence known as Manchuria.'[91] Koiso Kuniaki, Chief of Staff of the GA and later Prime Minister, outlined a forward-looking strategy: wartime self-efficiency in industrial production, raw materials and foodstuffs.[92] Japan would exhaust Manchuria's natural resources as the region was blessed with massive coal reserves. Called the 'capital of coal', Fushun was home to Asia's largest open pit.[93] The Japanese Army took over the Fushun coal mines during the Russo-Japanese War and refused to return them to their owners.[94] The first railway extension was from Fengtian to Fushun. By 1912, its daily output reached 5,000 tons, versus the 360 tons produced before the Japanese take over.[95] The *Herald* proclaimed in 1917 that the 'Fushun coal bed is indeed a rich one, and there is known to be enough workable coal to last for 400 years – at the present rate of excavating only 20 odd million tons per annum.'[96] The SMRC believed that Fushun has an estimated deposit of '1,200,000,000 tons of bituminous coal'; it was only one of many pits they mined.[97] The company publicised the total output as of March 1923, a staggering 4,392,000 tons.[98] At 1.5–2 *yen* per ton in Manchuria, the SMRC sold its coal in Japan at 6.8 *yen* per ton, seriously threating the domestic coal industry.[99] Pauley's report details the growth of the Japanese coalmining (see Table 6.2).[100]

Manchuria's coal powered the railways; it also fuelled industries back in Japan, such as in the extraction of fuel oil.[101] Iron ore played a similar role, and a large deposit was found in Anshan, near Fushun.[102] By 1923, Anshan Ironworks produced 18 per cent of the pig iron used for steel production in Japan and 13 per cent of that used for casting.[103] As the conquest of China and Southeast Asia progressed, Japan relied on resources from Manchuria to sustain its production of weapons.[104] But they took much more than just coal and iron ore. Manchuria was and still is China's grain basket. Henry

Table 6.2 Production and export of Japanese controlled coal in Manchuria, 1929–44

Year	**Production** (metric tons)	**Export** (metric tons)
1929	9,960,000	4,010,000
1930	9.020,000	3,830,000
1931	9,540,000	4,260,000
1932	8,830,000	3,360,000
1933	9,063,000	3,738,000
1934	10,338,000	4,657,000
1935	12,166,000	3,559,000
1936	13,874,000	3,845,000
1937	14,410,000	3,420,000
1938	15,000,000	NA
1939	19,000,000	NA
1940	21,200,000	NA
1941	23,800,000	NA
1942	24,168,670	1,931,400
1943	25,320,425	2,173,143
1944	25,626,704	2,294,005

Kinney believed that 'against a total arable area of about 64,500,000 acres', over 30,500,000 acres were under cultivation by 1927.[105] The GA took at least 6,000,000 tons of grain in 1934 and 33,306,000 tons in 1940, according to the *Jilin Provincial Gazetteer*.[106] This was not all.

Labelled the 'bean empire', the most lucrative trade in Manchuria was soybean.[107] 'Whoever ruled the soy trade would acquire a major influence in shape the future of the territory.'[108] Newchwang was the centre of this trade, which explains the British rationale for establishing control there and for financing a branch line from the Barrier railway.[109] Instead of being rewarded for supporting the Japanese during the Russo-Japanese War, the British had to watch the 'heavy profit' of soy trade being diverted from Newchwang to Port Arthur by the Japanese.[110] Matsusaka points out that by 1910, 'soy products alone had come to account for 47% of the railway's gross freight revenue'.[111] The SMRC acknowledged that beans and bean products 'comprise about half of the value of all exports', that is 115,859,000 out of 186,634,000 tons in 1922.[112] Manchuria produced 220 million tons of soybeans, 'nearly two-thirds of the world supply' in 1929.[113] The Japanese had destroyed the British soy trade and its stronghold Newchwang. The SMRC also monopolised the cotton and opium trade.[114] Its total earnings multiplied (*yen*):

3,667,272 in 1907–08
23,599,189 in 1917–18
53,643,744 in 1922–23.[115]

Ramon Myers believed that its total revenue reached 185 million *yen* in 1923.[116] Emer O'Dwyer also concludes that profits for the railways alone jumped from 3,667 (unit: 1000 *yen*) in 1907 to 75,766 in 1933 and 231,288 in 1943.[117] Japanese industries and assets in the region included far more than what are discussed here.[118]

The railways and the resources of Manchuria enriched Japan and enabled it to expand. But the Japanisation of Manchuria cannot be measured in commercial, industrial or monetary terms alone. Goto Shinpei explained what he meant by 'military preparedness in civil garb' as Ito Takeo remembered:

> We have to implement a cultural invasion with a Central Laboratory, popular education for the resident populace, and forget other academic and economic links. Invasion may not an agreeable expression, but [language] aside we can generally call our policy one of invasion in civil garb.[119]

This cultural invasion can best be seen from the augmentation of the Japanese population in the leased territory and the railway zones, where 'company towns' and 'station towns' emerged. Their populations were intimately linked to the SMR; this can be seen from the number of SMR employees (all Japanese):

13,621 (September 1908)
18.852 (spring 1915)
37,685 (autumn 1924)[120]

This increase in the workforce led to the increase of Japanese population by 1922:

82,145 Japanese (out of 682,173 in the leased territory)
67,726 Japanese (out of 177,813 in the railway zone)[121]

Henry Kinney estimated the number of Japanese employees at 20,801 out of a total of 35,397 in 1927, whereas Edwin Pauley believed that the number reached 170,000 by 1939.[122] The discrepancy in numbers is an indication that the subject matter deserves more research. The Japanese population continued to grow after 1931, when they ousted the Russians from north Manchuria.[123] These Japanese settlers were scattered along and among railway towns/cities, like Dalian, which Emer O'Dwyer has studied. Labelled the 'Japanese second city', the port city had a population of 121,611 by 1933.[124] Bill Sewell has examined the making of Changchun, the capital of the puppet Manchukuo. Its population grew rapidly after the Mukden Incident, reaching 600,000 by 1942.[125] The SMRC advertised Manchuria, especially the warm weather and unspoiled beaches of Dalian and Lüshun (see Figure 6.3):

> Hoshigaura, or Star Beach, is a seaside resort twenty minutes by motor from Dairen. A modern summer hotel and picturesque bungalows attract many visitors who enjoy the excellent bathing, tennis and golf of this charming spot.
>
> Ogandai (Port Arthur) is a beautiful beach resort, with pine-clad hills, a historical battlefield, cozy bungalows and a comfortable hotel. An old fishing village adds interest, and little tea houses perched on the cliffs enchant the eye. [126]

The Japanese renamed everything – another way to claim Manchuria. Many came to visit, including poet Yosako Akiko in 1928:

> In Lüshun Stationmaster Ito Shin'ichi was waiting for us. We rode off together in a horse-drawn carriage ... The acacias of Huangjin Mountain were especially beautiful, like snow. In an elegant room of the Yamato Hotel at Huangjin Mountain, we smelled the aroma of these flowers and were treated to lunch by Nishida ... After lunch we relaxed for a while at Ito's official residence on the side of Baiyu Mountain, ... After our respite, Ito put us on a small steamship at the station, and we made a tour of the harbor and a bit outside the harbor area ... We made the trip back from Lüshun by bus. The level road covered with asphalt provided the stitching between the hills, and on our right we frequently saw the green of the Yellow Sea. It was extremely pleasant.[127]

Figure 6.3 Yamato hotel in Dalian (Image courtesy of the Charles Wheeler collection and University of Bristol library, https://hpcbristol.net/visual/Wr-s122, Identifier Wr-s122)

She enjoyed the visit and felt at home in the conquered land. What characterises the Japanisation of Manchuria was the emphasis on Japanese settlement to dilute the local population. Japan saw a massive population increase in the 1920s, when it reached 60 million.[128] While Japanese migrants faced discrimination in North America and 'white' Australia, Manchuria, the 'new paradise', was seen as the perfect solution to population explosion.[129] So much was this the case that *Shen Bao* called Manchuria a 'second Korea'.[130] The SMRC labelled it the 'land of opportunities' and produced propaganda photographs, films and posters to promote immigration.[131] The Japanese were not alone – the Portuguese also used photographs to promote European settlement after it built railways in Angola and Mozambique.[132] Goto Shinpei's plan was to settle 'half a million or more immigrants' in the region within a decade.[133] The ultimate goal was to see 5,000,000 migrate in twenty years.[134] These immigrants were well organised, assigned different jobs and received military training. Once in Manchuria they lived and worked in garrison-style and served as reserves for the GA.[135] The programme also recruited Japanese women between the ages of seventeen and twenty-five, who were referred to as 'continental brides'. Historians believe that between 270,000 and 330,000 Japanese emigrated from 1932 to 1945.[136] Manchuria not only sustained Japan's industrial and military expansion but also absorbed and fed its excess population. But the consequences for the locals were dire as many lost their land, while others died.[137]

Little did Goto know that his policy would have devastating consequences in August 1945 when the GA abandoned its citizens, including women and children. Many died and thousands were left behind; most of the children were adopted by Chinese parents. Many of them were unaware of their backgrounds until after the normalisation of Sino-Japanese relations in 1972.[138] Unlike the Russians and the Germans who had no intention to assimilate the Chinese, the Japanese aimed at turning the Chinese/Manchurians into Japanese. The best tool was education. Japanese became the official language of Manchuria; all schools and universities had to adopt a Japanese curriculum, and everyone was required to speak Japanese. Not only did they indoctrinate, they also controlled the Manchurians using the latest technology: dactylography, as Issac Tan detailed.[139] The acquisition of a railway enabled Japan to build a colonial empire in Manchuria, prepare for the conquest of China Proper and Southeast Asia, and sustained decades of war. The SMRC, in the words of Edwin Pauley, had been 'the bulwark of Japanese political and economic interest in Manchuria'.[140] But Manchuria was not the only ground where the Japanese played the game of railway imperialism.

The Shandong bridgehead

One might be excused for thinking that Japan had enough on its plate by 1914 with a multitude of railways in southern Manchuria and colonies like Taiwan and Korea, but its appetite for expansion was growing. The Japanese saw an opportunity in Shandong at the outbreak of WWI, when Germany recalled its overseas troops, leaving the Jiaoji railways and Qingdao vulnerable. The British were more than happy to help. Formed in 1902, the Anglo-Japanese alliance was renewed in 1905 and again in 1911 for ten years.[141] They handed the Germans an ultimatum, demanding their surrender.[142] The *Herald* justified the Japanese revenge on the 'Triple intervention', even as it questioned Japan's misuse of the alliance for its own gain. On 23 August 1914, Japan declared war on Germany.[143] Vice-Admiral Alfred Meyer-Waldeck, the last governor of Qingdao, summoned all the German troops stationed along the Jiaoji railways and other parts of Shandong to Qingdao. He also mobilised German civilians and drafted the Chinese into service. The Bismarck Fortress was finally put to use. However, the Japanese did not attack Qingdao from the sea. Instead, 20,000 troops landed in Longkou on the north side of Shandong, in the British sphere and were met by Major-General Nathaniel Walter Barnardiston and his soldiers.

They planned to take Qingdao on land, not from the sea. It is ironic that the exact same tactic would be employed by the Japanese to attack the British in Singapore 28 years later, in 1942.

With the help of their British allies, the Japanese forces began their assault by capturing the Jiaoji railways, the stations, especially the terminus Jinan, and the coal mines. They starved the Germans held out in Qingdao. Then they laid siege to the city.[144] After nearly three months of heavy bombardment and fierce fighting, the Germans surrendered in November.[145] The former German colony became a city 'under three flags'.[146] Procuring the railways and Qingdao was only the beginning of a wider campaign. A month later, Japan put the *Twenty-one Demands* to the Yuan Shikai regime in January 1915 with an 'ultimatum' set for 7 May.[147] They demanded to inherit 'all the rights, interests, and concessions' of the Germans in Shandong, not just those in Qingdao.[148] And they demanded more railways:

> The Chinese Government agree to Japan's building a railway connecting Chefoo or Lungkow with the Kiaochow-Tsinanfu Railway.
>
> The two Contracting Parties mutually agree that the term of the lease of Port Arthur and Dairen and the term respecting the South Manchurian Railway and the Antung-Mukden Railway shall be extended to a further period of 99 years respectively.
>
> In order to help the development of the Nanchang-Kiukiang Railway, with which Japanese capitalists are closely identified, and with due regard to the negotiations which have for years been pending between Japan and China in relation to the railway question in South China, China to agree to give to Japan the right of constructing a railway to connect Wuchang with the Kiukiang-Nanchang line, and also the railways between Nanchang and Hangchou and between Nanchang and Chaochou.
>
> In view of the relations between the Province of Fukien and Formosa and of the agreement respecting the non-alienation of that province, Japan is to be consulted first whenever foreign capital is needed in connection with the railways, mines, harbour works (including dockyards) in the province of Fukien.[149]

Aside from connecting the Jiaoji with Chefoo (Yantai) in British-controlled north Shandong and extending the lease for lines in south Manchuria, Japan was preparing for expansion to south–central and southern China through at least five new lines. These railways fanned out from Nanchang, the capital of Jiangxi province, and travelled to various places including Jiujiang and Wuchang (Hubei province), both on the Yangzi River, as well as Hangzhou (Zhejiang) and Chaozhou (Guangdong), two coastal cities. They demanded a monopoly in Fujian province. These new lines would enable the Japanese to acquire more territory, station more troops, exploit more natural and human resources, and colonise the mid and lower Yangzi River valley. This would put them in direct conflict with the British, who had claimed the entire Yangzi River valley in 1898. The Japanese conquest of southern China had begun by 1915 – railways exposed their designs and strategies. Japan now considered east Mongolia as its exclusive sphere.[150] China would have to consult Japan if it wished to build railways or to open mines in the region. The *Demands* exposed the Japanese desire for not just east and southern China but also for Mongolia, putting them in direct conflict with Russia and the United States.[151] The US reacted quickly, stating that they:

> cannot recognise any agreement or undertaking which has been entered into or which may be entered into between the Governments of China and Japan impairing the Treaty rights of the United States and its citizens in China, the political or territorial integrity of the Republic of China, or the international policy relative to China commonly known as the Open Door Policy.[152]

WWI may have ended in 1918, but the Jiaoji railways remained firmly in Japanese hands. Shandong was discussed at the Peace Conference in Versailles, where the Japanese delegation requested that the Allies legitimise its possessions in Shandong, which they had held since 1914. Japan was now a power to be reckoned with, and the victors acquiesced to their demands, despite the fact that China had been their ally with thousands from the Chinese Labour Corp dying on the battlefront and trenches of France, the majority of them from Shandong.[153] Article 156 dealt with the Shandong railway exclusively:

> Germany renounces, in favour of Japan, all her rights, titles and privileges – particularly those concerning the territory of Kiaochow, railways, mines and submarine cables – which she acquired in virtue of the Treaty concluded by her with China on March 6, 1898, and of all other arrangements relative to the Province of Shantung.
>
> All German rights in the Tsingtao-Tsinanfu Railway, including its branch lines, together with its subsidiary property of all kinds, stations, shops, fixed and rolling stock, mines, plant and material for the exploitation of the mines, are and remain acquired by Japan, together with all rights and privileges attaching thereto.[154]

When this news reached Beijing in early May 1919, students and ordinary citizens protested on Tiananmen Square on 4 May. This was followed by nation-wide anti-Japanese demonstrations. The Chinese delegates to the conference did not sign the Treaty of Versailles, rendering it illegitimate, at least on paper. This meant continued Japanese occupation, known as the 'Shandong problem', remained unresolved. The ROC regime asked the League of Nations for help, calling the situation 'most unjust and unfair'.[155] Chung Hung Liu, a student at Harvard University, placed the blame squarely on Britain and America:

> Under the palmy protection of her British ally, Japan is sure to grow stronger and more active. Increasing strength and activity, prompted by selfish motives, will in all probability end in clash with others. As Great Britain and her kindred America are the two among other Western powers to have the largest interest in the Far East, they are above all else the more liable in the clash. Then the erstwhile friend will be a terrible enemy, and England may have to sigh that this enemy was of her own fostering.[156]

Liu was right. Japan would challenge the US and defeat Britain in WWII not just in Hawaii, Hong Kong and Singapore, but also in Malaya and Burma. With Russia in internal turmoil, Germany defeated, Britain and France preoccupied with post-war reconstruction, the US came to lead world affairs. President Woodrow Wilson called for 'New Diplomacy' in East Asia. The US proposed a new mechanism to check Japanese ambition and constrain its expansion – a banking consortium that included the US, Britain, France and Japan, with jurisdiction over foreign loans, especially those on railways.[157] They were of the opinion that:

> The interests of the Chinese can in existing circumstances best be served by the cooperative action of the various banking groups representing the investment interests of their respective countries in procuring for the Chinese government the capital necessary for a programme of economic reconstruction and improved communications.[158]

Unsure whether this applied to their existing initiatives, the Japanese representative M. Odagiri tried to ascertain by writing to American representative Thomas W. Lamont on 18 June 1919:

> I wish to communicate to you that we have been instructed by our principals in Japan that all the rights and options held by Japan in Manchuria and Mongolia, where Japan has special interests, should be excluded from the arrangements of pooling provided for the proposed agreements. This is based on the very special relations which Japan enjoys geographically, and historically with the regions referred to ...[159]

Although Japan was ready to abandon two new railways in Shandong in order to protect what they had gained in Manchuria and Mongolia, its hopes were dashed by the Americans. Lamont replied on 23 July: 'Mongolia and Manchuria are important parts of China, and any attempt to exclude them from the Consortium must be inadmissible. The "special interests" to which you allude have, in our opinion, never had to with economic matters.'[160] The 'inadmissible' position was reinforced by a firm letter from Robert Lansing, the US Secretary of State, on 30 July and another from the British Foreign Office on 11 August:

> Manchuria and Mongolia are important provinces of China and any attempt to exclude them from the scope of the Consortium would constitute a direct negation of the principles the Consortium is based, would provoke the revival of similar claims on the part of others nations and thus perpetuate the very difficulties which the Consortium is designed to obviate.[161]

Thomas Lamont also wrote to the *Herald*, exposing this issue to a much larger audience.[162] The agreement for the banking consortium was signed by the four countries on 15 October 1920.[163] The agreement forced Japan to dig deeper into Manchuria and Mongolia, since expansion to other areas was now constrained. But the Shandong problem remained unresolved. Discussion resumed during the disarmament conference in Washington DC where the Washington Naval Treaty was signed on 6 February 1922.[164] The cardinal principle was stipulated in Clause One, where the participants agreed 'to respect the sovereignty, independence, and the territorial and administrative integrity of China'.[165] Clearly, this was directed at not just Japan but also Russia.[166] Japan agreed to 'restore to China the former German leased territory of Kiaochow' and 'transfer to China the Tsingtao-Tsinanfu Railway and its branches, together with all other properties appurtenant thereto, including wharves, warehouses and other similar properties'.[167] Bruce Elleman has written expertly about the complexity of Shandong's return.[168] The good news came at a price – China had to reimburse Japan the value of the railways built by Germany:

> The actual value to be so reimbursed shall consist of the sum of 53,406,141 gold marks (which is the assessed value of such portion of the said properties as was left behind by the Germans), or its equivalent, plus the amount which Japan, during her administration of the railway, has actually expended for permanent improvements on or additions to the said properties, less a suitable allowance for depreciation.[169]

Just as Qing China had had to pay an extra 30,000,000 silver taels, on top of the 200 million indemnities, to see the return of the Liaodong Peninsula as a result of the 'Triple Intervention' in 1895, the Republic of China had to pay Japan for the return of the Jiaoji railways in 1922. Things had not changed at all. Japan agreed to withdraw its troops, especially those stationed along the railways and its terminus Jinan.[170] It could not afford to alienate the United States at a time when the Anglo-Japanese alliance was not renewed. Japan might have lost the Jiaoji railways and Qingdao, but it received the Caroline and Marshall islands from Germany.[171] Despite signing the Treaty, large numbers of Japanese stayed behind in Shandong: 12,070 Japanese subjects in Qingdao, 2,160 in Jinan and hundreds more scattered along the railway and mining towns.[172] This would have consequences, among them the Jinan Massacre in May 1928.[173] Jiang Jieshi had consolidated power in south China and established the Nanjing government in 1927. He then marched north to defeat the various warlords in order to unify the country. This had nothing to do with the Japanese, but in the name of protecting its citizens and commercial interests, thousands of Japanese troops rushed in from Nagoya, Port Arthur, Tianjin and Qingdao, reoccupying the Jiaoji railways, its stations and the provincial capital.[174] Chinese sources put the number of troops at 28,000, while the *Herald* counted 35,000.[175] On 3 May, the Japanese attacked Jinan, killing 6,123 Chinese, and wounding more than 1,700, according to Chinese sources.[176]

Where there are railways, there is competition and ultimately conflict. Region by region, railways spearheaded the Japanese conquest of China. Ever since the Meiji restoration, Japan had been travelling on a very different path from late Qing China. From 1874, when they made their first attempt on Taiwan, the Japanese were persistent and relentless in achieving their aims. Railways enabled them to acquire territory in Manchuria and Shandong. They were able to stations troops in the name of protection, using China's resources to feed a growing army and sustain decades of war, and colonise through station towns and immigration. Railways would soon drive their expansion into south China and Southeast Asia. But railways would also limit them as they became dependent upon them. This would become most apparent during WWII. Railways allow us a glimpse into what lay behind Japan's rise. The scramble for China manifested in the scramble for railways. Railways turned Manchuria and Shandong into 'cradles of conflict' as western powers and Japan jostled with each other. They geopoliticised not just railway development, but also the trajectory of modern China. But northeast and north China were not the only regions over which foreign powers fought for railway concessions. They also converged on southern and southwest China.

Notes

1 Takeo, *Life along*, 13–15.
2 'Assassination of Prince Ito', *Herald*, 30 October 1909: 228.
3 Headrick, *Tentacles*, 49.
4 Matsusaka, *Japanese Manchuria*, 38.
5 Matsusaka, *Japanese Manchuria*, 21.
6 Steven Ericson, *The sound of the whistle: railroads and the state in Meiji Japan* (Cambridge: Council on East Asian Studies, Harvard University, 1996), 6.

7 Kimitaka Matsuzato, 'Introduction: Russia as the decisive factor in the modern history of northeast Asia', in *Russia and its northeast Asian neighbors*, vii–xv.
8 Xue, *Hulujun*, 112–13.
9 MacMurray, *Treaties*, vol 1, 326–29.
10 'Russia's hold on Manchuria firmer', *Times*, 6 June 1903: 1; 'Latest telegraphic intelligence', *Herald*, 29 January 1904: 158.
11 'News from Europe', *Herald*, 19 February 1902: 308; Ian Nish, *Origins of the Russo-Japanese war* (New York: Longman, 1985), 192–257.
12 Adam Gower, *Jacob Schiff and the art of risk: America's financing of Japan's war with Russia* (New York: Springer International, 2018).
13 'The war: fighting at Port Arthur', *Herald*, 18 February 1904: 322 and 'Latest telegraphic intelligence', 4 March 1904: 410.
14 Christian Wolmar, *Engines of war: how wars were won and lost on the railways* (New York: Public Affairs, 2010).
15 'Japan: peace concluded', *Herald*, 15 September 1905: 604 and 'The treaty of peace', 3 November 1905: 266.
16 MacMurray, *Treaties*, vol 1, 522–28 (523).
17 Ramon H. Myers, 'Japanese imperialism', in *Informal empire*, 101–32; 'The Russo-Japanese conventions', *Herald*, 23 August 1907: 438.
18 Matsusaka, *Japanese Manchuria*, 1.
19 'Japan papers: the south Manchurian railway company', *Herald*, 17 August 1906: 420.
20 Myers, 'Japanese imperialism', 101–9 and 125–26.
21 South Manchuria Railway Company (SMRC hereafter), *Manchuria: land of opportunities* (New York: South Manchuria Railway Company, 1924), 8 and 1.
22 O'Dwyer, *Significant soil*, 8.
23 MacMurray, *Treaties*, vol 1, 549–67.
24 Alvin D. Coox, 'The Kwantung Army dimension', in *Informal empire*, 409–12; Lao Chongmin, ed., *Mantie dang'an ziliao huibian* (15 vols, Beijing: Shehui kexue wenxian chubanshe, 2011), vol 2, 208–75.
25 'Dongbei zhimi', *Shen Bao*, 2 July 1932: 13; Matsusaka, *Japanese Manchuria*, 90–91.
26 Matsusaka, *Japanese Manchuria*, 1.
27 Matsusaka, *Japanese Manchuria*, 4.
28 Tony Ballantyne and Antoinette Burton, 'Empires and the reach of the global', in *A world connecting, 1875–1945*, ed. Emily S. Rosenberg (Cambridge: Harvard University Press, 2012), 285–434 (306) and Myers, 'Japanese imperialism', 121.
29 MacMurray, *Treaties*, vol 1, 552–67.
30 'The Antung-Mukden railway', *Herald*, 14 August 1909: 389.
31 MacMurray, *Treaties*, vol 1, 552.
32 'The Kirin-Kuangchengtze railway', *Herald*, 11 September 1909: 646.
33 MacMurray, *Treaties*, vol 1, 554.
34 MacMurray, *Treaties*, vol 1, 552 and 'Youchuanbu zoujieshou Xinfeng tieluzhe', *Shen Bao*, 22 November 1907: 10.
35 MacMurray, *Treaties*, vol 1, 554–55.

36 'Ambassador Wright to the Secretary of State, American Embassy Tokyo, June 27, 1907', No. 6625/18, Office of the Historian, at https://history.state.gov/historicaldocuments/frus1907p2/d194, accessed 7 July 2021.
37 Moore, *Defining and defending*, 53–83.
38 'American interest in China', *Herald*, 22 May 1901: 1002 and 'American railroad enterprise in China', 22 April 1904: 856.
39 Maury Klein, *The life and legend of E. H. Harriman* (Chapel Hill: University of North Carolina Press, 2011), 288.
40 'The Fakumen railway', *Herald*, 28 February 1908: 461 and 'Manchurian railways', 1 May 1908: 291.
41 Chang, *International controversies*, 42–53.
42 Myers 'Japanese imperialism', 126–27.
43 'Chinese railway schemes', *Herald*, 8 February 1907: 277, 'The Tsitsihar-Chinchou railway', 24 December 1909: 714 and 'The Chinchou-Tsitsihar line', 4 Feb 1910: 245.
44 Bruce Elleman, *International competition in China, 1899–1991: the rise, fall and restoration of the Open-Door policy* (London: Routledge, 2015), 29.
45 'Russia turns down Knox's other plan', *Times*, 5 March 1910: 6.
46 'Railways in Manchuria: action by the China Association', *Herald*, 29 July 1910: 234; Luo Wenjun *et al.*, *Diguo zhuyi lieqiang qinhua tielu shishi* (Beijing: Tiedao chubanshe, 1998), 399.
47 MacMurray, *Treaties*, vol 2, 1054–55; 'Japanese railway concessions', *Herald*, 20 December 1913: 921 and 'Japan's demands in Mongolia', 27 March 1915: 926.
48 'Manmeng kaibutan', *Shen Bao*, 7 March 1915: 6 and 'Manmeng wulu', 21 December 1915: 6.
49 'The Tumen railway dispute', *Herald*, 18 November 1922: 428.
50 'Japan's plan in Manchuria', *Herald*, 16 February 1929: 263.
51 'Dongbei zhimi', *Shen Bao*, 2 July 1932: 13.
52 Matsusaka, *Japanese Manchuria*, 298.
53 Matsusaka, *Japanese Manchuria*, 299; 'Riren xiang Fengfang yaosuo luquan', *Shen Bao*, 21 October 1928: 10, 'Manmeng tielu jihua', 9 October 1930: 5 and 'Riren xuyi zuwo kaifa Manmeng', 13 December 1930: 6.
54 Kate McDonald, 'Asymmetrical integration: lessons from a railway empire', *Technology and Culture*, vol 56, 1 (January 2015): 115–49.
55 Danny Orbach, *Curse on this country: the rebellious army of imperial Japan* (Ithaca: Cornell University Press, 2017); David Hunter-Chester, 'Imperial Japanese army culture 1918–1945: duty heavier than a mountain and death lighter than a feather', in *The culture of military organisations*, eds Peter Mansoor and Williamson Murray (Cambridge: Cambridge University Press, 2019), 208–25.
56 MacMurray, *Treaties*, vol 1, 526.
57 'Events of Monday June 4', *Herald*, 9 June 1928: 407; 'Dongsansheng pairi shenglie', *Shen Bao* 19837, 8 June 1928: 6.
58 Alvin D. Coox, 'Kwantung Army', in *Informal empire*, 395–428.

59 Matsusaka, *Japanese Manchuria*, 368.
60 Matsusaka, *Japanese Manchuria*, 369.
61 'Japanese clash with Chinese', *Herald*, 9 June 1931: 334 and O'Dwyer, *Significant soil*, 277–78.
62 'Japanese occupy Manchuria', *Herald*, 22 September 1931: 398; 'Japanese takes all towns on Manchurian railroad', *Times*, 20 September 1931: 1.
63 Takeo, *Life along*, 135.
64 Stacie E. Goddard, *When right makes might: rising powers and world order* (Ithaca: Cornell University Press, 2018), 149.
65 Sadako N. Ogata, *Defiance in Manchuria: the making of Japanese foreign policy 1931–32* (Berkeley: University of California Press, 1964).
66 'The fighting in Manchuria', *Herald*, 24 November 1931: 261.
67 Nakagane Katsuji, 'Manchukuo and economic development', in *Informal empire*, 135–46.
68 'What League expects of its members', *Herald*, 22 March 1932: 424 and 'The Manchurian problem', 21 June 1932: 446.
69 League of Nations, *The situation in Manchuria: report of the Lytton commission of enquiry* (Geneva: League of Nations, October 1932), Library of Congress, at www.loc.gov/item/2021666890, accessed 30 July 2020 and 'Guolian diaochatuan baogao', *Shen Bao*, 3 October 1932: 3–4 and 6–8.
70 'Japanese attack Chapei', *Herald*, 2 February 1932: 146, 'Stout Chinese resistance' and 'Japanese bombard Woosung forts', 9 February 1932: 182 and 185.
71 'Japanese military supervision', *Herald*, 8 November 1933: 211.
72 'Kwantung governorship abolished', *Herald*, 15 August 1934: 235.
73 'Another railway for Manchuria', *Herald*, 31 October 1934: 172.
74 Su Weidong, ed., *Qinghua rijun dongning yaosai jiemi* (Beijing: Zhongguo wenshi chubanshe, 2005), 14–15.
75 Shi Ding, *Riben guandongjun qinghua zui'eshi* (Beijing: Shehuikexue wenxian chubanshe, 2005), 602.
76 Takeo, *Life along*, 60, 114–18, 160–64 and 180–81.
77 Tsuneishi Keiichi, 'Unit 731 and the Japanese imperial army's biological warfare program' and Suzy Wang, 'Medicine-related war crimes trials and post-war politics and ethics: the unresolved case of unit 731, Japan's bio-warfare program', in *Japan's wartime medical atrocities: comparative inquires in science, history and ethics*, ed Jing-Bao Nie, *et al.* (London: Routledge, 2010), 23–31 and 32–58.
78 David Tucker, 'Labor policy and the construction industry in Manchukuo: systems of recruitment, management and control', in *Asian labour in the wartime Japanese empire: unknown histories*, ed. Paul Kratoska (London: Routledge, 2015), 25–57 (27).
79 Matsusaka, *Japanese Manchuria*, 143; Xie Xueshi, ed., *Mantie dang'an shiliao huibian* (15 vols, Beijing: Shehui kexue wenxian chubanshe, 2011), vol 7, 529–44.
80 'A note on travel coolies tramping North', *Herald*, 8 April 1904: 736 and 'The coolies', 8 March 1924: 360.

81 Sewell, *Constructing empire*, 131.
82 Xie Xueshi, ed., *Mantieshi ziliao: meitiepian* (Beijing: Zhonghua shuju, 1987), 476; Qingdaoshi dang'anguan, *Tieti xiade zui'e: Riben zai Qingdao lueduo laogong shimo* (Beijing: Dang'an chubanshe, 2003), 21.
83 Köll, 'Tianjin-Pukou', in *Manchurian railways*, 123–42 (137).
84 Li Binggang *et al.*, *Riben zai Dongbei nuyi laogong diaocha yanjiu* (Beijing: Shehuikexue chubanshe, 2009), 11, 33–40, 400–25, 473–541 and 37.
85 Su, *Dongning*, 15.
86 Su, *Dongning*, 57, 173 and 215; Xie, *Mantie dang'an*, vol 7, 544–79.
87 Su, *Dongning*, 181; Li Binggang, *Nuyi laogong*, 483.
88 Tucker, 'Labor policy', in *Asian labour*.
89 Edwin Pauley, *Report on Japanese assets in Manchuria* (Washington DC: USGPO, July 1946), 74.
90 'Chinese coolies in Japan', *Herald*, 20 September 1907: 681; Zhu Miaochun *et al.*, *Laogong xueleishi* (Shanghai: Sanlian shudian, 2005), 1.
91 Matsusaka, *Japanese Manchuria*, 181.
92 Matsusaka, *Japanese Manchuria*, 214–18.
93 Henry W. Kinney, *Modern Manchuria and the south Manchurian railway* (Dairen: Japan Advertiser Press, 1928), 69–70; Tsu-yu Chen, 'The South Manchurian Railway company and the mining industry: the case of the Fushun coal mine', *Cross-currents (Honolulu, Hawaii)*, vol 4, 2 (2015): 630–57.
94 'Jueyi buhuan Fushun meikuang', *Shen Bao*, 18 June 1907: 4 and 'Fushun meikuang jiaoshe', 3 August 1910: 6.
95 SMRC, *Land of opportunities*, 30.
96 'The Fushun mines', *Herald*, 20 January 1917: 122 and 'China's huge coal resources', 11 September 1935: 421.
97 'Japanese mining in Manchuria', *Herald*, 23 May 1914: 644; 'Zhuandian', *Shan Bao*, 26 July 1921: 10.
98 SMRC, *Land of opportunities*, 31.
99 Matsusaka, *Japanese Manchuria*, 137.
100 Pauley, *Report*, 72–74.
101 'Fuel oil from coal', *Herald*, 3 November 1932: 172, 'Coal-oil industry near Mukden', 19 August 1936: 311 and 'Oil monopoly', 27 April 1938: 141.
102 'Nanman tielugongsi jingying shiye', *Shen Bao*, 5 October 1924: 5–6 and 'Ri Mei jingzheng zhongzhi Manmeng dili', 14 February 1928: 9; Kinney, Modern Manchuria, 72–73.
103 Matsusaka, *Japanese Manchuria*, 305 and Koji Hirata, 'Made in Manchuria: the transnational origins of socialist industrialisation in Maoist China', *The American Historical Review*, vol 126, 3 (September 2021): 1072–1101.
104 'Mining progress in Manchoukuo', *Herald*, 1 January 1941: 19.
105 Kinney, *Modern Manchuria*, 58.
106 'Riben guandongjunde xingwang', Zhishuo at Jilin difangzhi, http://dfz.jl.gov.cn/jldywh/dbkl/201701/t20170110_5217618.html, accessed 14 December 2021.
107 'Bean empire falls on evil days', *Herald*, 13 September 1939: 441.

108 Matsusaka, *Japanese Manchuria*, 139. See also the works of Miki Masafumi.
109 'The romance of the soy bean', *Herald*, 17 June 1911: 726 and 'The competition for Manchuria', 27 May 1930: 335.
110 'South Manchurian railway', *Herald*, 26 June 1909: 236 and 'Trade figures', 12 July 1933: 50.
111 Matsusaka, *Japanese Manchuria*, 133 and 146.
112 SMRC, *Land of opportunities*, 49; Myers, 'Japanese Imperialism', 117–18.
113 'South Manchurian railways', *Herald*, 3 July 1926: 10.
114 'Riben duzhan dongsansheng meihuo shichangtan', *Shen Bao*, 4 September 1922: 13.
115 SMRC, *Land of opportunities*, 71.
116 Myers, 'Japanese imperialism', 114.
117 O'Dwyer, *Significant soil*, 132, Table 4.1.
118 Pauley, *Report*, 35–53.
119 Takeo, *Life along*, 14.
120 Myers, 'Japanese Imperialism', 115; Matsusaka, *Japanese Manchuria*, 56–57, Appendix B 414–15.
121 SMRC, *Land of opportunities*, 84.
122 Kinney, *Modern Manchuria*, 32; Pauley, *Report*, 126.
123 'Continued support for Manchukuo', *Herald*, 23 August 1933: 288.
124 O'Dwyer, *Significant soil*, 38–39; Myers, 'Japanese Imperialism', 116; Kinney, *Modern Manchuria*, 45.
125 Sewell, *Constructing empire*, 134.
126 SMRC, *Land of opportunities*, 91.
127 Yosano Akiko (translated by Joshua A. Fogel), *Travels in Manchuria and Mongolia: a feminist poet from Japan encounters pre-war China* (New York: Columbia University Press, 2001), 131.
128 'Our Japan letter', *Herald*, 25 June 1927: 550.
129 Sandra Wilson, 'The "new paradise": Japanese emigration to Manchuria in the 1930s and 1940s', *The International History Review*, vol 17, 2 (May 1995): 249–86.
130 'Ribenzhi manzhou jiaotongwang jihua', *Shen Bao*, 24 June 1929: 9.
131 SMRC, *Land of opportunities*; Jie Li, 'Phantasmagoric Manchukuo: documentaries produced by the South Manchuria railway company, 1932–1940', *Positions: East Asian Culture Critique*, vol 22, 2 (2014): 329–69.
132 Hugo Silveira Pereira, 'The camera and the railway: framing the Portuguese empire and technological landscapes in Angola and Mozambique, 1880s–1910s', *Technology and Culture*, vol 64, 3 (2023–07): 737–59.
133 Matsusaka, *Japanese Manchuria*, 143.
134 Gao Lecai, *Riben Manzhou yimin yanjiu* (Beijing: Renmin chubanshe, 2000), 5.
135 'Shi Manzhou wanquan zhiminhua', *Shen Bao*, 16 May 1933: 9, 'Disanci wuzhuang yimin jihua', 7 April 1934: 6 and 'Yimin Manzhou fangzhen', 11 May 1936: 7.
136 Meng Yueming, *Riben kaituotuan ruzhi shimo* (Shenyang: Liaoningdaxue chubanshe, 2012); Anke Scherer, 'Japanese emigration to Manchuria: local

activists and the making of the village-division campaign' (Unpublished dissertation, Ruhr-Universitat Bochum Fakultat fur Ostasienwissenschaften, 2006).

137 'Rixun huiyi', *Shen Bao*, 13 June 1932: 8, 'Yiyongjun yu Riben yimin', 6 August 1932: 3 and 'Chen Youren tan Meiguo jiying xiezu Zhongguo', 27 March 1933: 7.

138 Some have returned to Japan, see The Support and Communication Centers for People Returning from China, at www.sien-center.or.jp, accessed 23 June 2023.

139 Issac C. K. Tan, 'Science and empire: tracing the imprint of dactylography in Manchuria, 1924–1945', *Japan Forum*, vol 32, 4 (2020): 531–54.

140 Pauley, *Report*, 126.

141 'News from Europe', *Herald*, 19 February 1902: 308, 'The new Anglo-Japanese alliance', 29 September 1905: 708 and 'Late telegraphs', 5 August 1911: 318.

142 'The Japanese, ultimatum', *Herald*, 22 August 1914: 566.

143 'The war in Shantung', *Herald*, 10 October 1914: 99 and 'Late telegraphs', 22 August 1914: 616.

144 'The siege of Tsingtao', *Herald*, 5 September 1914: 778 and 'Part played by the British fleet in Tsingtao', 26 June 1915: 908.

145 'The fall of Tsingtao and tributes to Japan', *Herald*, 19 December 1914: 862.

146 Wilson L. Godshall, *Tsingtau under three flags* (Shanghai: Commercial Press, 1929).

147 MacMurray, *Treaties*, vol 2, 1231–35 and 1234–35; 'Japan's demands of China', *Herald*, 13 March 1915: 771 and 'Great Britain and Japan's demands upon China', 3 April 1915: 21.

148 MacMurray, *Treaties*, vol 2, 1231.

149 MacMurray, *Treaties*, vol 2, 1231 and 1233.

150 MacMurray, *Treaties*, vol 2, 1232.

151 'The amendments of demands', *Herald*, 1 May 1915: 346.

152 'Reservation of rights by American government, May 13, 1915', in MacMurray, *Treaties*, vol 2, 1236.

153 'The Shantung question', *Herald*, 6 September 1919: 593.

154 'Treaty of Peace with Germany', Library of Congress, at www.loc.gov/item/19026725/, accessed 27 July 2020.

155 'The case for Shantung', *Herald*, 23 October 1920: 219.

156 Chung Hung Liu, 'The Anglo-Japanese Alliance: its past and present', *The Chinese Students' Monthly*, vol XVII (November 1921 to June 1922): 130–36 (135).

157 'Allied banking consortium', *Herald*, 7 June 1919: 622.

158 Carnegie Endowment for International Peace, *The Consortium: the official text of the Four Power Agreement for a loan to China* (Washington DC: The Endowment, 1921), 67–72 (68).

159 Carnegie Endowment, *The Consortium*, 19–20.

160 Carnegie Endowment, *The Consortium*, 20–21 (21).

161 Carnegie Endowment, *The Consortium*, 28–30 (30).

162 'Japan and the consortium', *Herald*, 20 August 1920: 559 and 'Japan's demand of the consortium', 18 September 1920: 740.

163 Carnegie Endowment, *The Consortium*, 67–72; 'Japan and the 21 demands', *Herald*, 18 February 1922: 427 and 'China and the new consortium', 4 June 1921: 653.
164 'The Shantung agreements signed in Washington', *Herald*, 11 February 1922: 355 and 'Integrity of China and open door', 18 March 1922: 731.
165 MacMurray, *Treaties and agreement with and concerning China 1919–1929* (Washington: Carnegie Endowment for International Peace, 1929), 89–93 (91).
166 'Russian aggression in China', *Herald*, 20 August 1927: 322.
167 MacMurray, *Treaties 1919–1929*, 80–88 (83).
168 Bruce Elleman, *Wilson and China: a revised history of the Shandong question* (London and New York: Routledge, 2015).
169 MacMurray, *Treaties 1919–1929*, 80–88 (83).
170 MacMurray, *Treaties 1919–1929*, 100–1.
171 These islands were sold by Spain to Germany in 1899. The US claimed them during WWII.
172 Zhang Yan and Sun Yanjing, eds, *Jinan can'an* (Beijing: Guojia tushuguan chubanshe, 2014), 62–65.
173 'Riben yingfu luhuo quanze', *Shen Bao*, 6 May 1928: 13–14; 'Two Chinese protests to Japan', *Herald*, 12 May 1928: 213–14.
174 Takeo, *Life along*, 114–15.
175 Zhang and Sun, *Jinan*, 120–22 and 'Japanese reinforcements', *Herald*, 5 May 1928: 174.
176 Shehui kexueyuan jinshisuo, *Riben qinhua qishinian shi* (Beijing: Shehuikexue chubanshe, 1992), 281–82.

7

'Civilisation follows the locomotive': French variation[1]

> England has strongly positioned itself on the entire coastline of India; it has seized Rangoon, Pegou, Martaban, Maulmein and the three keys to the strait: Pou-lo-Penang, Malacca and Singapore. France, since the fall of its Asian domination, no longer had any position in the Far East; she felt the need to create a new one, and found it in An-nam: in 1860 and 1867 she conquered the six southern provinces of this state ...
>
> All the questions can be reduced to these two terms: What is the shortest trade route from China to Europe? Who should control it?[2]

Paul Eugene Louis Deschanel complained about British gains in Burma and Malaya in 1883. The future president of the French Republic provided a clear rationale for the conquest of Vietnam – the most direct route to China. His vision would be realised with the construction of a railway that would link Haiphong in Vietnam with Kunming in Yunnan province. French manufacturer Decauville and M. Thévenet had made efforts at introducing and building railways in China. Government-led entry into the 'hunt for railway concessions' came after the Sino-French War. This chapter investigates the making of the Yunnan to Vietnam or the Dianyue [滇越] railroad. It exposes the similarities as well as the differences between French railway imperialism and the other cases discussed in the preceding chapters, exposing the ferociousness of Anglo-French railway rivalry in southwest China.

Eyes bigger than stomach

The most important article of the Li-Patenotre Treaty signed at the end of the Sino-French war in 1885 begins with this:

> The object of France and China entering into this Treaty being to secure friendly commercial relations between neighbouring States, it is now desired to duly realise this object by France opening up roads in Tongking and

promoting the early construction of railways; and it is agreed that if in the future China desires to lay down railroads, the advice of French experts will be available.[3]

This was the first time a foreign power had demanded railway rights in a treaty – a signal that, by the mid-1880s, the game of empire was adapting to new technologies at the disposal of imperialists. This was only the beginning. Auguste Gerard, French minister in Beijing, imposed a series of agreements on the Qing in 1887, including the *Supplementary Commercial Convention*, which also included railway rights.[4] More agreements followed in the wake of the 'Triple Intervention', which gave France another opportunity to make demands. Article V of the 1895 *Additional Convention* to the above *Supplementary* stated that: 'It is agreed that railways, either those already in existence or those projected in Annam, may, after mutual agreement, and under conditions to be defined, be continued on Chinese territory.'[5] The French colonial regime in Indochina had been building a railway to Đồng Đăng on the Sino-Vietnam border and sought to extend the line to Longzhou, across the border in China's Guangxi province. Hardly had construction on this railway begun, the French pestered the Qing again in June 1897. Auguste Gerard was so demanding and persistent that Weng Tonghe referred to the Frenchman as 'disgusting' in his diary.[6] The French were hardly content with the Longzhou extension: 'as soon as the Đồng Đăng to Lung-chou shall be finished, a request will of necessity be made it to continue the said line in the direction of Nan-ning and Pe-se (Baise)'.[7] This extension was to run from Longzhou to the heart of the Guangxi (see Figure 7.1).

Like the Japanese, the French had an insatiable appetite for railways. A similar character to Sergei Witte, Alfred Tirpitz and Goto Shinpei, the new governor-general of Indochina, Paul Doumer, had a vision for railway development and was determined to implement it.[8] An 'ardent imperialist', Doumer planned to construct 3,200 km of railways in the region and into China.[9] He dispatched missions dominated by military officers rather than engineers, to survey southwest China and gather intelligence. He intended to turn Yunnan into a horse remount for the Indochina army that could be used to occupy and ultimately annex the region. The French ironclad battleship, *Bayard*, sailed into Guangzhouwan (Zhanjiang) on the Leizhou Peninsula in west Guangdong province, which they renamed Port Bayard.[10] Like the Russians and the Germans, the French undertook field surveys without seeking Chinese permission, trusting that they would inevitably procure what they had desired on paper. Their opportunity soon came in November 1897, when the death of two German missionaries in Shandong led to more French gains.

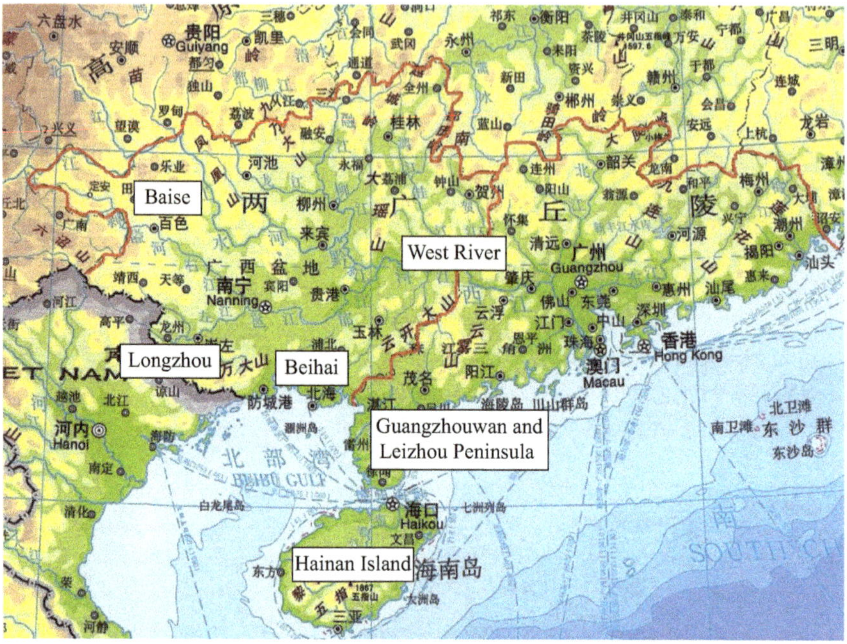

Figure 7.1 Map of Guangxi, Guangdong, Leizhou Peninsula and Hainan (Baidu public domain)

The French leapt at the opportunity, quickly rectifying earlier agreements to better suit their interests as well as securing new deals, including the *Agreement in Regard to a Concession to Build a Railway from Tongking to Yunnan*, and the *Convention for the Lease of Kwang-chow Wan*, in April and May 1898.[11] They helped the French lay claim to Chinese provinces bordering Vietnam: Yunnan and Guangxi. The *Agreement* finalised the railway into Yunnan:

> The Chinese government grant to the French Government, or to the French Company, which the latter may designate, the right to make a railway from the frontier of Tong-king to Yun-nan-fu; the Chinese having no responsibility (*charge*) but to furnish land for the road and its dependencies. The route (*trace*) of this line is actually surveyed (*etudie*), and will be fixed later on in agreement with the two Governments.[12]

This led to the birth of the Dianyue railway from Haiphong to Hekou on the Sino-Vietnam border, terminating in Kunming, capital of Yunnan (see Figure 7.2). Less than a month after this agreement was signed, the French Foreign Minister, Gabriel Hanotaux, tasked the new French Minister to China, Stephen Pichon, with another job – to 'ask the Chinese

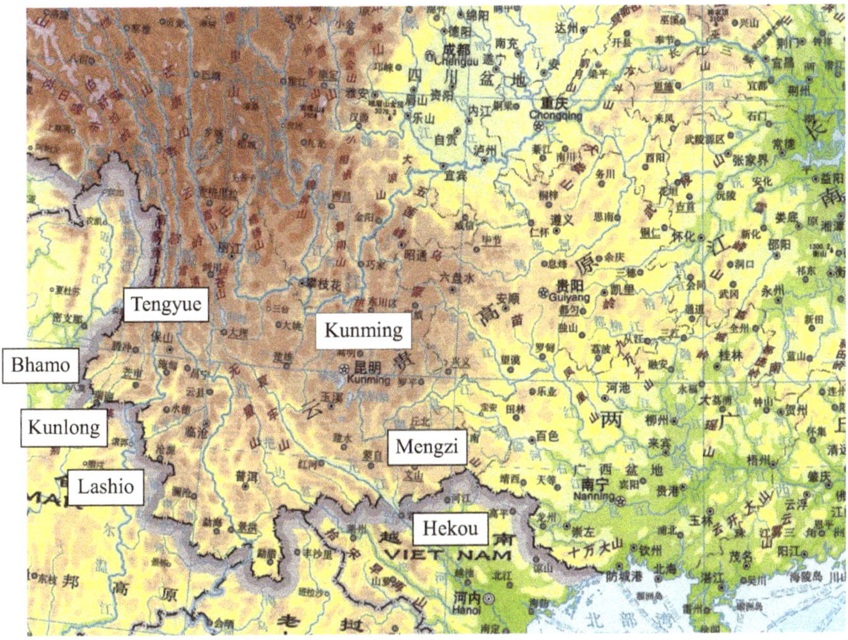

Figure 7.2 Map of Yunnan, border with Vietnam and Burma (Baidu public domain)

government for the concession, to a French Company, of a railway destined to connect the port of Pakhoi (Beihai) with a point to be fixed upon on the course of the West River'.[13] Pichon's reply dated 28 May confirmed Chinese consent – the Qing had responded within days, indicating the extent of French aggressiveness as well as the Qing's weakened position. Beihai was an important coastal city in western Guangdong, and the West River is the largest subsidiary of the Pearl River – this would put the French in direct conflict with the British, who considered old Canton and the Pearl River Delta their exclusive sphere. Unsatiated with railways into Yunnan, Guangxi and west Guangdong, the French demanded more lines on the Leizhou Peninsula under Guangdong's jurisdiction in the *Convention*:

> The Chinese Government authorizes France to construct a railway on the bay of Kuang-chow Wan, by Lei-chou, with a point to be designated on the west coast of Lei-chou, in the neighbourhood of On-pu, ... China will give the land ... and see the protection of the railway and the stock ...[14]

This would effectively be a cross-peninsula line. Leizhou was in a strategic location, the last peninsula in China not yet occupied by a foreign power.[15]

Positioned between the South China Sea and the Gulf of Tokin, the Peninsula overlooks the route to Hong Kong, commanding the Qiongzhou Strait and Hainan Island, which the French had already obtained with the *Declaration Concerning the Non-alienation of the Island of Hainan* in 1897.[16]

The French rationale was similar to that of the Japanese: more railways would ease their penetration into China and exclude their competitors. They demanded more railways than they could build and without knowing the difficulties involved. In fact, none of the railways into Guangxi and Guangdong would ever be built. This was the same in Africa as the French government endorsed the Trans-Saharan railway in 1879, but it was shelved until 1890; only a small portion was built by 1944.[17] The only line the French managed to construct in China was the 468 km narrow-gauge Dianyue line. The French sphere of influence more than doubled with the inclusion of southwest China and Hainan Island, an area that surpasses the size of Vietnam, Cambodia, Laos and French Siam combined. But constructing railways in a region known for its extremely challenging geography would not be easy, as highlighted by the companies that conducted route surveys.[18] The same could be said of the French's ability to wield influence on a land inhabited by not just Chinese and Muslims but also a multitude of indigenous tribes in deep valleys and impenetrable jungles who had resisted Han Chinese rule and assimilation for hundreds of years. This delicate situation did not escape the eyes of Francis Garnier, who explored the region in the 1860s, and those of missionaries who operated in the province.[19]

'La pénétration française au Yunnan'[20]

French railway imperialism bore many similarities to the model discussed in Chapters 4, 5 and 6, but with notable variations. The 1898 *Agreement* stipulated that the Chinese had 'no responsibility (*charge*) but to furnish land for the road and its dependencies'. Unlike the Russians, the Germans and the Japanese who seized land on their own initiative, the French forced the Qing authorities to proffer land fit for their purpose. Such a demand was an unwanted financial and administrative burden for local governments, the consequences of which soon manifested when preparations for the construction of the Dianyue were underway. Yunnan was inhospitable terrain for railways. Archibald John Little, a Manchester merchant, travelled through the province in the early 1900s:

> The Yunnan plateau is nothing but an endless succession of small isolated oases – cuvettes or basins – some filled with deep water lakes, others partially occupied by shallow meres, – dotted about amidst a sea of rugged mountains ... the few small rivers that flow above ground have cut out deep narrow

defiles in the limestone ... while their gorges form a practically impassable barrier to intercommunication.[21]

Little identified the biggest challenge for the railway in terms of 'how best to climb the wall-like ascent of 5000 feet from the Red River valley on to the plateau'.[22] One of Doumer's surveying party proposed a 'western route', also called the 'Guillemoto route'.[23] Starting from Hekou on the Sino-Vietnam border, this line would pass Xinxianhe, Mengzi, Honghe and Jianshui before reaching its destination Kunming. Doumer had favoured this route, but the people of Yunnan resisted it. They had battled the Chinese for centuries and now they would exacerbate the French. Resistance against the French erupted in Mengzi immediately after Doumer's visit in June 1899. The rioters attacked the well-fortified town but were repelled by Qing forces. They soon turned to the Imperial Maritime Customs, controlled by foreigners, and set it ablaze.[24] The situation in Kunming was even more dire. The French surveying party based themselves in the local pagoda – a sacred place, similar to what the Germans did to the Confucius Temple. They were, as Robert Lee put it, preparing to 'establish the machinery of the Indo-Chinese government in Yunnan'.[25] The Yunnanese laid siege to the pagoda, offering Doumer the perfect pretext to send in troops as he had long advocated for territorial expansion, against the wishes of the French government.[26] Like the Japanese Army, Doumer acted on his own initiative and was looking to secure a *fait accompli*.

To prevent a crisis from materialising, diplomat Auguste François was dispatched to Kunming in October 1899.[27] This led to a standoff between François, with Paris' backing, and Doumer, which continued into 1900, when the Boxer movement spread to Yunnan.[28] The French community of Kunming was attacked and besieged for a few days.[29] French lives and property would remain vulnerable for years to come.[30] After the rebellion, the 'western route' was abandoned and replaced with plans to find another route.[31] However, this new orientation was disastrous for Ding Zhenduo, the governor-general of Yunnan-Guizhou provinces:

Ever since Guillemoto and others arrived in spring 1898 to conduct route survey, I have instructed local officials and gentry to work with them, and sent soldiers to protect them. The survey began in Hekou, passing Honghe, Xinxianhe, to arrive at Mengzi. From here it goes to Lin'an, Jianshui, Tonghai, Xinxing, Chenggong, etc and other places to arrive at Kunming. At Honghe and Xinxianhe, we purchased land and built a small road, a warehouse and a hospital. Because the original agreement says we have to provide land, I had to move funds from other places to pay for this. But now the Company has changed the route from Nanxihe to Mengzi, from here it moves to Mile, Yiliang and then the capital. What we had purchased and built were wasted. The government has bought from the locals and it would be difficult to retreat

> and recover the funds, not to mention the local officials and gentry who helped, and the expenses spent on soldiers and horses.[32]

His regional government had purchased land for the 'western route', not knowing the French would abandon it. That aside, he was forced to immediately raise further funds to acquire land for the new 'eastern route'. He sought help anywhere he could: from the Qing court to powerful officials like Zhang Zhidong. He also coerced local merchants. The French made the Chinese pay for the land needed for railway construction, saving them money and the trouble of dealing with locals. Purchasing land for the French railway was only the beginning of the governor-general's nightmare as construction began in 1904.

Unlike the Russians, the Germans and the Japanese who used railways to justify their military presence and occupation, the French would make the Yunnan regime provide and pay for their protection. But why was this necessary? First, France was constrained by the *Declaration with regard to the Kingdom of Siam and other matters (advantages in Yunnan and Szechuan)*, signed with Britain in 1896. Article IV of this document stipulated that 'all privileges and advantages of any nature are to be shared by the two powers'.[33] In other words, any move by the French to send in large numbers of troops could be replicated by Britain. This was why the French government opposed Doumer's scheme to annex Yunnan. Second, the French military was involved in fierce competition with the British in Southeast Asia, and were preoccupied with the ebb and flow of conquering and reconquering in multiple locations.[34] Protecting the French and the railway in hostile territory wrought financial havoc to many local officials, evidenced in their memorials to the Qing court. They protested about the vast expenses they were expected to endure. Governor-general Ding summed up their complaints: 'From Hekou to Kunming, there are more than one thousand kilometres. Yunnan has not been exposed to the outside world and the people are rough. In addition to funds needed to purchase land, we also have to station soldiers to protect everything. This is another gigantic expense.'[35]

Ding asked the *Zongli Yamen* and the Qing Treasury for 1,000,000 taels (silver) as a matter of urgency but this was not all. As construction progressed, the newly completed segments, stations, building material, as well as the mines, also needed protection. The *Mining Concession* introduced a solution:

> [I]n order to obtain security for the Europeans and all the persons in the mines, in order to ensure peace and tranquillity and to prevent complications, the Syndicat du Yunnan, Limited, may, on giving notice to the local authorities, recruit near its mines native soldiers, who shall be placed under the command of an officer chosen for that purpose ... [36]

Using local resources to protect foreign lives and property was cheap for the French, but it pitted the local population against each other and armed potential rebels. This was particularly dangerous in a region where Han-ethnic relations were extremely delicate and often led to rebellions, like the Muslim uprising of 1856. As construction progressed, 'two companies of French troops' arrived in Mengzi, the largest and most important station outside of Kunming, to serve as 'railway guards'.[37] They were hardly enough for a line that stretched further into the interior. The *South China Morning Post* explained the problem in February 1908 as completed segments began to operate:

> Robbers, brigands, and bad characters of all sorts formed the police force, which soon became a public danger. Instead of protecting French residents, they robbed and murdered at will. All complaints were disregarded by the Chinese authorities, so that, last year, robbery and murder had free play on the railway line, with Chinese officers and soldiery as prime movers and abettors. ... Countless robberies, attacks on the trains, and attempts to get the train off the rails, have been committed with impunity.[38]

This was the worst-case scenario. The locally recruited police force had outsmarted the French, while local officials turned a blind eye either because they had been bribed or were conniving with the 'police force' to undermine the French. The Indochina colonial authority responded in April 1908 by sending in more troops, approximately '11 battalions of 250 men each, or about 3,000 men'.[39] But these forces were not enough to guard a railway that climbed to 6,000 feet above sea level and passed through areas inhabited by indigenous tribes who were keen to protect their territories against any encroachment. This did not include the various French communities scattered in the region. One such community was found by Archibald Little in Yiliang, which he described as a 'French colony of railway people – thirty-five foreigners living in a pretty little enclosure all to themselves'.[40] Little also counted 'some sixty Europeans' living in Mengzi, with a French consulate and the Imperial Maritime Customs which was largely staffed by foreigners.[41] Arming locals and using them to protect the railway would quickly ignite anti-Qing activities and give rise to well-armed warlords, whom Wilbur Burton called a 'military oligarchy', that would plague southwest China for half a century to come.[42]

In addition to the railways, the numerous mines in the region also needed protection. The Yunnanese had been mining for more than 1,000 years.[43] The region's mineral wealth had been noted by French colonial officers including Francis Garnier, who located stores of copper, gold, silver, mercury, iron, lead and zinc, among others.[44] British explorer Arthur Davenport saw mines extracting white copper, blue copper, red copper, silver and iron.[45]

The region was also endowed with other resources such as coal, tin, cadmium, indium, thallium, antimony, crocidolite and sulphur. The French had demanded mining rights in both the 1887 *Supplementary* and the 1895 *Additional Convention*. A French newspaper reported the income generated by Yunnan's minerals in 1900:

> The Commissioner of Customs uses the word 'phenomenal' in his report to characterize them. They are due almost exclusively to the considerable rise in tin. ... The year 1899 was therefore very brilliant for the Mong-tzé trade. It was, moreover from our point of view, one marked by tax income.[46]

The future looked bright for the Indochina colonial regime – the Dianyue railway opened a vast region for French industry and commerce. The 1902 *Agreement for Yunnan Mining Concession* signed away the mining rights in seven locations to a French–British Syndicate.[47] Article I insisted that the mines in these areas were 'abandoned' by the 'State of China', rendering them free for the French to take. These included 'copper, gold, silver, coal, iron, etc', and 'platinum, tin, as well as the deposit of petroleum, precious stones, and cinnabar'. The French also sought to appropriate private mines, especially the profitable ones, as stated in Article II:

> Apart from the State mines whose working will be started in the above-mentioned districts, if the Syndicate wishes to work mines belonging to private persons, it will apply to the High Provincial Authorities, who will order an investigation to be made; and if the said investigation does not bring to light any objection the local officials will be authorised to negotiate with the proprietor for hiring the mine or the land ...[48]

In making these demands, the French were following the precedent established by the Russians, the Germans and the Japanese. They were entitled to open new mines too, as stipulated in Article III. These Articles gave the Syndicate a mining monopoly and they dictated how revenue was to be distributed:

> The Imperial Government's share 25%
> The Provincial Government's share 10%
> The share of the Syndicate for distribution among its shareholders 65%

The Yunnan government handed the French state-owned mines, allowed them to take over private ones, paid for the land needed for the railway, and protected French lives, yet they only received 10 per cent of the revenue while the Syndicate received the lion share. Once again, it was the people of Yunnan, not the government, who rejected the *Mining Concession*, as yet another uprising against the French broke out in May 1903. The 20,000 strong rebel force tried to prevent foreigners from taking over the mines and constructing the railway. They swept through several towns, including Gejiu, Mengzi, Tonghai and Kaiyuan. Sadly, their uprising was put down by their own government, not the French.[49] Similar to their sustained demands for railways, regardless of whether they were able to build them or not, the French asked for more mines than they could operate.

As was the case for the Russians, the Germans and the Japanese, more mines meant more railways, more expansion and more profit. Railways and mining work in tandem, a defining example of economic imperialism. Article IV of the *Mining Concession* stipulated:

> On waste lands situated near the mines, the Syndicat au Yunnan, Limited will be able to build railways, roads, canals, all means of communication which it shall deem necessary ...
>
> Since the joining up of the railroads of the Syndicat with the railway has as its object the facilitating of the marketing of its mineral products and the transport of material, as soon as the principal trunk-line shall be constructed, special rules will be drawn up by common agreement ... and the work of joining up the lines may be commenced.[50]

The trunk-line mentioned was the Dianyue, completed in April 1910. All mining lines would connect with it, turning it into a rail network linking up with those in Vietnam. Among them was a line from Gejiu (Kochiu) to Mengzi.[51] Nicknamed 'China's tin capital', Gejiu was also rich in copper, zinc, tungsten, beryllium, bismuth, gallium, germanium, cadmium, silver, gold and nepheline.[52]

The *Herald* called the Dianyue railway a 'remarkable feat of railroad engineering'.[53] However, it could not have been constructed without the labour of Chinese coolies. J. P. Daughton has argued that 'the construction of the Congo-Océan railway was an act of extraordinary violence'.[54] This was exactly the case with the Dianyue. As construction began, the real cost that characterised its construction became apparent, which was perfectly expressed by one mainland scholar: 'one drop of blood for every single nail, one life for every piece of sleeper'.[55] It was the most challenging line built on Chinese soil thus far, traversing limestone formations and karst landscapes with deep gorges flanked by jagged mountains with near-vertical peaks and unnavigable rivers (see Figure 7.3). Added to this terrain was the hot damp weather that bred tropical diseases and indigenous tribes who often attacked intruders. Yunnan's geography can be seen from the drawings of Francis Garnier.[56]

The first difficult task involved transporting building equipment and material to construction sites where there were no roads and where the land elevated gradually to as high as 6,000 feet. Initially, the French recruited labourers locally and from neighbouring provinces but the rate of desertion

Figure 7.3 Looping railway track and tunnels in Yunnan (Image courtesy of the Cottrell family and University of Bristol library, HPC, https://hpcbristol.net/visual/Co-s118, Identifier Co-s118)

was high due to abuse, disease and exhaustion. Pierre Ibos, a colonial officer who worked on the railway described that 10,000 coolies were recruited in 1904, but the majority fled.[57] George Marbotte, an engineer who worked on site, recalled that 60,000 coolies from different regions were recruited between 1903 and 1906.[58] Rang-ri Park-Barjot put the figure at 60,700, whereas Virginia Thompson estimates it was 80,000.[59] The railway company Fives-Lille also drafted workers from north China, especially those who had laboured on railways in Manchuria. He Zongzhang, a local official, reported their plight in Yunnan:

> First recruits from Tianjin, Shuntian, Zhili and Jinan, about 6000 coolies, arrived by sea and were sent down to the various segments. Their pay was 6 cents per day but that was embezzled at every level. They had to pay for their own food. In early spring, the northerners still wore their padded trousers and robes, but in Yunnan the miasma makes the summer really hot. They would set up a shack, covering three-pronged wood strips with straw and sleep on it, very damp. ... after a few days, they began to die. Only one or two out of ten could walk. Locals were so sick of this that they did not ask whether they were still alive, they burnt the shacks and buried them. For those dying on the roadside, they were kicked down to the gully below ... those who were not sick, they deserted. In a few days, they were all gone.[60]

Marbotte was so concerned that he tried to find out what had happened to those who had been kicked down to the gully – to his horror, they had been eaten by wild dogs.[61] To attract new labourers, the French promised better pay, with wages ranging from 50 cents to 75 cents per day as Selda Altan details in an excellent article.[62] Although this rate of pay was advertised in *Shen Bao* for a lengthy period, the working conditions of the coolies did not improve. The Qing government dispatched more officials to investigate the situation, Shen Zuyan pointed to the main problem:

> The foreign contractors were very harsh, the coolies are called and checked twice a day, if they are found resting, a day's salary would be deducted, and if they did not perform well or showed signs of fatigue, they would be whipped. All kinds of maltreatment, as if they are not human beings.[63]

These foreign contractors were mostly Italian, some French and other Europeans. According to Archibald Little, there were 1,000 of them in Kunming alone, where they directed 10,000 coolies.[64] Selda Altan's research into the French archives reveals that French diplomats knew that the contractors treated the coolies inhumanely, 'as if the life of a poor coolie is not worth a pipe of tobacco'.[65] Swiss engineer Otto Meister wrote letters home, telling his family that 'coolies die like flies' and that 'the railway has become their grave and they lay there one by one, just like in the aftermath of a war'.[66] Amaury Lorin and Vatthana Pholsena approximate that 12,000 of the over 60,000 labourers died of pollution, hunger, disease and maltreatment.[67] This kind of French brutality can also be seen in the construction of the Congo-Océan railway where 'between 15,000 and 23,000' forced labourers died.[68] The *Herald* tried to estimate the number of casualties in Namti valley, the most difficult location to reach:

> The vile climate of Namti Valley has levied a heavy death toll on those who have dared to open up its primeval jungles and gullies. The death rate among the coolies imported from various parts of the Empire and put to work in this dreaded valley, may, without exaggeration, be estimated at 5,000, or 70 per cent of the total number employed on that particular section of the line.[69]

The terrain was treacherous. Frederick Tablot described the Namti valley as a 'deep, wide, V-shaped fissure, one side dropping down perpendicularly for several hundred feet. The line pierced its way through one bluff, and had to jump across the rift to enter the opposite wall of rock.'[70] This rift necessitated the construction of a bridge of at least 200 feet across and 300 feet above the river below. Georges Bodin designed a bridge that looked like an inverted V, resembling the Chinese character *ren* [人]. But assembling the bridge at such height where there is no path to the cliff was a life-and-death challenge. The bridge was composed of two pre-manufactured parts. To begin with, the coolies needed to open a path and carry the equipment and material to the cliff. Talbot believed that 'mules could handle weights up to 600 pounds, and that the natives could struggle along with loads varying between 200 and 300 pounds, but neither man nor beast could cope with anything exceeding 7 feet in length'.[71] Next, the coolies had to drill, suspended off the side of the cliff, cutting out niches and extracting shelves on both sides, upon which winches could be erected and the bridge anchored (see Figure 7.4). This was an extremely dangerous operation; some were literally crushed to death in mid-air by huge, unstable hammers while others fell to the valley below. Then came the assembling process:

> Large gangs of coolies were disposed of Indian file 7 feet apart, and the chains were trailed over their shoulders like a giant serpent. In this way, they wound up crags, climbed steep bluffs and threaded narrow defines for some 13 miles. These chains were passed around the winches and the outer ends were attached to the upper point of the bascules. ... Gangs of coolies were stationed at each winch under the supervision of a French engineer and at the word of command the chains were slowly paid out, causing the bascules to heel over towards one another ...[72]

Claudia R. Plens compared railways in Latin America to 'scars in the Atlantic Rainforest'.[73] The Dianyue, the Namti viaduct in particular, is undoubtedly a 'scar' on the Yunnan plateau. Labelled the 'death valley', the viaduct

Figure 7.4 Namti valley and railway bridge today (Baidu public domain)

became synonymous with the Dianyue railway, not because of its engineering achievement but because of its high mortality.[74] French official Maurice Casanave was sent to investigate the casualties. He concluded that many perished in Namti as a result of the unhealthy climate and lack of basic medical facilities; he put the blame on the contractors and recommended new legislations to govern Asian labour.[75] One solution was to bring workers from Vietnam, who were regarded by the French as being more loyal and manageable than the Chinese. The other solution was to recruit nationwide, even in Hong Kong.[76] But this did not resolve the problem as the construction of the railway would continue to suffer from labour shortages and desertion – a headache for Consul Auguste François, who inspected many times and left detailed reports.[77] Whether in Manchuria or Shandong, the Americas or Australia, Chinese coolies suffered the same fate – they were synonymous with cheap labour and exploitation.[78] But the congregation of large number of labourers on railways and in mines also spelt trouble. In 1908, the order of things was upended by the Hekou Uprising, instigated by Dr Sun Zhongshan and his allies, and the precursor of the Nationalist Revolution to come.[79] Railways became the site of revolution.

The construction of the Dianyue railway would transform Yunnan. British traveller Arthur Davenport noted that various foreign goods, from Russian cloth to Dutch camlets, made their way to Yunnan.[80] From its opening as a Treaty-port in 1889, Mengzi saw the introduction of many foreign goods, including coffee. With the expansion of trade came more foreign merchants. The *South China Morning Post* reported that they were 'mostly French and Italian', and they established 'a small club, the Cercle du Commerce, with the billiard table and tennis court'.[81] The number of foreign residents would grow as Germany, Italy, the United States and Japan set up consuls and as foreign companies established branches. They included Compagnie Lyonnaise Indo-Chinois, Asiatic Petroleum, and Crédit Agricole Corporate and Investment Bank. Decauville was finally engaged in successful business as they supplied all the engines and coaches for the railway – a reward for their decades of 'palm-greasing'.[82] Other successes included British American Tobacco and Nanyang Brothers Tobacco. They reaped huge profits, turning Yunnan into the capital of the cigarette industry that continues today.[83]

Kunming, too, saw its social and cultural fabric change with the arrival of the railway as it 'proved to be a picturesque *tour de force* rather than an economic miracle'.[84] It saw the emergence of French-style architecture which made the multi-ethnic city even more exotic than it already was.[85] In fact, the Dianyue would turn Yunnan into a holiday destination as many Europeans escaped the heat of Hong Kong, Singapore and other parts of Southeast Asia. They came to Yunnan for the cool breeze, the majestic limestone forests, karst landscape, lakes and exotic culture. 'Mengtze offers the foreign resident good riding, pleasant walks and capital shooting', as the *Herald* advertised.[86] Archibald Little knew what appealed to foreigners, commenting that Yunnan's climate was 'the most equable in the world', with its 'perpetual sunshine' in winter, and where one could live 'with open doors and windows all the year round as in the tropics, and enjoy the fresh air minus tropical heat and damp'.[87] Yunnan was acclaimed as a 'summer resort' by 1914.[88] American diplomat John Davies was sent there; he enjoyed sports in *Le Cercle Sportive Francais* and riding.[89] Kunming shared its fate with Harbin, Qingdao and Dalian in this regard. Railways had made them accessible and turned them into tourist destinations. This certainly was not limited to China.[90] Scarborough in Britain and the ski resorts in the French Alps are examples. Where there are railways, there will be holiday destinations.

But Yunnan was also home to malaria and bubonic plague due to its warm and damp weather. They were a major concern for the Indochina colonial regime as the railways could spread them quickly. Vigilant, they despatched physicians, largely military doctors, to Mengzi, Simao and Longzhou where they had consulates. Florence Bretelle-Establet authored an excellent article about this French undertaking. She notes of Dr Barbezieux who was sent to Mengzi where he complained that 'Chinese towns are dirty, and the inhabitants' hygiene is deplorable, that every house is active or latent seat of epidemics, that ignorance is everywhere.'[91] Like the Germans, he thought the dirty Chinese needed better sanitation, and his approach was similar:

> In the area where we live, in the European quarters, let's pave the streets, install electricity, build healthy houses, let's make drains and build sewers, let's draw drinking water, incinerate corpses, disinfect centers of epidemics, let's organise special sanitation boards and a town policy of assistance … only then will we have done something good and enduring which will be understood, appreciated and copied.[92]

The military doctor put his words into action and founded a consular hospital in Mengzi. Initially, the hospital only served French and foreign nationals who worked at the Imperial Maritime Customs, on the railways and in the mines. From 1914, as Che Lin has illustrated, it began to accept locals.[93] It seems his experience had changed Dr Barbezieux's opinion. The province converted many foreigners; not just doctors like him, but also missionaries like Père Jean-Marie Delavay and explorers like Joseph Rock.[94] He founded another hospital in Kunming, where he dedicated himself to educating Chinese doctors and nurses. His efforts were much appreciated by the Governor-general Xi Liang, who showered him with titles and gifts. His superiors, however, were angry, as the hospitals were funded by the French. This would soon change, as the hospitals opened to the public and came to be financially dependent on patients. Bretelle-Establet believes that the French used medicine to not just keep epidemics at bay but also to win over the Chinese, and that the consular hospitals 'proved to be not a marginal phenomenon but a real beginning of a large-scale development of modern, scientific medicine' in Yunnan.[95]

The success of the hospitals may seem to suggest that the French presence and the Dianyue railway were of an overall benefit for China. This line of reasoning was, in fact, held by many, including Gaston Doumergue, an MP and later president, who believed that 'the Chinese were the main beneficiaries of this extraordinary French investment'.[96] Selta Altan notes that the Dianyue railway 'never realised the expansionist dreams of French colonialists in Indochina and therefore has been studied as a failure of French imperialism'.[97] Park-Barjot concurs with this, blaming insufficient profitability.[98] Jean-François Rousseau identifies three 'unmet expectations' for the Dianyue enterprise.[99] First, it did not 'serve the anticipated military purpose', as France did not conduct any military operations in Yunnan after the railway was completed in 1910. Second, it failed to 'create wealth in an undifferentiated manner' in the places it traversed. Third, it was a 'financial bottomless pit', draining French resources. But these were not the French aims to begin with. The lead French newspaper, *Colonial Despatch*, had explained the rationale behind the Dianyue back in 1901: 'The Laokai-Yunnan line opens Yunnan to us: this is indisputable and if even the first years of operation were very expensive, the political and strategic interest we have in penetrating this province must take precedence over any financial consideration.'[100] The railway was built for strategic and political purposes to exclude its arch rival – Britain. There is little doubt that the Gallicisation of Yunnan through railways might have been more aggressive had it not been for constraints imposed on the French by the *Declaration* they had signed with Britain in 1896 and the resistance put up by the people of Yunnan.

The Yunnan bridgehead

Political and strategic interests took precedence over financial considerations. The French were not alone as it was also the British rationale. Historian D. McLean summed up: 'British financial interests in China, since 1895, had been closely linked with political and strategic considerations.'[101] Railways had turned Manchuria and Shandong into bridgeheads, and they would do so to Yunnan. The French had been competing with the British for colonial strongholds around the globe.[102] Richard S. Gundry, editor of the *Herald* (1865–78), complained: 'Where England goes, all the world is free to follow: Burma is as much open as to French trade as to English – to Rouen and Lille as to Manchester and Bradford – but where France goes nothing English is admitted, if protective duties can keep it out.'[103] Railways made Anglo-French competition more fierce as they tried to outmanoeuvre each other. For instance, Lord Salisbury built the Uganda railway 'with the idea of frustrating a chimerical French threat to British control over the Suez route to the East'.[104] His French counterpart would do the same when their global rivalry took them to Asia.[105]

Robert Lee's research reveals that French diplomats fought hard for the right to finance or supply the Barrier and the Jinlu railways.[106] Yet, they lost the battle to the British. This pushed the French to invest in the Russo-Chinese bank and to ally with Belgium, who was awarded the contract to finance and build the Luhan line.[107] Lee believes that the French bribed Sheng Xuanhuai for this contract.[108] Wherever they made inroads, the French were always met with British competition. This developed in tandem with their entanglement in Southeast Asia. When the French moved to annex north Vietnam (Tokin) in the early 1880s, the British marched north to conquer upper Burma. They both raced to reach the Chinese border with the intention of expanding further.[109] Seeking some form of limitation to their competition, they signed the aforementioned *Declaration* in 1896. Article IV read:

> The two Governments agree that all commercial and other privileges and advantages conceded in the two Chinese provinces of Yunnan and Szechuen either to Great Britain or France, in virtue of their respective Conventions with China of the 1st March, 1894 and the 20th June 1895, and all privileges and advantages of any nature which may in the future be conceded in these two Chinese provinces, either to Great Britain or France, shall, as far as rests with them, be extended and rendered common to both Powers and to their Nationals and dependents ...[110]

This Article forced the two powers to share rights and gains in Yunnan and Sichuan. This would take their rivalry to new heights, beginning with the French procurement of the Đồng Đăng to Longzhou railway in 1895. The more the French gained, the more they incited the British to stake its claims, instigating a frenzy for concessions. Furious at French gains, the British demanded compensation and forced the Qing to open the West River for trade, which transpired in February 1897. This spurred the French to demand their own compensation, which led to the 1897 *Agreement*, in which the French sought to extend the Longzhou line to Nanning and Baise. The desire to outsmart the British led the French to commit to projects that, in Robert Lee's assessment, 'would never benefit French commerce', but would instead, ironically, 'allow British and Indian goods to reach Chinese markets more expeditiously'.[111] The French realised that and abandoned the Longzhou initiative, turning their attention to Yunnan where Britain had already forced an agreement on the Qing in 1894 that allowed them build railways and 'connect them with the Burmese lines'.[112] The race to build railways into Yunnan had begun.

British merchants had been discussing the possibility of a railway from Burma to Yunnan since the 1850s.[113] Some put pressure on their MPs; Edward Baines (Leeds), for example, spoke for 'several chambers of commerce' who recommended 'the construction of a railway from Rangoon through British Burmah to Western China'.[114] The following decades saw many surveys and multiple proposals. Captain Richard Sprye suggested 'a cheap railway from Rangoon, across Pegu and Burma or Ava, to Western China' and believed this was within the government's 'easy power'.[115] Following an expedition to the region, Major Sladen concluded that a railway from Bhamo in Burma to Momein or Tengyue (see Figure 7.2) in Yunnan 'would effectively tap the resources of Yunnan and put us in direct communication with the wealth and resources of South Western China'.[116] Augustus R. Margary led another mission, before being killed by indigenous tribes in 1875 and Archibald R. Colquhoun, then executive engineer of Indian Public Works, was sent for another survey in the early 1880s. He reiterated the importance of such a railway:

> There are not days left in which we can afford to lose any opportunity. The international fight is too keen to permit this. With the French in Tonquin making persistent efforts to be before us in securing the trade of southern China, it is necessary that we should neglect no chance to gain the command of this market and retain the place we have held so long in the mercantile position of nations.[117]

Beating the French was the British priority. Conquhoun proposed a line 'passing through Zimme and Kiang-hung to Ssu-mao', to which he added that 'there is now no suzerain power to prevent that from being worked.'[118] Lack of political control in the Sino-Burmese border region would give the British free reign. There were other proposals: one from Kunlong (see Figure 7.2) to Yunnan, another from Magaung to Yunnan and then northward towards Sichuan.[119] Major Henry R. Davies was commissioned to survey routes twice in the 1890s. He maintained that 'The eventual object of any railway scheme now proposed must be not merely the encouragement of local trade across the Burmese frontier, but the attainment of a through route from India to Ssu-chuan and to Eastern China.'[120] His vision for Britian reached far beyond Yunnan, aiming at Sichuan or the Upper Yangzi, and eastern China or the Lower Yangzi – they should seek to control the entire Yangzi River valley from head to tail. The purpose was, as Davies stated, 'twofold':

> To draw the trade of Yun-nan towards Burma, and thereby prevent the trade of Western Yun-nan being taken from us by the French.
>
> To reach the rich province of Sse-ch'uan, there to connect in the future with the projected Hankow to Ch'eng-tu railway, thus forming a thorough route between India and Shanghai – a link in the great trunk line which will doubtless some day connect Calais with eastern China via India.[121]

Excluding the French was, once again, the priority. Davies had grand ambitions, similar to those of Cecil Rhodes, who dreamt of an African railway from Cape to Cairo – neither was ever completed. Davies envisioned connecting southwest China with Shanghai via the Chengdu to Hankou railway. As for connecting eastern China with Calais, this remained an unfulfilled dream. Davies estimated that the 1,000-mile Kunlong to Chengdu line would cost at least £15 to 20 million and take at least 10 years to build, excluding the 1,100-mile journey from Chengdu to Hankou. Like the French, the British ambition for opening up southwest China was prompted by a desire for shorter routes and exclusive access to south-central and eastern China. Davies was at least realistic about the challenges, remarking that 'the construction of the whole line from Kunlong to the Yangtze is too vast a project to be undertaken at once'.[122] But he also highlighted the problem with indifference:

> But if we are content to do nothing in the way of railway extension into Yunnan the French will in the future be in possession of the greater part of the line which will in some day connect India with eastern China and will also be in a position to take from us the trade of western Yunnan.[123]

British indifference would allow the French to exploit the situation, although the latter soon found themselves taking on an enormous and prohibitively expensive project. Another exploration was led by Lieutenant Colonel C. C. Manifold whose 1903 report seemed to be the final nail in the coffin:

> There is no doubt of the advantages that would accrue from such a railway if it could be constructed at a reasonable cost, but great difficulties have been put in the way of such a line of communication by Nature, who has thrown off from the high Tibetan plateau into Yunnan a great mountain mass filling up most of the province, forming a plateau land which she has cleft into great arms by rivers, whose line of flow is from North to South. Any line from Burma through Yunnan to the Upper Yangtze Valley must therefore be carried over this difficult country in right angles to the axis of these mountain ranges.[124]

The great mountain mass described were the Hengduan Mountains, which straddle southwest China, Tibet and north Burma, and the rivers mentioned were the Nujiang (Salween), the Lancang (Mekong) and the Yangzi (Jinsha locally). Eager to do business in the Yunnan–Sichuan region, Archibald Little reflected: 'it is doubtful if a private company will be found to build it'.[125] Little was exact in his judgement that a private company would not have the resources to build a line through such a challenging landscape. Only during times of crisis might governments consider such a costly undertaking. Manifold reminded his readers that the French line under preparation was a government initiative. In a paper read before the Royal Colonial Institute in April 1904, Fred Shelford delivered a verdict on the 'French scheme for railways in West Africa', that they would 'entirely dwarf the British constructed railways'.[126] The same can be said of the case for Yunnan. After countless missions and five decades, connecting Burma with Yunnan and Sichuan remained little more than a talking point, 'imagining' in the words of Yin Cao.[127] Little lamented:

> while we have been talking the French has been acting. Our Indian Government commenced a railway, which was to 'tap' Western Yunnan, in a half-hearted way, and then stopped short one hundred miles from the Yunnan frontier; and so the Mandalay-Kunlong line now runs one hundred and seventeen miles north-east of Mandalay, and there ends in the jungle and has, of course, little or no traffic.[128]

The British had not given up and were close to completing this line in 1908, but the project had to be abandoned due to the extreme difficult terrain as it approached Yunnan. The British mission to speed into Yunnan on a railway ended with a French victory, as the Dianyue reached Kunming in April 1910. The Anglo-French railway battle for Yunnan might have been over, but their race for Sichuan had only just begun. To compound British misery, the French continued negotiations with China and acquired the extension to Sichuan in January 1914.[129] The *Herald* was quick to report on the deal:

> This forms an exceedingly important strategic link as it effects a junction at Yunnanfu with the French railway from Haiphong to Chengtu, and with the Franco-Belgium railway from Tatungfu in the North of Shanxi, which in turn joins the trunk-line from Lanchowfu in Kansu to Haimenting on the coast. Thus, this concession has given French, Russian and Belgium interests an immerse railway system, north, south, east and west over China.[130]

The *Herald* was exaggerating. The French might have outdone the British in Yunnan, but that did not constitute success in linking Kunming with not just Sichuan but also railways in north–central and northwest China. The Sichuan extension, like many other lines the French sought, was never commenced. Would their scramble for railway concessions have continued had they not been distracted by the outbreak of WWI? Britain appears to have lost its large appetite for China after the war, whereas the French never gave up their dream of build railways into Sichuan, even securing a cross provincial line between Chengdu and Chongqing in 1936.[131]

The French case teaches us unique lessons about railway imperialism. Like the German Navy and the Japanese Army, Paul Doumer drove a military wedge into his railway vision. But he failed. This speaks to the vigour of the French foreign service as it maintained control, at least in the case of Yunnan. The French forced the Yunnan government to provide land and protection, but this strategy backfired – they were outmanoeuvred by the Yunnanese. They schemed hard and planned well, yet were unable to fully exploit the railway and the mines, which came up against the tide of the Railway Rights Recovery Movement in the 1900s. They did not succeed in Gallicising Yunnan but propelled the province onto the path of medical modernisation. There is no doubt that the French, like the Russians, the Germans and the Japanese, took advantage of the Qing's weakness, but the Dianyue railway, just like the CER, the Jiaoji and the SMR, left behind unintended consequences. At the cost of the Qing's sovereignty, environment and resources, the Dianyue opened up Yunnan, modernised its transportation and laid the foundation for industrialisation. Archibald R. Colquhoun called railways the 'best of pacificators'.[132] Yet their ability to pacify was limited. Railway imperialism exposed the ordinary people and educated Chinese leaders in the 1900s when the iniquities of imperialist-built railways would become a rallying cry for patriotism around the country and a catalyst for revolution.

Notes

1 Lorin, 'La civilisation suit la locomotive'.

2 Paul Deschanel, *Le question du Tonkin* (Paris: Berger-Levrault et cie, 1883), 10 and 28.

3 'The Franco-Chinese Treaty', *Herald*, 19 June 1885: 705; 'ZhongFa huanyue', *Shen Bao*, 10 December 1885: 1 and 'ZhongFa tongshang xinyue', 12 August 1886: 11; Wang Tieya, *Jiuyuezhang*, vol 1, 466–68.

4 MacMurray, *Treaties*, vol 1, 26–32.

5 MacMurray, *Treaties*, vol 1, 30.

6 Weng, *Riji*, vol 6, 2891, 2906, 2922 and 2923–25, 2929 and 2939.

7 MacMurray, *Treaties*, vol 1, 31.

8 Colin Dyer, 'Singapore in 1897 as seen by Paul Doumer', *Journal of the Malaysian Branch of the Royal Asiatic Society*, vol 90, 313 (2, 2017): 113–16; Paul Doumer, *L'Indo-Chine francaise souvenir* (Paris: Vuibed and Noney, 1905).

9 Lee, *Exploitation*, 181 and T. H. Trang Phan, 'Paul Doumer: aux origines d'un grand projet: le chemin de fer indochinois', *Historie, Economie et Société*, vol 30, 3 (September 2011): 115–40.
10 Van Dijk, *Pacific strife*, 337–40.
11 MacMurray, *Treaties*, vol 1, 123–25 and 128–30.
12 MacMurray, *Treaties*, vol 1, 124.
13 MacMurray, *Treaties*, vol 1, 125.
14 MacMurray, *Treaties*, vol 1, 129–30.
15 Bert Becker, 'French Kwang-Chow-Wan and British Hong Kong: politics and shipping, 1890s and 1920s', in *British and French colonialism in Africa, Asia and the Middle East*, ed. James R. Fichter (Cham: Springer, 2019), 181–221.
16 MacMurray, *Treaties*, vol 1, 98.
17 T. W. Roberts, 'The Trans-Saharan railway and the politics of imperial expansion, 1890–1900', *Journal of Imperial and Commonwealth History*, vol 43, 3 (2015–06): 438–62.
18 Rang-ri Park-Barjot, 'Le patronat français des travaux publics et les réseaux ferroviaires dans l'empire français: l'exemple de chemin de fer au Yunnan (1898–1913)', *Publications de la Société française d'histoire des outre-mer*, vol 6, 1 (2008): 653–70.
19 Francis Garnier, *The French in Indochina: with a narrative of Garnier's explorations in Cochin-China, Annam and Tonquin* (London: T. Nelson, 1884), 130–31; Jean Michaud, *'Incidental' ethnographers: French Catholic missions on the Tokin-Yunnan frontiers, 1880 to 1930* (Leiden: Brill 2007).
20 Pierre Ibos, *Le chemin de fer du fleuve rough et la pénétration française au Yunnan* (Paris: Henri Charles-Lavauzelle, 1906); 'Les chemins de fer de pénétration en chine', *La Dépêche Coloniale*, 9 April 1901: 1.
21 Archibald J. Little, *Across Yunnan: a journey of surprises* (London: Sampson Low & Marsten, 1910), 85.
22 Little, *Across Yunnan*, 86.
23 Ibos, *Pénétration française*, 54.
24 'Jingbao quanlu', *Shen Bao*, 18 May 1903: 12.
25 Lee, *Exploitation*, 213–14.
26 Lee, *Exploitation*, 190–99.
27 Désirée Lenoir, *Le consul qui en savait trop: les ambitions secrètes de la France en Chine, 1886–1904* (Paris: Nouveau Monde, 2011) translated by Xu Tao *et al.*, *Dianyue tielu yu Yunnan wangshi* (Nanjing: Yilin chubanshe, 2019), 156–61.
28 Bretelle-Establet, 'Resistence and receptivity'.
29 Lee, *Exploitation*, 232–33.
30 *Bulletin du Comite de L'Asia française*, 1 janvier 1912: 37; 'Tongking Yunnan railway', *Herald*, 8 August 1914: 432.
31 'Yuedian tielu', *Shen Bao*, 29 March 1903: 2 and Ibos, *Pénétration francaise*, 54–55.
32 'Diandu Ding zou Dianyue tielu xufei shenju qingzhi buokuanzhe (1904)', Yunnan jiyi at Yunnan dang'an guan, www.ynda.yn.gov.cn/html/2015/shiliaoyanjiu_0903/5176.html, accessed 15 March 2023.

33 MacMurray, *Treaties*, vol 1, 54–55.
34 Van Dijk, *Pacific strife*, 201–26 and 227–44; 'Regulation for mixed police on Sino-Annamite frontier', in MacMurray, *Treaties*, vol 1, 32.
35 'Diandu Ding', at Yunnan dang'an guan, www.ynda.yn.gov.cn/html/2015/shiliaoyanjiu_0903/5176.html, accessed 15 March 2023.
36 MacMurray, *Treaties*, vol 1, 913
37 'French troops in Yunnan', *Herald*, 22 March 1907: 650; 'Zhongfa cebingzhi jiaoshe', *Shen Bao*, 29 April 1907: 11.
38 'Yunnan railway', *South China Morning Post*, 29 February 1908: 11.
39 *Bulletin du Comite*, 1 Avril, 1908: 27.
40 Little, *Across Yunnan*, 91.
41 Little, *Across Yunnan*, 115.
42 Wilbur Burton, 'French imperialism in China', *Current History*, vol 39, 4 (1 January 1934): 428–31.
43 Ulrich Theobald and Jin Cao, eds, *Southwest China in a regional and global perspective (c. 1600–1911): metals, transport, trade and society* (Leiden: Brill, 2018), 87–123 and 184–223.
44 Garnier, *French in Indochina*, 132.
45 Arthur Davenport, *Report on the trading capabilities of the country traversed by the Yunnan mission* (London: Harrison and Sons, 1877), 13–25.
46 'Chine et Tonkin', *La Mayenne*, 2 September 1900: 2.
47 MacMurray, *Treaties*, vol 1, 911–14; 'Dianhui yiping', *Shen Bao*, 2 July 1900: 2.
48 MacMurray, *Treaties*, vol 1, 911.
49 'Yunnan feiluan shixian Lin'an', *Shen Bao*, 24 May 1903: 1.
50 MacMurray, *Treaties*, vol 1, 911–12.
51 'The Kochiu mining district of Yunnan', *Herald*, 16 June 1937: 481.
52 Richard R. Tawney, *Land and labour in China* (London: Allen and Unwin, 1932), 129.
53 'Tongking-Yunnan railway', *Herald*, 25 September 1909: 729.
54 J. P. Daughton, *In the forest of no joy: the Congo-Ocean railroad and the tragedy of French colonialism* (New York: W. W. Norton, 2021), 145.
55 Luo Zhiyuan, 'Yike daoding yidixue, yigen zhenmu yitiaoming [一颗道钉一滴血一根枕木一条命]', *Chuangzao*, 2 (2010): 108–9.
56 Garnier, *French in Indochina*, 109, 113 and 137.
57 Ibos, *Penetration francaise*, 60.
58 Pierre Marbotte, *Un chemin de fer au Yunnan: l'aventure d'une famille francaise en chine* translated by Xu Tao, *Yige Faguo jiating zaiZhongguode jinli* (Kunming: Yunnan meishu chubanshe, 2010), 49–50.
59 Park-Barjot, 'Le patronat français' and Thompson, *French Indo-China*, 208.
60 Yunnansheng dang'anguan, *Dianyue tielu shiliao huibian* (Kunming: Yunnan chuban jituan, 2013), 34, 37–38.
61 Marbotte, *L'aventure d'une famille*, 78–79.
62 Altan, 'Politics of life and labor' and *Chinese workers of the world: colonialism, Chinese labour and the Yunnan Indo-china railway* (Stanford: Stanford University Press, 2024).

63 Yunnansheng, *Dianyue shiliao*, 9 and 37.
64 Little, *Across Yunnan*, 65.
65 Altan, 'Politics of life and labor'.
66 Sylvia Meister and Wang Jin (translator), *Didang yougu: 1903 zhi 1910 yiwei Sulishi gongchengshi qinlide dianyue tielu* (Kunming: Yunnan renmin chubanshe, 2018), 49 and 76.
67 Lorin, 'La civilisation suit la locomotive', 47 and Pholsena, 'Technology and empire', 552.
68 Daughton, *Congo*-Océan *railroad*, 8.
69 'The railway of Yunnan', *Herald*, 27 July 1906: 213.
70 Frederick A. Talbot, *The railway conquest of the world* (London: William Heinemann, 1911), 303.
71 Talbot, *Railway conquest*, 302.
72 Talbot, *Railway conquest*, 303–4.
73 Claudia R. Plens, 'Railways: landmarks and scars in the Atlantic Rainforest', in *Archaeologies of the British in Latin America*, ed. Charles E. Orser Jr. (Cham, Switzerland: Springer, 2019), 201–18.
74 Thompson, *French Indo-China*, 208.
75 Altan, 'Politics of life and labor'.
76 'The Yunnan railway recruiting coolies', *South China Morning Post*, 3 December 1904: 5.
77 Xu Tao and Zhang Xinzi, *Lingshi Fang Suya* (Nanjing: Yilin chubanshe, 2019), 309–18.
78 Karuka, *Empire's tracks*, 82–101.
79 Fan Dewei, *Wushennian (1908) Hekou qiyi yanjiu* (Beijing: Shehui kexue chubanshe, 2016), 11–30.
80 Davenport, *Report*, 26.
81 'Mengtze, Yunnan', *South China Morning Post*, 3 April 1906: 2.
82 Xu and Zhang, *Fang Suya*, 319.
83 Yang Shouchuang, *Yunnan yancao fanzhanshi* (Beijing: Shehuikexue wenxian chubanshe, 2018), 30–45.
84 Thompson, *French Indo-China*, 208.
85 Jiahui Hu and Kosuke Matsubara, 'Development of French inspired urban architecture with the construction of the Yunnan-Vietnam railway', *Journal of Asian Architecture and Building Engineering*, vol 22, 4 (2023): 1915–50.
86 'Lungchow and Mengtze', *Herald*, 16 May 1890: 590.
87 Little, *Across Yunnan*, 67–68.
88 'Development of Yunnan-fu', *Herald*, 14 July 1914: 22.
89 John Paton Davies, *Dragon by the tail: American, British, Japanese and Russian encounter with China and one another* (New York: Norton, 1972), 161.
90 Geoffrey R. Bird and Michael B. Conlin, eds, *Railway heritage and tourism: global perspectives* (Blue Ridge Summit: Multilingual Matters, 2014).
91 Bretelle-Establet, 'Resistence and receptivity'.
92 Bretelle-Establet, 'Resistence and receptivity'.
93 Che Ling, 'Jiaohui, tielu, gonglu yu jindai Yunnan xiyide fazhan', *Qujing Shifan Xueyuan Xuebao*, vol 32, 1 (2013): 5–9.

94 Michaud, *Incidental ethnographers.*
95 Bretelle-Establet, 'Resistence and receptivity'.
96 Lee, *Exploitation*, 248.
97 Altan, 'Politics of life and labor'.
98 Park-Barjot, 'Le patronat français'.
99 Rousseau, 'Imperial railway failure'.
100 La *Dépêche Coloniale*, 'Utilité économiques de la ligne le Yunnan', 6 May 1901: 2.
101 McLean, 'Townley Agreement'.
102 Fichter, *British and French colonialism.*
103 R. S. Gundry, *China and her neighbours: France in Indo-China, Russia and China, India and Thibet* (London: Chapman and Hall, 1893), introduction.
104 Robinson, 'Introduction', in *Railway imperialism*, 3.
105 Cady, *French imperialism*, 136–59; Keay, 'Mekong exploration commission'; Kennedy, 'Anglo-French rivalry'; Chandran, *Burma-Yunnan railway*; Didac Cubeiro Rodriguez, *The pearl of the East: the economic impact of colonial railways in the high age of imperialism in Southeast Asia* (Cham: Palgrave Macmillan, 2023), 51–81.
106 Lee, *Exploitation*, 41–49.
107 MacMurray, *Treaties*, vol 1, 65–66 and 79–80.
108 Lee, *Exploitation*, 77.
109 Valdemar Hoskiaer, *Les routes commerciales du Yunnan: province choinoise au nord du Tonkin* (Paris: E. Plon, Nourrit, 1883), 9–15.
110 MacMurray, *Treaties*, vol 1, 54–55.
111 Lee, *Exploitation*, 171 and 177,
112 'Agreement modifying the convention of March 1, 1894, relative to Burmah and Tibet', in MacMurray, *Treaties*, vol 1, 94–98.
113 'Domestic intelligence', *Bradford Observer*, Thursday 5 October 1865: 3; 'British direct trade with the inland west of China', *Exeter and Plymouth Gazette*, Friday 15 December 1871: 8; 'London and the provinces', *The Standard*, Tuesday 24 April 1883: 3.
114 Mr Baines, 'Observations question', *Hansard Parliamentary Debates*, vol 198, Friday 30 July 1869.
115 Richard Sprye, *Correspondence between Captain Richard Sprye and William-Ewart Gladstone* (Whitefish: Kissinger Publishing, 2010), 16 and 73.
116 E. B. Sladen, *Official narrative of and papers connected with the expedition to explore the trade routes to China via Bhamo* (Rangoon: British Burma Press, 1869), 238 and 'Major Sladen's journey to western China', *Herald*, 25 August 1870: 315.
117 A. R. Colquhoun, *Across Chryse: being the narrative of a journey of exploration through the south China border lines from Canton to Mandalay* (2 vols, London: Sampson Low, 1883), vol 2, 191.
118 Colquhoun, *Across Chryse*, vol 2, 46.
119 'The trade routes to Szechuan', *Herald*, 9 October 1896: 632.

120 H. R. Davies, *Yun-nan: the link between India and the Yangtze* (Cambridge: Cambridge University Press, 1909), 7 and *Report on the part of Yunnan between the Bhamo frontier and the Salween* (Rangoon: Government Printing, 1894).
121 Davies, *Yun-nan*, 8.
122 Davies, *Yun-nan*, 15.
123 Davies, *Yun-nan*, 16.
124 'An address on recent journeys in China', *Herald*, 23 October 1903: 859.
125 Archibald J. Little, *The Far East* (Oxford: Clarendon Press, 1905), 124.
126 Fred Shelford, *The development of West Africa by railways* (London: Royal Colonial Institute, 1904), 20.
127 Yin Cao, 'The Yunnan-Burma railway, 1860s–1940s: imagining, planning and rejecting a railway that was never built', *Journal of Southeast Asian Studies*, vol 54, 2 (June 2023): 298–315.
128 Little, *Across Yunnan*, 69.
129 'French railway concession', *Herald*, 21 February 1914: 560; MacMurray, *Treaties*, vol 2, 1099–1108.
130 'French railway concession', *Herald*, 31 January 1914: 336.
131 'Chengtu-Chungking railway $34,500,000 loan', *Herald*, 23 December 1936: 500.
132 Archibald R. Colquhoun, *The mastery of the Pacific* (New York: Macmillan, 1902), 385.

Part III

Instrument of nation building

8

The 'quest for railway autonomy': China mobilises[1]

A satirical cartoon titled 'Lantern convention of different countries' appeared in *Minhu Ribao* or *People's Cries Daily*, a Shanghai newspaper, in the summer of 1909 (cover image). A short, bald Chinese man carries what appears to be a globe. Behind him is a British man holding a railway engine in the shape of a lantern, who is in turn followed by a French man, a German, an American, a Russian and a Japanese, each holding one coach of the train. The smoke emitted above reads 'current situation with Chinese railways'. The message is clear: Chinese railways were in the hands of foreign powers. This conclusion was evident in the gauges used in Manchuria alone:[2]

Russian section of CER (1088 miles) on 5′ gauge
Japanese controlled SMR (508 miles) on 3½′ gauge
Newchwang branch of the Barrier line (300 miles) on 4′ 8½″ gauge
Chinese controlled Hsinminting to Fengtian on 3½′ gauge
Fengtian to Antung (128 miles) on 2½′ gauge
Korean trunk (600 miles) linking with Antung on 4′ 8½″

The existence of five different gauges in one region speaks to the complexity of railway politics in the final decade of the Qing. The lines partitioned the country into separate spheres of foreign control. Railway imperialism educated ordinary people and mobilised local officials as railway nationalism emerged. It was driven by the 'Railway Rights Recovery Movement' which would culminate in the Nationalist Revolution of October 1911. Focusing on the specific railways that fuelled the flame of the movement, this chapter challenges the conventional wisdom on railway nationalism. It instead highlights regionalism as an important driver and argues for a re-examination of railway finance and shareholders.

'An industrial achievement of unique difficulty'[3]

In 1903, the Qing court launched its eleventh-hour effort at reform. Dubbed the 'New Policy', this attempt aimed to build a Western-style military called the New Army, introduce Western-style universities/education, reduce the bureaucracy, institute modern laws and encourage private enterprise. Weipin Tsai's description of this period as the 'Qing's last flowering', albeit in reference to the Post Office, is a fitting assessment for railway development.[4] This last-minute effort was impromptu, too little and too late. It would only hasten the Qing's downfall by encouraging the Chinese to build their own railways and to buy back the rights of lines financed, built and controlled by foreign powers.[5] The Qing court published the *Concise Regulation for Railways*.[6] They tried to prevent further foreign encroachment by calling upon the Chinese to invest and build railways. The *Regulation* attached several conditions:

> When Chinese entrepreneurs apply to construct a railway, if there are foreign shares, in addition to asking for permission from this Ministry (Commerce), they should also ask the Ministry of Foreign Affairs to investigate the foreigners before returning to this Ministry where a final decision can be made.
>
> When collecting shares, the Chinese must always have the dominant majority. If they must have foreign shareholders, these cannot exceed the Chinese portion. The exact number of foreign shares must be made absolutely clear at the time of application and no other foreign loan is allowed in such cases. If anyone fabricates, deceives or obfuscates their license to construct will be revoked.[7]

Chinese control and dominance were fundamental. It also specified that those who had already procured the licence to build must not contract foreigners or use the railway as collateral for a foreign loan, under threat of government takeover.[8] This policy marked a significant juncture. The right to initiate and construct railways had stayed in government hands: from the Tangxu, the Barrier and Luhan to those sanctioned by treaties with foreign powers. The *Regulation* essentially privatised railway development and opened up a new avenue for contentious politics. A Department of Posts and Communications (DPC) was established to administer railway enterprising. The DPC drew up an ambitious plan consisting of four trunk lines: an eastern rail system that paralleled the SMR-CER and terminated at Aihui on the Sino–Russian boundary, a northern one that stretched to Kyakhta across the Qing border with Russia, a western one to Ili that neighboured Kazakhstan and a southern system that called at Guangzhou before terminating at Hong Kong. The late Qing had finally developed a national vision by 1907, but it would remain lines and dots on paper as the Qing's days were numbered.[9]

Like a call to arms, the *Regulation* motivated many Chinese across the country and even beyond. Between 1903 and 1910, private railway companies and regional projects sprang up.[10] Some were initiated by provincial governments or local merchants through various schemes like taxation and public share. Guangdong province marched at the forefront. The Chaoshan [潮汕], or Chaozhou to Shantou railway, was one example. Funded and built by Zhang Yinan (Chang Yu-nan), an overseas Chinese who made a fortune in the Dutch East Indies, the 42 km railway was completed in 1906.[11] It linked two important provincial cities and facilitated regional economic development. A second example was the 133 km Xinning [新宁] line between Xinning and Xinhui, built by Chinese American Chen Yixi (Yi-hsi).[12] These local lines were successful because the province was home to overseas Chinese who had managed to procure both the capital and expertise needed to ensure their enterprises paid off. This contributes to the key problem raised earlier: the unorganised, piecemeal and regional approaches to railway development.

Encouraging local governments and mobilising the private sector did help the Qing to frustrate foreign encroachment. The best example was the Beijing to Zhangjiakou/Kalgan, or Jingzhang [京张], railway. To understand its origin, we go back to the 1902 'Exchange of Notes between China and Russia' where Prince Qing reassured the Russians by dismissing rumours of British involvement, promising to build it without resorting to foreign aid.[13] The Notes ended:

> The Chinese government agrees that, if any additional railroads are built in the future to the north of Peking or to the northeast to the direction of the Russian frontier, and unless they are built with Chinese capital by Chinese officials, if the intention is manifested in holding conference in view of turning the construction to a foreign nation, such intention should be primarily discussed with the Russian government or a Russian syndicate for the purpose of granting the concession (for the work); and the concession shall by no means be granted to any other nation or a syndicate of any other nation.[14]

Clearly, the Russians had put pressure on the Qing to exclude the British.[15] We also know that Russia rejected the Knox initiative and threatened to build a line from Kyakhta to Zhangjiakou if the Americans went ahead with their Jinzhou-Aihui project. The Russians had reason to wish for a line to Zhangjiakou, a 'thriving bustling city of from 50,000 to 60,000 inhabitants doing a busy trade with Kiakhta on the north, Sining near Lake Kokonor on the west, and Peking on the south'.[16] Situated 180 km to the northwest of Beijing, Zhangjiakou served as a gateway to Russia where Chinese traders and caravans gathered before they set out for Kyakhta, the hub of the Sino–Mongol–Russian trade (see Figure 8.1). Furthermore, Kyakhta could be connected with the Trans-Siberian line, while a branch line to Zhangjiakou would cut a shorter and more direct route to Beijing.[17]

To undercut both the Russians and British, Yuan Shikai informed the Russians that his local government would build the Jingzhang.[18] Like his

Figure 8.1 Beijing, Zhangjiakou and Kyakhta (Baidu public domain)

predecessor Li Hongzhang, Yuan was resourceful and effective. He tapped into the profits made by the Barrier railway and private investments.[19] He recruited Zhan Tianyou, one of the first boys sent to study in America in 1872.[20] Zhan had gained a great deal of experience by assisting Claude Kinder on the Tangxu railway and by constructing the Western Tombs line for the Empress Dowager. Construction of the Jingzhang began in May 1905.[21] The 200 km railway posed huge challenges, as it had to traverse the Yanshan Mountain, upon which the Great Wall sits. Zhan had to be flexible and innovative in understanding what the natural environment would permit. Laying tracks meant tunnelling, and since the Qing did not possess the modern equipment needed, Zhan resorted to using the old Chinese method of explosives and drilling from the side or the top. He had to build bridges between the mountains, which presented another technical challenge. Since the route travelled upwards, the difficulty was unprecedented. Frederick A. Talbot was impressed with the result:

> The road is excellently built, and the engineer displayed his ingenuity in coping with the problem of carrying the line through the Nankow Pass. The pass guards the entrance of the main road through the Great Wall, and to overcome the obstruction a gradient of 1 in 30 had to be introduced for a distance of 13 miles. ... The alignment of the railway up this pass is noteworthy. The road clings for the most part to the side of the mountains, crossing deep rifts and

wide clefts, as well as cutting through spurs and humps and compassing massive crags.[22]

To facilitate the ascent, Zhan equipped both ends of the trains with engines to pull and push at the same time. The *Herald* did not hide its surprise:

The intervention of the foreigner in extending the conquest of the steel rails over this historic route might have savoured of vandalism: in the hands of the Chinese the line becomes an eloquent expression of the dispossession of the old by the new, the outward and visible sign of the unmistakable forces now at work as much on the inner conscience of the people as on the material condition of the empire.[23]

It seemed that the Chinese people had finally risen to the challenge of railways. Its completion in October 1909 was celebrated with fanfare.[24] The *Times* called it an 'industrial achievement of unique difficulty' because:

The project was pronounced entirely impracticable for any Chinese engineering and constructive talent, and extremely difficult for the best foreign engineers. It has been accomplished in a thorough fashion at a lower cost than that of some of the forging-built roads in the level Yangtze plains.[25]

Talbot delivered a resounding conclusion:

The completion of the work, however, offers convincing testimony that the Chinese under competent supervision, are perfectly capable of building railways without the aid of foreigners, and that no fault can be found so far as solidity and durability are concerned. For centuries the Chinese have been famed for their prowess in matters pertaining to civil engineering.[26]

This prowess could also be seen in the constructions of the Great Wall more than two thousand years before. Talbot knew Chinese history better than the likes of Jardine Matheson, Sergei Witte and Paul Doumer, who were only interested in their own gains.

Hailed as the first railway designed, financed and built by the Chinese themselves, the trains and engines were, nonetheless, all imported, as indicated by the presence of Mallet locomotives.[27] Its significance lay in the fact that it validated the idea that the Chinese could build their own railways without foreign loans and supervision. Its success encouraged others to take up the challenge. Officials in Mongolia (then part of the Qing) asked for permission to extend the line to Kulun (Ulaanbaatar).[28] Merchants in Zhejiang asked for a Zhejiang to Guangdong line, those in Shanxi proposed a line from Datong to Puzhou, while Yunnan officials applied for a Yunnan to Sichuan railway.[29] Turning to the private sector aided the Qing in frustrating foreign powers. However, they would also generate problems. First, it contributed to the unorganised and regional nature of railway development. This was also the case with mining and telegraph lines, which again point to the disorganised nature of late Qing reform. Second, working with the private sector came with problems which the Qing had no idea how to tackle. Third, it encouraged regionalism, which Elisabeth Kaske labels as 'provincialism', at a time when the Qing were trying to keep the empire together as anti-Manchu organisations/activities proliferated.[30]

Catalyst for revolution

The fervour of railway nationalism spread to the Sichuan basin. Called 'heavenly country', the province is rich with natural resources, populous and strategically located as it neighbours Tibet to the west and Yunnan to the south, hence access to South–Southeast Asia. To the north and northwest lay the Shaanxi–Gansu region, the gateway to Central Asia, Mongolia and Russia. But the chains of mountains that separate it from the rest of the country make it difficult to access, posing huge challenges to construction (see Figures 8.2 and 8.3).

This was not the only problem. Sichuan was coveted by both Britain and France. A. F. Legendre, a French army physician, saw that the ultimate aim of the Haiphong to Kunming railway was to reach Sichuan, 'the largest and the richest province'.[31] Henry R. Davis came to the same conclusion: 'Beyond Yunnan lay a far richer and more populous province – that of Ssu-ch'uan', Davis fantasised a line into the upper Yangzi River valley, that would form 'a thorough route between India and Shanghai'.[32] Keith Neilson and T. G. Otte have rightly articulated that 'the profit motive was fused with strategic calculations' when foreign powers competed with each other by laying railway tracks in the race for empire.[33] Hankou would be central to the British fantasy. Stephenson had seen the importance of Hankou in the 1850s and proposed five lines fanning out from the city: northward to Beijing, southward to Guangzhou, eastward to Shanghai, westward to Sichuan and southwest-ward to Yunnan.[34] If the British could secure a railway from Sichuan to Hankou, their march to the lower Yangzi River valley and Shanghai would be made easier and, above all, stop the French northward advance from Yunnan.[35] But Britain and France were not the only foreign powers who were interested in the Sichuan–Hankou region. Both Belgium and the United States had a stake in such a line. The Belgian-financed Luhan would terminate in Hankou where the Hankou to Canton railroad, awarded to American Chinese Development Company, would begin.[36]

Where there were railways, there were inevitably scrambles. Like Manchuria, Shandong and Yunnan, Sichuan was becoming a 'bridgehead'

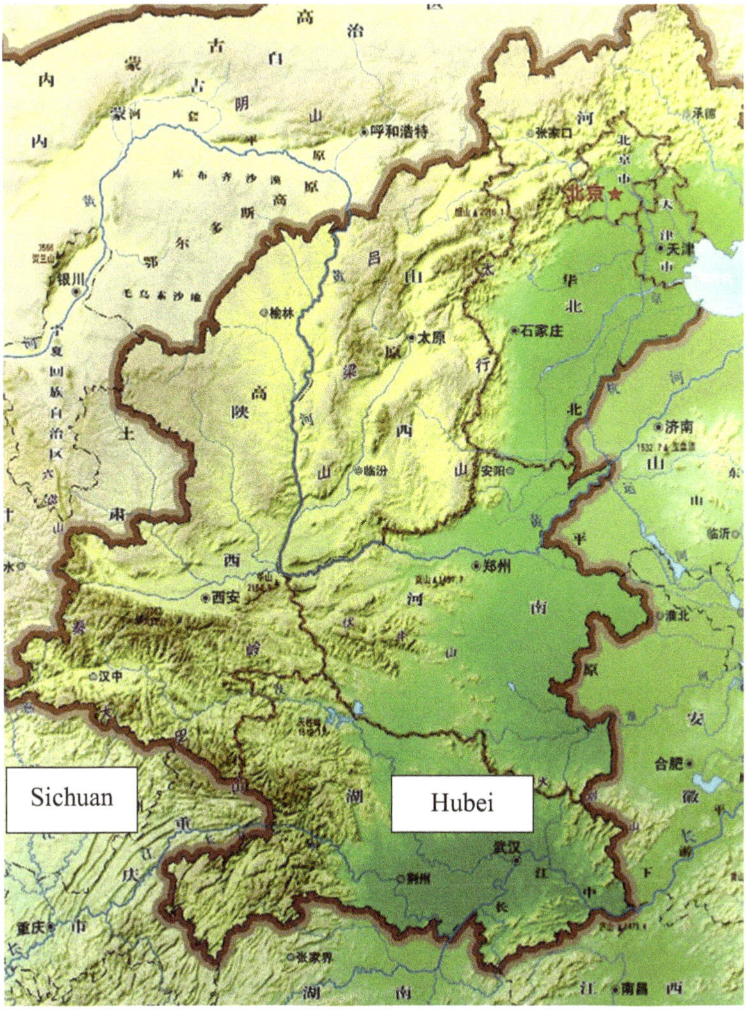

Figure 8.2 Hubei and Sichuan geography (Baidu public domain)

by the early 1900s. The two British consortiums, the Peking Syndicate and the British and Chinese Corporation, lobbied so hard that the Foreign Office intervened. D. McLean believed that 'the struggle for control of the Szechuan extension was the predominant consideration' and that its acquisition was a 'matter of considerable political importance'.[37] The Foreign Office put enormous pressure on the Qing until the Townley Agreement was signed in 1903. Although they did not secure the right to finance and build a line into Sichuan, the two consortiums secured three railways in

Figure 8.3 Western Hubei topography: Enshi Grand Canyon (Baidu public domain)

north–central China: one from Daokou to Zezhou that crossed the Luhan; another from Xuzhou to Kaifeng, which connected the Jinpu with the Bianluo or the future Longhai; and a third from Pukou to Xinyang that linked Nanjing with the Luhan. D. McLean claimed that Prince Qing promised to the British that 'if no native capital was available when China desired the line to be built, then application should be made to British and American financiers', and 'in any negotiations for the contract he would instruct the Chinese delegates to show British capitalists more favour than their American associates'.[38]

To pre-empt the foreign powers' advance into Sichuan, Xi Liang, the governor-general of Sichuan, decided his local regime would build the Sichuan to Hankou line, known as the Chuanhan [川汉] railway. He memorialised the court in 1903:

> Foreigners have wanted it for so long and have been strategizing. Some Chinese have already networked with them and tried to raise funds. This has led to many problems. We must take initiative, set up a government company to attract shares and to protect our own interest. This cannot wait. Some suggest we borrow from foreigners and allow them to build. But this would allow foreigners to take over and leave us with no right.[39]

With his leadership, the Chuanhan Railway Company (CRC) was founded in 1904 and the Sichuanese would finance and build the line themselves. This was a government-initiated public share company, with no borrowing from foreigners. To retain Chinese autonomy, the guidelines made clear that foreigners could neither buy nor hold shares. This directive was printed on the shares issued: 'In accordance with the printed regulations, this certificate is neither negotiable nor transferable to any person or persons other than Chinese; if these conditions are not observed this certificate shall become null and void.'[40] This declaration is indicative of what Arthur Barry termed the 'growing intolerance of foreign guidance and assistance' that dominated railway construction in late Qing China.[41]

The CRC shares would come from three key sources: private purchase, taxation and local government investment.[42] The first is clear: any Chinese could invest by buying shares. The second, called *chou gu* [抽股], literally 'draw share' when translated, was a scheme to raise funds through levies, with the word 'draw' denoting the compulsory nature of share allocation. For example, for each *dan* of grain harvested, three *sheng* would be 'drawn' to serve as shares.[43] This varied in different localities and was extended to profitable commodities like salt and tea. In Baxian, for example, this was levied on opium sale and consumption.[44] The third source consisted of investments by various local government agencies. Many forward-looking Sichuanese invested in the future as they believed that they were sure to see good returns with a government-led project.[45] Even Sichuan students studying in Japan bought shares.[46]

Also called 'rent share' and 'railway donation', the 'draw share' requires further attention.[47] It was the most complex form of fund-raising, as its sources were diverse and its methods of conversion complicated. Significantly, it meant that the ordinary people now owned shares in the CRC, even though they had never experienced a railway before. *Shen Bao* claimed that the CRC had managed to collect 6,000,000 taels by November 1905.[48] However, an audit conducted by the DPC in May 1911 provides a more accurate calculation (silver taels).[49]

Year	Bought	Government	Rent	Opium	Salt/tea	Total
1904–07						8,175,789.00
1908	69,420.58	37,375.00	1,519,259.94	250,098.64	56,660.53	1,932,814.69
1909	36,589.84	9,875.00	1,343,459.47	196,797.48	64,281.51	1,650,983.30
Total						11,759,586.99

The CRC managed to collect a staggering 8,175,789 taels in its early years. The largest was 'rent share', drawn from, some might say imposed upon, farmers and small businesses. In fact, some were unwilling or reluctant to contribute, this would have consequences.[50] It is vital that we understand this particular share component. En-han Lee and Elisabeth Kaske remain the few scholars who have paid attention to it.[51] The following pages delve into the details of Chuanhan finance and its shareholders to reveal how they led to the Sichuan Railway Protection Association and ultimately Nationalist Revolution in 1911.

Having secured enough funding, the CRC aimed to start construction in the spring of 1906. But the government-initiated project struggled from the very start, as the project was plagued by the mismanagement of funds. Local government and officials, including the CRC board of directors, misappropriated nearly 4,000,000 taels.[52] An investigation by the DPC revealed the depth of the situation: of the 8+ million taels collected, more than 2,000,000 were embezzled by the provincial Copper Bureau and nearly 1,560,000 were used on various expenses, including banquets.[53] Engineers were still in the surveying stage of their work by November; this led to protests from shareholders.[54] In April 1907, the CRC was changed into a private enterprise.[55] It hired Zhan Tianyou as chief engineer and even asked China's embassies to recruit students then studying in Europe and Japan.[56] But by October 1907, the CRC was still trying to finalise the route, which had hitherto changed several times, in an effort to minimise terrain challenge.[57] As Kaske notes, there were also disagreements among the engineers about where to start.[58] Construction finally began in December 1909 in Yichang, Hubei province. Although the CRC now had funding, expertise and popular support, it struggled to replicate the Jingzhang success. By July 1910, only 16 miles of track had been laid after more than 3,000,000 taels were spent.[59]

New problems also emerged to plague the Chuanhan at this time. Attentions were divided in Hubei with another line, the Hankou to Canton or Yuehan [粤汉] railway. April 1906 saw the completion of the Luhan line, which linked Beijing with Hankou. The right to finance and build the Yuehan was awarded to the American Chinese Development Company (ACDC) in 1900. Evidently, the Qing had adopted a strategy of preventing one foreign power from gaining too many railway concessions by sharing out the opportunities among them. However, their plans were soon thwarted when Qing officials discovered that the majority of shares in the ACDC were sold and now owned by Belgium.[60] Allied with France, Belgium also had ties with Russia, evidenced through the finance and construction of the Luhan and Zhengtai lines. Allowing these three powers into the Yuehan line would anger the British, who considered the Yangzi River valley and southern China their sphere, as the Yuehan would terminate in Guangzhou where the Canton to Kowloon railway was under construction. This worried the Qing court; they promptly cancelled the ACDC contract in 1905 despite strong protest from the company, American diplomats and even President Roosevelt.[61] For this decision, the Qing had to pay the ACDC '$6,750,000 indemnification covering the rail properties and bonds sold by the company'.[62]

With no money at his disposal, Governor-general Zhang Zhidong had to borrow from the Hong Kong government in order to start the indemnification payment. Needless to say, there was no money left for the Yuehan. At this point the project was salvaged by the gentry and merchants of Hubei, Hunan and Guangdong provinces, who stepped in to finance the Yuehan.[63] The government and people of Hubei province thus had two railways on their hands.[64] The Yuehan was afflicted with issues, from peasant resistance to encroachment on their land to unwillingness to fund segments outside one's locality. Some were hostile towards American engineers surveying routes.[65] It was also plagued by similar problems that beset the Chuanhan: corruption and mismanagement, which En-han Lee has written about.[66] The most egregious was the shortage of funds. Funding two railways was an expensive endeavour. Elisabeth Kaske rightly argues that Zhang Zhidong believed the 'project was not feasible without a foreign loan'.[67] The only problem was that borrowing from foreigners was exactly what local investors opposed.[68]

Sheng Xuanhuai was brought in to tackle the issues aggravating the Chuanhan and the Yuehan railways. A beneficiary of the 'government supervised commercial management' business model, Sheng made both a name and a fortune by shuttling between Qing officials and foreign companies, managing reform projects. He participated in the negotiations for the Wusong and was involved in contracting the Luhan and the Jinpu while heading the All-China Railway Company and overseeing the construction of the Western Tombs line.[69] Labelled the 'father of Chinese enterprises', his philosophy, like that of his old patron Li Hongzhang, was to utilise foreign capital and expertise to industrialise China. He approached Britain, Germany, France and the United States, and drew up an agreement for a loan of £6 million, which was ready to be signed by August 1909.[70] This angered many officials with some calling to impeach Zhang who had authorised it.[71] We don't know whether this contributed to Zhang's sudden death in October 1909, but his passing would forever change the fate of not just the two railways but also the Qing, now caught between domestic resistance and foreign pressure.[72] The *Herald* anticipated that 'signing the Loan Agreement may produce a serious commotion'.[73] This would materialise in 1911.

Disasters do not travel alone. The Yuehan problem came at a time of ever-greater crisis for the Chuanhan. The CRC's chief financial officer, Shi Dianzhang, had been sent to Shanghai to invest in order to acquire more funding. He was entrusted with 3,500,000 (silver taels) from Sichuan shareholders.[74] The most profitable investment in the late 1900s was rubber, such that the *Times* and the *Herald* referred to the boom as 'mad' and 'tornado'.[75] Growth in car manufacturing and speculation saw the value of rubber skyrocket. Shi invested nearly 2,000,000 in rubber shares owned by several Chinese banks and a further 850,000 owned by a company founded by the Scottish businessman George McBain.[76] But the rubber stock market crashed in the summer of 1910, triggering a global sell-off, wiping out the CRC investment. In addition, Shi himself had spent 120,000 on his own expenses, which meant that at least 2,970,000 of the 3,500,000 entrusted to him, the majority of the CRC's investment largely made up of 'draw share', was gone.[77] The number offered by the DPC is higher – they claimed that more than 3,200,000 was lost.[78]

The rubber stock market crash did not just hit the Chuanhan railway, it jeopardised banks in Shanghai as they all bought rubber shares.[79] The Qing government had to step in to bail them out as they were responsible for paying foreign indemnities. To solve the problem with the Chuanhan and Yuehan railways, Prince Regent Zaifeng, father of the three-year-old Puyi Emperor, promoted Sheng Xuanhuai to the post of minister for the DPC in May 1911. Sheng advocated not only signing the aforementioned agreement with foreign banks but also nationalising the two railways. This was endorsed by the Prince Regent who believed that the central government could better coordinate and manage these cross-province projects.[80] Sheng

Xuanhuai did his best explaining to local officials the benefits of nationalisation.[81] Nationalising meant that the DPC would take over the Chuanhan project. After Shi Dianzhang was imprisoned and the CRC's chief executive, Qiao Shunan, was dismissed from office, the CRC still had about 7,000,000 taels left in the hands of Li Jixun, its last director in charge of construction on site in Hubei.[82] Sheng lured Li to Beijing with the promise of promotion to chief executive, and Li surrendered the remaining funds in his care.[83]

Sheng acted quickly, signing the *Final Agreement for the Imperial Government Hukuang Railway* with British, French, American and German banks in late May.[84] This loan of £6,000,000 was obtained by signing away various local levies, the salt tax of Hunan and Hubei provinces for example, as collateral.[85] The four powers divided up the two lines by appointing their own engineers:

> Germany appoints the engineer for the first 800 kilometers.
> The United States of America appoints the engineer for the next 400 kilometers.
> England and France appoint the engineer for 600 kilometers apiece of the remaining 1,200 kilometers of the Hankou-Szuchuen line.
> United States appoints further a sub-engineer for the last 200 kilometers of the 800 kilometers allotted Germany.[86]

The four powers made sure that each of their shares matched those of the others. Had this materialised, they would have carved out south–central, southern and southwest China with railways. While the foreign powers clamoured for their own interests, the Prince Regent and Sheng had more difficult matters on their hands. They seemed to have forgotten that the people of these provinces had invested in these railways precisely because they had been against borrowing from foreigners.

It is vital for us to follow the trail of money and shareholders because they provide an important corrective to the prevalent account which has stressed the emergence of nationalism as an explanation for the movement and the revolution in 1911. Nationalising the Chuanhan and the Yuehan railways would have consequences, not least because the government had to pay off their shareholders. The losses from years of corruption and the stock market crash became a thorny issue. The shareholders of the two lines were treated differently. Those of the Yuehan railway, investors from Guangdong, Hubei and Hunan provinces, were eligible to receive government shares almost at the full value of their original investment.[87] But the Sichuanese investors in the Chuanhan line would only receive compensation for what was left in their account – the 7 million taels its last director had handed over to Sheng. This amounted to roughly 63 per cent of their investment, according to the DPC's own figures.[88] The Qing government, in other words, declared itself not responsible for funds lost to corruption and in the stock market crash. Why such discriminatory schemes? One answer could lie in the fact that the shareholders of Hubei, Hunan and Guangdong were powerful officials and merchants with political connections, whereas Sichuan shareholders were mostly ordinary people. This disparity would be exploited by political organisations; it would soon lead to consequences that neither the Prince Regent, Minister Sheng nor the foreign powers had anticipated.

News of the *Final Agreement*, nationalisation and the discriminatory compensation scheme began to spread; it outraged Sichuan shareholders. They congregated at the CRC's headquarter in Chengdu; their meetings led to the birth of the Sichuan Railway Protection Association in June 1911, with Pu Dianjun as Head and Luo Lun as his deputy. Born into an influential family, Pu obtained the *jinshi* degree in 1904, after which he was sent to study law in Japan. Upon his return in 1909, he became Speaker of the Sichuan Assembly. Another native of Sichuan, Luo obtained the *juren* degree in 1902. A well-respected public intellectual, he edited newspapers and taught in local academies before being elected to the Assembly. This so-called Assembly was a local political consultative group set up at the height of the 'New Policy', when the Qing court toyed with the idea of a constitutional monarchy. Leading the Assembly meant that Pu and Luo were, in theory, constitutional monarchists. For them, protecting Sichuan's railway rights was the perfect platform to campaign for a devolution of political power to the province.[89] They were also CRC shareholders and were angered that the Qing court had made decisions without consulting them. For the Qing, constitutional monarchy was merely a mechanism to continue the status quo and the Assembly was little more than an empty name.

This summer of discontent saw the shareholders of the CRC, whose investment had disappeared and whose compensation was short-changed, stage protests and strikes, with some refusing to pay taxes.[90] This brought the province's economy to a standstill, and the new governor-general, Zhao Erfeng, called for a crackdown. At an emergency shareholder meeting on 5 September, the Association publicised a 'Letter on the deliberation of Sichuan self-preservation':

> The situation in China now is not too late to remedy. We cannot wait for our luck, but must salvage what we can. All those important facilities, such as military harbour, commercial port, iron production, customs duty collection, frontier land, steam ship, railway, postal service, arms manufacture, personnel and administration – all the fundamentals that concern the foundation of the country and life of ordinary people have been signed away in treaties to foreigners by the government. ... Today the government has snatched away

our funding for railways and handed them to foreigners. This has agitated and woken up our 70,000,000 compatriots.[91]

It is not true that the government handed the funding to foreigners; they had borrowed from the foreign powers in order to continue both the Chuanhan and the Yuehan, but clearly this was manipulated to incriminate the Qing. The letter called for a stop to strikes that were seen as counter-productive to the local economy. On 7 September, Zhao Erfeng invited leaders of the Association for talks in his office. When they arrived in good faith, he had them arrested, including Pu and Luo. As news of the arrests spread, hundreds gathered and tried to storm the government office before soldiers opened fire.[92] At least twenty-six were killed in what would thereafter be known as the 'Chengdu Blood Incident'.[93] *Shen Bao* devoted a series entitled 'Drops of blood for Sichuan railway' to the struggle.[94] The Sichuanese also mobilised their fellow compatriots around the country to spread the news and solicit nationwide support.[95] The *Herald* called this a 'crisis' of 'unknown quantity'.[96] The newspaper had acutely predicted the outcome of this agitation.

The 'letter' spoke the language of insurrection and separatism – it reads like the work of political activists, rather than ordinary shareholders. In fact, none of this could have happened without the work of the United League (UL), an anti-Qing revolutionary party founded by Dr Sun Zhongshan and Huang Xin in 1905, active among overseas Chinese and foreign-educated students/intellectuals. The situation in Sichuan was a perfect storm for the UL in its struggle to overthrow the Qing. The author of the 'letter', Zhu Yuansheng, was a UL member.[97] Throughout the summer of 1911, UL members in Sichuan worked assiduously towards their goal: using the railway movement to overthrow the Qing by agitating shareholders, organising militia, encouraging uprisings in cities and separatist movements in the countryside, and by attacking the capital Chengdu.[98] They mobilised women, organising a Women's Railway Protection Association.[99] They targeted workers, rickshaw coolies, soldiers, police and the ordinary.[100] This finally materialised on 25 September when UL member Wu Yuzhang managed to oust government troops from Rongxian county. Wu joined the UL while studying in Japan. His declaration of Rongxian's independence from the Qing marked the dawn of the Nationalist Revolution. The Sichuan shareholders' fight for lost railway investment and compensation had been successfully hijacked by a political party for its revolutionary cause.

How did foreign educated students and intellectuals like Wu, manage to defeat government forces? The answer lies in local militias in the form of organised crime. The so-called Nationalist Revolution comprised a complicated web of relations which enabled the UL and its allies to capture small towns like Rongxian and establish their own regimes. The struggle of the Chuanhan railway shareholders was not just exploited by constitutional monarchists but also infiltrated by UL revolutionaries and organised crime syndicate called *pao ge* [袍哥], or 'gowned brothers' when translated literally.[101] Extremely effective in defending local communities and interests, many of them had martial skills and used modern weapons; they were also CRC shareholders. Although different from the constitutional monarchists and UL revolutionaries, they had a common enemy – the Qing dynasty that had ruled China since 1644. Hatred for the Manchus had united three different political forces. The ordinary shareholders of the Chuanhan railway and their grievances had become footnotes to history. With more localities following the example of Rongxian by late September, it looked as though the Qing were about to lose Sichuan as it was ready to declare its independence.

Governor-general Zhao resorted to a crackdown and all available military forces in Sichuan had been dispatched to suppress pockets of rebellion. The size of the mountainous province with its scattered towns/cities meant that more troops were needed. The closest place to call for help was neighbouring Hubei province. Its military force was also dispatched to Sichuan, leaving a vacuum in the province and thus another opportunity for the UL. The Hubei government was on high alert.[102] The UL planned an uprising in Hankou for 16 October, but a bomb exploded prematurely on 9 October in the Russian concession of the city. This exposed their scheme, leading to the arrest of its leaders and a curfew, thus impeding UL members from meeting to re-strategise. Yet, it could not prevent the UL members hidden within Hubei's New Army from organising themselves. The UL had infiltrated the New Army – a key platform of the Qing's 'New Policy', its rank and file were filled with young Chinese who had studied in Japan and secretly joined the League. They carried out an uprising in Wuchang, hence the label Wuchang Uprising, on 10 October, by taking over the provincial government and declaring Hubei's independence. Soon other provinces followed suit, declaring their independence from the Qing.

Since its foundation in 1905, the UL had staged nine failed uprisings, but it was a railway that led them to eventual success in 1911.[103] The Chuanhan railway had united the people of Sichuan, but it also began to tear the vast province limb from limb. This latter point has been ignored in existing scholarship due to the efforts to highlight the success of the revolution. On 21 November 1911, the Great Han North Sichuan Military Government [大汉蜀北军政府] was declared in Guang'an. This military regime chose to associate itself with the Han dynasty in a bid to delegitimise the Qing. The next day, the Sichuan Military Government [蜀军政府] was established in Chongqing.[104] Soon, seventy-five counties declared their independence. On 27 November, Chengdu proclaimed its independence and the Great Han

Sichuan Military Government [大汉四川军政府] was founded with Pu Dianjun, leader of the Sichuan Assembly and Railway Protection Association, installed as governor.[105] On the same day, the Manchu Governor-general Duanfang was murdered, ending the Qing's rule in the province. There were now at least three military governments in Sichuan.[106] The warlord era that would define the Republic of China had begun.

Imagined foes

Historians of modern China have studied the Railway Rights Recovery Movement and its pivotal role in the Nationalist Revolution of 1911. Aside from En-han Lee and Elisabeth Kaske, Madeleine Chi examined the efforts to buy back the Shanghai–Hangzhou–Ningpo line.[107] Mary Rankin analysed how emotionally charged rhetoric turned into revolutionary violence.[108] Xiaowei Zheng studied the transformation of elite consciousness before/during the revolution.[109] There is no shortage of scholarship that has emphasised the importance of the UL and Dr Sun. But there are also critiques. Joseph Esherick questioned whether the 1911 uprising could be classed as a revolution, and Sun's legacy with the UL dominated narrative.[110] The centennial has generated more interest.[111] Rana Mitter has called it an 'unanchored revolution'.[112] Li Danke drew attention away from the frequently studied elite and highlighted the importance of popular culture.[113] Heung Shing Liu saw 1911 as the culminating point of decades of internal rebellion, foreign aggression and political decline.[114] Whereas Arif Dirlik described it as the 'high peak in a revolution of long duration that is yet to be completed'.[115] It is clear that politics of nationalism and revolution, rather than Chuanhan railway finance and the struggle of its shareholders, have dominated the research.

This chapter has exposed blind spots in existing scholarship and has discussed the acute issues that have been neglected. It is not entirely incorrect to label the efforts of Chuanhan shareholders as the 'Railway Rights Recovery Movement'. Many Chinese, not just the Sichuanese, were angry at foreign encroachment and supported the call to recover railway rights, as well as those of mining and shipping as Anne Reinhardt has discussed.[116] But the label is misleading. The struggle of Sichuan shareholders had less to do with 'rights' and more to do with share loss and unfair compensation; it had little to do with nationalism and more with regionalism. The majority of 'draw' shareholders were ordinary people – what mattered to them was the promised return of their investment. The label does not do justice to their movement as their fight was hijacked by constitutional monarchists and UL revolutionaries, who surfed the tide of anti-Manchu sentiment and whose goal were to replace the Qing dynasty. The ordinary shareholders, many of them illiterate, did not leave behind any writing. Historians have largely dwelt on the anti-Qing material left behind by constitutional monarchists and UL revolutionaries.

The genuine grievances of Sichuan railway shareholders – share loss and discriminatory compensation, have been forgotten. Letting them take centre stage has exposed the less noble and more humble origin of the movement. It has shed new light on the regional, rather than national, factors that quickened the collapse of the Qing. This has encouraged me to advance a new proposition: the real foes of the Chuanhan railway and its shareholders were neither the imperialists nor the Qing court, they were corruption, risky investment, lack of leadership, regionalism and above all the hijacking of their struggle by political parties with the help of organised crime. They are topics that deserve much more research and discussion than the few paragraphs here.

Henry R. Davies concluded, after his extensive survey in Yunnan and Sichuan, that railways in China could be built by the Chinese themselves, but 'two things are necessary – a sufficient number of skilled engineers, and some honest men to be in charge of the finance. Time will doubtlessly supply the former but will honest men ever be found?'[117] His understanding was astutely perceptive. The main issue that plagued the Chuanhan was corruption. Official records showed that the CRC secured over 11 million taels between 1904 and 1909, enough to kickstart the project. But this money was mis-spent, not by the shareholders but by the CRC's own board of directors and the local government. Why didn't the shareholders direct their anger at the CRC leadership? The ultimate course of events indicates a degree of manipulation by the constitutional monarchists. Instead of taking responsibility and addressing the problem, they appropriated the concerns of shareholders in their effort to devolve power and consolidate their political hold on the region. But they were soon eclipsed by the more powerful UL, who skilfully directed the anger of Chuanhan shareholders at the Qing court and foreign powers – neither of which had anything to do with share collecting or CRC spending.

Adding insult to injury, collected shares were squandered in risky investments. A Confucian-trained scholar–official who served in mid-level provincial posts in the interior, Shi Dianzhang had no training in international finance nor knowledge of stock-market trading. Neither a banker nor a speculator, he was also new to Shanghai, a tempting metropolis where he was entrusted with more than 3 million taels. The result was hardly surprising: like his superiors, he also misappropriated the funds entrusted to him. The old imperial examination system was abolished in 1905, but it would take time for the new education system to produce graduates with the skills required for fiscal responsibility. This alludes to a larger issue. The CRC board members themselves were also classically trained local elite and scholar–officials. Sheng Xuanhuai's letters reveal that scholars from the Hanlin Academy were sent to serve as engineers on railway projects – no wonder the CRC failed.[118] The project enjoyed local government support, but its leaders had no experience in running a private enterprise. The loss of funds through corruption and risky investments were further compounded by the compensation scheme. The Qing court was not responsible for the money lost to corruption and stock market crash. Sheng Xuanhuai could not afford to alienate powerful officials and merchants from southern China but had no cause to fear the small shareholders from remote Sichuan. He had miscalculated – it was precisely these small shareholders who had sealed his fate and that of the Qing.

The CRC lacked leadership and accountability. Unlike Li Hongzhang and Yuan Shikai who were hands on and effective, its board members and directors were not held accountable for mismanagement and misconduct. Strong characters like Li and Yuan were essential in managing large-scale cross-provincial infrastructure projects that required oversight and coordination. Leadership at a cross-regional or national level was necessary, as the railways traversed different provinces where many saw no reason to fund segments outside their own locality, neither did they understand why outsiders should interfere with segments in their area. Hindsight seems to suggest that Prince Regent and Sheng Xuanhuai had been right about the need for central government direction. This can be seen in the fact that the Yuehan railway was not completed until 1936 and the Chuanhan in 2010. In both cases, the Republic of China and post-Mao regimes poured financial, technological and human resources into completing these lines: in the former case, to prepare for Japanese invasion; and in the latter, to finally realise the dream of the Sichuan people, in time for the centennial of the Nationalist Revolution.

The insurrection inspired by the Chuanhan movement has been labelled 'nationalist', but I would argue that regionalism was the primary motivation, particularly with regards to the Sichuanese who fought for their shares, compensation and 'self-preservation'. Regionalism was evident in the slogan that circulated in response to nationalisation: 'The CRC is finished. Sichuan is finished. China is finished.'[119] Mary Rankin was correct in referring to these types of proclamations as 'emotion-laden rhetoric' and 'rhetorical violence'.[120] We see similar language in the Mining Rights Recovery Movement. Foreign powers coveted the rich coal reserve in Shanxi, and the regionalist slogan was likewise 'If Shanxi's coal is lost, then Shanxi is lost!'[121] What motivated the Sichuanese to invest was the promised return for their investment and benefit to Sichuan, and it was the same motivation that mobilised them in the event of their loss. Regionalism became more acute as the Qing began to disintegrate by the 1900s. Railway development had thus far been a regional affair, this would shape the post-revolution decades where warlords would take railway regionalism to new heights. Developing this regionalism interpretative framework sheds new light on the Railway Rights Recovery Movement and the so-called Nationalist Revolution.

It would appear that the framework fashioned by Benedict Anderson fits perfectly with the case here. Sichuan constitutionalists, the United League, organised crime and ordinary shareholders 'imagined' that they belonged to the same 'community', but they were in reality a motley crew. The constitutionalists were local gentry collaborating with the Qing in order to maintain their local powerbroker status while knowing the Qing's days were numbered. But they had a formidable rival: the ideologically driven UL whose goal was to overthrow the Qing and establish a new political order. These groups were, in fact, competing forces and strange bedfellows. The 'gowned brothers', on the other hand, often terrorised localities in the name of maintaining public order and enforcing justice. The ordinary shareholders were primarily concerned with their financial gain and loss. But their wrath was skilfully exploited by partisan agents who elevated it to political heights and knew how to exploit the situation. Decades of encroachment by foreign powers had wrought strong anti-Qing sentiments, the Qing became the scapegoat for the CRC's own failures. This was a perfect storm that served the political parties, which is evidenced from what they secured after revolution: the constitutionalists took control of Chengdu with Pu Dianjun as governor, whereas the UL secured Chongqing with the Sichuan Military Government. Anthony Reid has delineated the different shades of nationalism: imperialism-induced, outrage at state humiliation, state-sponsored and bottom-up.[122] The struggle of the Sichuan shareholders could be classified as a bottom-up regional movement that was hijacked by the UL who exploited general anger at imperialist encroachment and outrage at the Qing's humiliation, to advance its nationalistic goal.

Railways became a catalyst and triggered socio-political upheaval; this was not limited to China. Argentina also experienced railway nationalism and revolution in the early twentieth century as an 'increasing number of Argentines began to view the British-owned railways as instruments of foreign domination and as tokens of retrogression and strangulation of Argentina's economic development'.[123] Mexico went even further. The United States built the country's railroad, but when America suffered economic recession, Mexico also found itself in hard times. Like the late Qing court, the Mexican government tried to enforce central control, but this only further alienated anti-imperialistic agenda in the provinces and led to revolution.[124] South Africa underwent a similar process, where 'international capital, local railway politics, and imperial policies' significantly affected the course of railway development.[125] Sichuan shareholders propelled the UL into the seat of power as the Republic of China (ROC) was founded in January 1912. But the ROC regime did not solve any problems for them – their investments were lost forever; they were never compensated and the railway was never built during the Republican period (1912–49). The revolution the Sichuan shareholders had inspired made the subsequent story of railways in China even more politicised and complicated.

Notes

1 Lee, *Quest*.
2 Kent, *Enterprise*, 80.
3 'Light in China', *Times*, 26 November 1911: 14.
4 Weipin Tsai, 'The Qing empire's last flowering: the expansion of China's post office at the turn of the twentieth century', *Modern Asian Studies*, vol 49, 3 (May 2015): 895–930.
5 Lee, *Quest*, 50–84 and 85–148.
6 'Shangbu chongding tielu jianmin zhangchengzhe', *Shen Bao*, 31 December 1903: 2.
7 'Tielu jianmin zhangcheng', *Xuwenxian tongkao*, vol 364, 11087–89.
8 'Tielu jianmin zhangcheng', *Xuwenxian tongkao*, vol 364.
9 Zen Sun, *British interests*, 19.
10 Jin Chongji and Hu Shengwu, *Xinhai gemin shigao* (4 vols, Shanghai: Renmin chubanshe, 1980–91), vol 2: 281–89.
11 Yen Ching-hwang, 'Chang Yu-nan and the Chaochow railway (1904–08): a case study of overseas Chinese involvement in China's modern enterprise', *Modern Asian Studies*, vol 18, 1 (February 1984): 119–35.
12 'Changban Xinning tielu', *Shen Bao*, 1 April 1905: 3–4 and 'Xinning tielu', 1 February 1906: 4.
13 MacMurray, *Treaties*, vol 1, 335–36.
14 MacMurray, *Treaties*, vol 1.
15 'The proposed Peking-Kalgan railway', *Herald*, 24 July 1903: 190; 'E'shi ganshe jingzhang tielu', *Shen Bao*, 9 June 1905: 2 and 'E'shi qiangbi Zhongguo jiekuan', 10 June 1905: 2.
16 'The Manchurian Railway', *Herald*, 30 October 1899: 862.
17 'The Kiakhta–Peking railway', *Herald*, 13 November 1903: 1042.
18 'Yuanshuai dafu E'shi Jingzhang tielu guiZhongguo zizao', *Shen Bao*, 14 June 1905: 2.
19 'Jingfeng tielu zhiyuli', *Shen Bao*, 23 June 1908: 12 and 'Jingzhang tielu shanggu zhangcheng', 7 September 1903: 2.
20 'Zhidu zouzhun Zhan Tianyou zhuanban jingzhang tielu', *Shen Bao*, 21 July 1906: 2.
21 'Jingzhang tielu kaixiu', *Shen Bao*, 19 May 1905: 3.

22 Talbot, *Railway conquest*, 297.
23 'Dispossessed: the Kalgan railway', *Herald*, 15 May 1908: 425 and 'Kalgan: railway progress', 11 July 1908: 82.
24 'Jingzhang tielu kaiche jisheng', *Shen Bao*, 7 October 1909: 4 and 'Jingzhang tielu kaiche', 9 October 1909: 4; Jiaotongbu, *Zhan Tianyou xiansheng zhuchi jianzhu jingzhang tielu gongcheng zhaopianji* (Taibei: Kejiziliao zazhishe, 1974).
25 'Light in China', *Times*, 26 November 1911: 14.
26 Talbot, *Railway conquest*, 298.
27 'Jingzhang tielu goumai hongshan zuixin jiqi', *Shen Bao*, 9 March 1907: 4.
28 'Tiyi xiuzhu zhangku tielu', *Shen Bao*, 2 October 1906: 3 and 'Yizhu Kulun di Qiaketu tielu', 29 July 1906: 3 and 'The capital: the Kalgan railway', *Herald*, 31 October 1908: 262.
29 'Diandu zouqing xingzhu Dianshu tieluzhe', *Shen Bao*, 27 May 1905: 1–2, 'Quzhou yichou ziban Zheyue tielu', 26 June 1905: 4 and 'Kaiban Taisui tielu zhishuwen', 3 March 1906: 3; Lee, *Quest*, 42–44.
30 Kaske, 'Sichuan as a pivot', in *Southwest China*, 379–423.
31 A. F. Legendre, *Le Far West chinois: Deux Années au Setchouen* (Paris: Plon-Nourrit, 1905), 145.
32 Davies, *Yun-nan*, 8.
33 Keith Neilson and T. G. Otte, 'Railpolitick: an introduction', in *Paths of empire*, 1–15 (7).
34 Stephenson, *Report*, 3 and Kent, *Enterprise*, 5–6.
35 'YingFa chuixian chuansheng lukuang', *Shen Bao*, 16 June 1911: 4.
36 Braisted, 'American China development company'.
37 McLean, 'Townley Agreement'.
38 McLean, 'Townley Agreement', 159–60.
39 Dai Zhili, *Sichuan baolu yundong shiliao huizuan* (3 vols, Taibei: Zhongyanyuan jinshisuo, 1994), vol 1, 255; Wang Yanwei, *Qingji waijiao shiliao* (269 vols, Beijing: Shumu wenxian chubanshe, 1987), vol 173, 4–5.
40 Sichuansheng dang'anguan, *Baolu zhendang: Sichuan baolu yundong dang'an tuji* (Chengdu: Sichuan daxue chubanshe, 2011), qianyan (foreword).
41 Arthur J. Barry, 'Railway development in China', *Journal of Royal Society of Arts*, vol LVII, 2948 (21 May 1909): 541–58.
42 'Chuanhan tielu zonggongsi jigu zhangcheng', in *Tielushi ziliao*, vol 3, 1062–65.
43 'Chongqing Chuanhan tielu choukuan banfa', *Shen Bao*, 25 May 1905: 2 [一石抽穀三升].
44 Dai, *Shiliao huizuan*, vol 1, 283–84.
45 'Chuanhan tielu choukan', *Shen Bao*, 2 April 1905: 3 and 'Chuanhan tielu jigu', 17 April 1905: 2.
46 'Zigao Chuanhan lukuan', *Shen Bao*, 29 January 1905: 2.
47 Zheng Guanglu, *Sichuan baolu yundong lishi zhenxiang* (Chengdu: Sichuan minzu chubanshe, 2011), 122.
48 'Kaizhu Chuanhan tielu shouduan gongcheng', *Shen Bao*, 18 November 1905: 2.

49 Jiaotongbu, *Jiaotongshi luzhengbian* (18 vols, Shanghai: Jiaotongbu, 1935), vol 16, 46–48; 'Sichuan Chuanhan gongsi jigu chengji 1904–1909', in *Tielushi ziliao*, vol 3, 1096.
50 'Chouxing Chuanlu', *Shen Bao*, 29 Jan 1905: 2 and 'Chuanhan tielu zhaogu Chongqing', 14 February 1905: 3.
51 Lee, *Quest*, 85–141; Kaske, 'Sichuan as a pivot', in *Southwest China*, 379–423.
52 'Chuanhan tieluzhi dahuiyi', *Shen Bao*, 3 June 1907: 11.
53 Dai, *Shiliao huizuan*, vol 1, 379–82; 'Duzhibu wei Chuanhan tielushi cheng duchayuan', *Shen Bao*, 11 October 1907: 11 and 'Youchuanbu zou chaming Chuanhan tielu kuanmuzhe', 6 December 1907: 10–11.
54 'Chuanhan tielu jishi', *Shen Bao*, 26 November 1906: 3.
55 'Chuanhai tieluzhi neirong', *Shen Bao*, 16 April 1907: 4.
56 'Chuanhan tielu yonghua gongchengshi', *Shen Bao*, 29 April 1905: 3.
57 'Chuanhan tielu shending guixian', *Shen Bao*, 22 October 1907: 5.
58 Kaske, 'Sichuan as a pivot', 379–423.
59 Dai, *Shiliao huizuan*, vol 1, 378.
60 Moore, *Defining and defending*, 128.
61 'Cancellation of the American China Development Company's Canton-Hankau railway concession', No. 147 to 164, Office of the Historian, at https://history.state.gov/historicaldocuments/frus1905/ch36, accessed 31 May 2021 and 'The United States and China Agreement to sell the Canton-Hankou railway', in MacMurray, *Treaties*, vol 1, 519–22.
62 Braisted, 'American China development company'; Lee, *Quest*, 50–78.
63 Pang Guangyi, *Minguo shiqi Yuehan tielu lishi yanjiu* (Hefei: Gongye daxue chubanshe, 2014); Zhu Congbing, *Zhang Zhidong yu Yuehan tielu* (Hefei: Gongye daxue chubanshe, 2011).
64 'Hubei guanqianju chengban Chuanhan tielu', *Shen Bao*, 14 September 1906: 14.
65 Moore, *Defining and defending*, 127–35.
66 'Jiniang bianduan', *Shen Bao*, 27 May 1903: 2 and 'Lugong zishi', 22 December 1903: 1; Lee, *Quest*, 106–9.
67 Kaske, 'Sichuan as a pivot', 379–423.
68 'Wei Hubei Chuanhan tielu jiewaizaishi quangao', *Shen Bao*, 1 September 1907: 3 and 5 September 1907: 3.
69 Sheng Chengmao, *Sheng Xuanhuai yu jindai Zhongguo tielu jianshe* (Wuhan: Wuhandaxue chubanshe, 2022), 7–16, 39–41, 64–70 and 127–29.
70 'American enterprise in China', *Herald*, 10 July 1909: 73 and 'Notes on native affairs', 21 August 1909: 473.
71 'The Yuet-Han railway loan', *Herald*, 25 September 1909: 722.
72 'The Yuet-Han loan', *Herald*, 25 February 1910: 435.
73 'Notes on native affairs', *Herald*, 9 September 1910: 611.
74 Dai, *Shiliao huizuan*, vol 1, 426–31.
75 'British public mad over rubber stock', *Times*, 28 February 1910: 1; 'The foreign trade of China', *Herald*, 17 June 1911: 735 and 'Finance and trade in 1911', 13 January 1912: 119.

76 'Chuan jingguan chengqing chaozhui lukuan', *Shen Bao*, 20 October 1910: 4 and 'Chuanhan tielu baihuai qingxing', 8 November 1910: 26; York Lo, 'From No 1 on the bund in Shanghai to a kaolin mine in Cha Kwo Ling, Hong Kong–the century old China coast saga of the McBains', The Industry History of Hong Kong Group, 14 April 2022, at https://industrialhistoryhk.org/from-no-1-on-the-bund-in-shanghai-to-kaolin-mine-in-cha-kwo-ling-the-century-old-china-coast-saga-of-the-mcbains/, accessed 27 June 2023.
77 'Youbu chaban Chuanlu daokuan', *Shen Bao*, 5 June 1911: 5–6; W. A. Thomas, 'An intra-empire capital transfer: the Shanghai rubber company boom 1901–1912', *Modern Asian Studies*, vol 32, 3 (1998): 739–60.
78 Dai, *Shiliao huizuan*, vol 1, 384.
79 'Hudao pichi Chuanlu jinli', *Shen Bao*, 12 August 1910: 18–19 and 'Hudao duifu Shi Dianzhang', 18 August 1910: 18–19.
80 Liu Xuefeng, 'Zaifeng and late Qing railway policy', in *China: how the empire fell*, eds Joseph Esherick and George C. X. Wei (London: Routledge, 2011), 222–58.
81 Yang Wending letter to Sheng Xuanhuai on 19 May 1911, sxh44-0013 and cuhk:1743842-01/02, Sheng Xuanhuai Archive, Chinese University of Hong Kong.
82 Wang Ermin and Wu Lunni, *Sheng Xuanhuai shiye handiangao* (2 vols, Hong Kong: CUHK, 1993), vol 1, 760 (No. 471).
83 Sheng Xuanhuai telegraph to Ruicheng and Duanfang on 29 August 1911, sxh13-0221 and cuhk:1727709-01, Sheng Xuanhuai Archive, Chinese University of Hong Kong.
84 MacMurray, *Treaties*, vol 1, 866–99.
85 MacMurray, *Treaties*, vol 1, 870.
86 'Ambassador Hill to Secretary of State, American Embassy Berlin, May 20, 1910', No. 5315/927, Office of the Historian, at https://history.state.gov/historicaldocuments/frus1910/d254, accessed 31 May 2021.
87 Wang and Wu, *Sheng*, vol 2, 750 (No. 457); Dai, *Shiliao huizuan*, vol 1, 597–98.
88 Wang and Wu, *Sheng*, 764–66 (No. 477); Sheng Xuanhuai telegraph to Zhao Er'feng on 30 August 1911, sxh13-0222 and cuhk:1727711-01 to 04, Sheng Xuanhuai Archive, Chinese University of Hong Kong.
89 Zheng, *Zhenxiang*, 138–40.
90 'Sichuanren zhi zhenglure', *Shen Bao*, 17 July 1911: 10, 'Chuanren baolu zhixuelei', 29 August 1911: 5, 'Chuanren zhenglu zhi dianzougao', 6 September 1911: 5 and 'Chuansheng luanshi jiwen', 19 September 1911: 4–6.
91 Dai, *Shiliao huizuan*, vol 2, 1105; Ma Linghe, *Qingmo Minchu tielu waizhaiguan de yanjiu* (Shanghai: Fudan daxue chubanshe, 2004).
92 Dai, *Shiliao huizuan*, vol 2, 1133–34.
93 Zheng, *Zhenxiang*, 228–29; 'Zhao Erfeng zhizao Chengdu xue'an', *Sichuan dashi*, 233–34, at Sichuan shengqing wang, www.scsqw.cn/, accessed 30 October 2021.
94 'Chuanlu didixue', *Shen Bao*, 20 September 1911: 4.
95 Dai, *Shiliao huizuan*, vol 2, 1124–30.

96 'The Szechuan crisis', *Herald*, 23 September 1911: 772 and 'The Szechuan crisis', 7 October 1911: 6.
97 Dai, *Shiliao huizuan*, vol 2, 1109–10.
98 'Tongmenghui cehua quanchuan wuzhuang qiyi', *Sichuan dashi*, 230–31, at Sichuan shengqing wang, www.scsqw.cn/, accessed 30 October 2021.
99 Dai, *Shiliao huizua*, vol 2, 1022–27.
100 Dai, *Shiliao huizua*, vol 2, 1056–63.
101 Di Wang, *Violence and order on the Chengdu plain: the story of a secret brotherhood in rural China, 1939–49* (Stanford: Stanford University Press, 2018); Ouyang Enliang, *Xinan paoge yu xinhai gemin* (Beijing: Zhigong chubanshe, 2011).
102 'E'sheng yanfang gemindang cufazhi zhoumi', *Shen Bao*, 15 May 1911: 10–11.
103 'Dr. Sun Yat-sen', *Herald*, 11 November 1910: 319.
104 'Chongqing guangfuzhi zhenxiang', *Shen Bao*, 13 January 1912: 12, 'Sichuan dulihou zhihaojie', 16 January 1912: 3 and 'Shujun zhenfufu zuzhi dagang', 30 March 1912: 6.
105 'Gongdian', *Shen Bao*, 23 February 1912: 1 and 17 March 1912: 1.
106 Robert Kapp, *Szechwan and the Chinese republic: provincial militarism and central power, 1911–1938* (New Haven: Yale University Press, 1973).
107 Chi, 'Shanghai-Hangchow-Ningbo'.
108 Rankin, 'Nationalistic contestation'.
109 Zheng, *Politics of rights.*
110 Joseph Esherick, '1911: a review', *Modern China*, vol 2, 2 (April 1976): 141–84 and 'Reconsidering 1911: lessons from a sudden revolution', *Journal of Modern Chinese History*, vol 6, 1 (2012): 1–14.
111 R. Bin Wong, 'Centennial perspective on China's 1911 revolution', *China Information*, vol 25, 3 (November 2011): 275–82.
112 Rana Mitter, '1911: the unanchored Chinese revolution', *China Quarterly*, vol 208 (December 2011): 1009–20.
113 Li Danke, 'Popular culture in the making of the anti-imperialist and nationalist sentiment in Sichuan', *Modern China*, vol 30, 4 (October 2004): 470–505.
114 Heung Shing Liu, *China in revolution: the road to 1911* (Hong Kong: Hong Kong University Press, 2011).
115 Arif Dirlik and Roxann Prazniak, 'The 1911 revolution: an end and a beginning', *China Information*, vol 25, 3 (November 2011): 213–31.
116 'Lun Zhesheng jiyi zixing tielu yidu wairen lanquan', *Shen Bao*, 31 May 1905: 1, 'Jinpu tielu wenti', 5 July 1909: 4, 'Shouhui Zedao tieluquan', 8 May 1905: 4, 'Bingqiu xiangzu Zhegan tielu', 17 June 1905: 2, 'Nanxun tielu jugu huiyi', 23 June 1905: 3, 'Jiuguang tielu nengfou shefa jianchi jigu ziban', 24 June 1905: 2, 'Yueshen cheng E'du lizheng luquan', 19 February 1905: 3 and 'Meizhou liuxuesheng shouhui yuehan luquan tiaochen', 26 March 1905: 17; see also Reinhardt, *Navigating*, 183–87.
117 Davies, *Yun-nan*, 17.
118 Li Jingfang letter to Sheng Xuanhuai, 2nd, month not clear, sxh36-0111 and cuhk:1739511-01, Sheng Xuanhuai Archive, Chinese University of Hong Kong.

119 Zheng, *Zhenxiang*, 147.
120 Rankin, 'Nationalistic contestation'.
121 Roger Thompson, 'If Shanxi's coal is lost, then Shanxi is lost!: Shanxi's coal and an emerging national movement in provincial China 1898–1908', *Modern Asian Studies*, vol 45, 5 (September 2011): 1261–88.
122 Anthony Reid, *Imperial alchemy: nationalism and political identity in Southeast Asia* (Cambridge: Cambridge University Press, 2010), 1–24.
123 Wright, *British-owned railways*, 5; William J. Fleming, 'Profits and visions: British capital and railway construction in Argentina, 1854–1886', in *Railway imperialism*, 71–84.
124 William E. French, 'In the path of progress: railroads and moral reform in Porfirian Mexico', in *Railway imperialism*, 85–102.
125 Kenneth E. Wilbur, Jr, 'Engines of empire and independence: railways in South Africa, 1863–1916', in *Railway imperialism*, 25–40 (37).

9

He who controls railways controls China

> On the evening of May 5, 1923, I was traveling between Nanking and Peking, together with a few other newspapermen. ... Our train, consisting of first-, second-, and third-class coaches, was China's crack 'Blue Express,' the first train of all-steel coaches ever seen in the Orient, which had been purchased by the Chinese Railway Administration in the United States ...
>
> It was early spring and a bright moon was shining, making the barren rocky Shantung Mountains quite visible in the distance. We had raised the window so as to enjoy the warm breeze, and just before retiring I looked out the window and remarked to Berube that we were passing through 'bandit territory,' ...
>
> The train had just crossed the divide from Kiangsu into Shantung Province and was proceeding slowly, when there was a sudden grinding of brakes and the cars came to an abrupt halt – so abrupt that many passengers were tumbled out of their berths. There was a great deal of shouting and firing outside, and I stuck my head out of the window to see what was going on. I quickly withdrew it, however, when a bandit fired his rifle in the air within a couple of feet of my head, but I had time to see what looked like a small army of men swarming down the embankment, yelling and firing their rifles as they came. They climbed into the cars through the windows, ran along the corridors and began routing the passengers out of their berths while they ransacked the baggage. ...[1]

The excerpt presents a witness account of the 'Lincheng Outrage', an infamous hijack that took place on the Jinpu railway.[2] The first-class coaches carried Europeans and Americans, including businessman Carl Crow, journalist John B. Powell and Lucy Aldrich, the daughter of Senator Nelson Aldrich and sister-in-law to John D. Rockefeller, Jr.[3] Although the female hostages were released after a week, the rest were held for over a month while negotiations took place. The story was later dramatized in the 1932 film *Shanghai Express*. But this was not the first time the Jinpu had been attacked, nor would it be the last.[4] Although railways had proved a catalyst to revolution, they wrought new havoc in the Republic of China. Just as the foreign powers had used railways to spread their influence, so too would a host of new players, from bandits and warlords to the various ROC regimes. This chapter sheds light on how railways predetermined and characterised the history and politics of the warlord-divided and war-torn era.

'Strategy to Build a Nation'[5]

The ROC was founded on 1 January 1912, when Dr Sun was sworn in as its provisional president in Nanjing.[6] But in Beijing, the Qing dynasty carried on with the six-year-old Puyi Emperor on the throne. To bring Beijing and north China into the orbit of the ROC, Sun promised Yuan the presidency if he could engineer the abdication of the boy emperor, which Yuan achieved in March 1912.[7] Yuan's ROC regime is often referred to as the Beiyang government, a reminder of its origin in the Beiyang commissioner. Though different characters, Yuan and Sun had one thing in common – a passion for railways. Sun became director of the National Rail after he stepped down, with Zeng Kunhua serving as his chief adviser. A graduate of the Iwakura Railway School in Japan who began his career in the DPC, Zeng advocated an education system devoted to railway science and systematic management of rail networks, laying the foundation for China's railway industry to come.[8]

Railways were central to Dr Sun's imagining of a new China that would be *on a par with* the West.[9] In an interview with a newspaper in June 1912, he stated: 'Transportation is the mother of enterprise and railways are the mother of transportation. A country's wealth is measured by the numbers of its railways.'[10] Like fellow Cantonese and Taiping rebel Hong Rengan, Sun saw railways as a symbol and manifestation of a modern industrial nation – the more railways he could build, the richer and more powerful China would be. Railways were the most important instrument for nation building. Richard White and William G. Thomas have emphasised the role of railroads in promoting economic development, national integration and the making of modern America.[11] Thomas argues that the railroad became 'one of the most obvious, and the most prevalent, forms of symbolic technology'.[12]

Dr Sun set out to build a national rail system, publishing his ideas in a book titled *Jianguo Fanluo* or *Strategy to Build a Nation.*[13] C. Martin Wilbur discussed his vision briefly.[14] It is vital we engage with Sun's railway strategy in detail, not just because its national scope offer a contrast to the haphazard and regionally orientated approach of the Qing but because it has continued to inspire the post-Mao regime. He conceived of six railway 'systems' stretching across China like a giant spider's web: the Northwest, the Southwest, the Central, the Southeast, the Northeast and the Plateau. The systems fanned out from three ports: Tianjin, Shanghai and Guangzhou. The railways he proposed varied from 3,500,000 *li* to 200,000 *li*.[15] However, this would not be an easy feat for the young ROC. George E. Anderson highlighted the problem it faced: 'The railways of China have been constructed very largely without regard to a national system. Each road is more or less local, not only in its interests and support, but in its physical characteristics.'[16] Sun's plan had serious flaws, not least that it was so bold that even the post-Mao regime, with all its technological capabilities, resources and manpower, has struggled to realise it (see Figure 9.1).

Sun's plan was audacious, yet highly unrealistic. The Northwest system branched out from Tianjin to as far north as Mohe on the Qing–Russian boundary, and as far west as Ili, Kashgar and Yutian on the border with Kazakhstan, Kyrgyzstan, Tajikistan, Afghanistan, Pakistan and Kashmir.[17] Sultan Abdul Hamid II built the Hejaz railway, from Damascus to Medina with the goal of facilitating the Hajj and enhancing the image of the Sultanate in the Islamic world.[18] What did Sun aim to accomplish? Hugo Silveira and Anne Hardgrove discussed how the Portuguese and Trinidadian governments used railways to reach remote area in order to control the local

Figure 9.1 'Strategy to Build a Nation' railway map (Sun Zhongshan, *The international development of China* (Taipei: China Cultural Service, 1953), back matter)

population.[19] Perhaps Sun thought so, too. The proposed lines appeared like layers of wires separating China from its neighbours. Sun explained the need for such an elaborate system:

> This railway system sits at an important place and dominates the world. It will become the trunk line of a Eurasian railway system, linking the populations of Asia and Europe. For those who travel from the Pacific to Europe, this is the closest route. The branch line from Ili will connect with railways in India and Europe (those going to Baghdad, Damascus, and Cairo), they will become one chain. From the ports I envisioned, one can travel directly to the Cape of Good Hope.[20]

The phrase 'dominate the world' indicates just how much he had learnt from foreign powers. This system was to travel west along the old Silk Road, connecting China with Eurasia, before traversing Africa from head to tail – the dream of Cecil Rhodes. The northeast and northwest were so important to him that he elaborated further where railways are depicted as rising like fortresses, defending China along the borders of Manchuria, Mongolia and Xinjiang.[21] These numerous lines in sparsely populated areas of little commercial value would, in Sun's words, help 'to colonise Mongolia and Xinjiang'.[22] He planned to move 10 million people to these regions and considered it a matter of urgency in order to reduce overpopulation in central China. But who would migrant to regions with arid land and hostile weather? His plan on paper failed to reckon with the extremely challenging topographies on the ground: impenetrable mountains, Yellow River loess, Mongol steppes, vast grasslands, sinking basins and deserts which remain difficult to cross due to sandstorms that often bury not just rail tracks but entire settlements. No wonder Jia Benyi called his plan 'romantic'.[23]

Dr Sun appears to have been obsessed with the northwest and southwest frontiers. Three of the six systems were devoted to the vast region inhabited by Mongols, Uyghurs, Kazakhs, Tibetans and many other non-Han ethnic groups. The extent of his ambition can best be seen in the Plateau system, which was meant to link Lhasa in Tibet with Lanzhou to the north, Chengdu to the east and Dali to its southeast.[24] The Yutian (Hotan) to Lhasa line is a case in point. Starting in western Xinjiang near the border with Pakistan and Kashmir, this line was to traverse the Himalayas, the Ladakh Range, Karakorum, Mount Kunlun and the Pamirs. How was this to be achieved then, when it is not even possible today? What purpose would this line, located on the 'roof of the world', serve? Perhaps he thought it would deter foreign invaders. If so, the topography of the 'roof' itself surely provided a more formidable deterrent to attacks than any railways could.

Connecting dots on paper, Sun's plans completely disregarded the technical and material difficulties, and failed to account for the expertise needed.

Wilbur justifiably articulates that he had 'very limited personal knowledge of China's terrain'.[25] This is most visible in the designs of the Northwest and Plateau systems; it can also be seen in the Southeast system.[26] Adopting British ideas, he conceived a direct line from Shanghai to Chongqing. Setting out from Shanghai, the railway passed Hangzhou, Mount Tianmu and Shexian in the Yellow Mountain before entering Jiujiang on Lake Boyang in Jiangxi province. From there, it followed the Yangzi River west to Hubei province, passing the Hankou region and dipping down to Lake Dongting in Hunan province. Then the line abruptly turned north, charging back into Hubei. From here, it followed the route of the old Chuanhan railway to Yichang and Wanxian.[27] The challenge in building this line was enormous: not only would it have to cross the tempestuous Wu and Yangzi rivers but also the wetlands in Boyang and Dongting lakes, as well as five jagged mountains with karst landscape: Wudang, Daba, Wuling, Wushan and Dalou, which separate Hubei from Sichuan. These are treacherous terrains, even in the twenty-first century.

Bridges would be needed to cross the rivers, lakes and wetlands, and bridges required knowledge, equipment and material that the young ROC did not possess. Yet Sun was keen to persist with this project despite the prohibitive costs and technical challenges for the exploitation of natural resources, expressed when discussing the Guangzhou to Chongqing line.[28]

> Nine hundred miles long, this line passes through rich mining and agricultural areas. In north Guangdong, they have discovered coal deposits, iron, antimony, and tungsten in Lianzhou. In the southwest of Hunan, there is tinstone, antimony, coal, iron, copper, silver. In Sichuan, the Youyang area has antimony and mercury. The agricultural produce along the line includes sugar cane, peanut, cannabis, kalo nut tree oil, tea, cotton, tobacco, raw silk, grain. It is also endowed with many bamboos, forests and related products.[29]

This was one of seven lines he conceived for the Southwest system, justifying their construction with the opportunities provided by mineral and agronomic resources along the routes.[30] Sun envisioned a new and prosperous China built through his rail systems. Yet he clearly favoured Guangdong, his birth province, as the Southwest system fanned out from the maritime hub in seven directions, linking old Canton with not just central China but also with Vietnam and Burma through French and British built or planned lines. This can also be seen from the Southeast system that departed from Shanghai, Nanjing, Fuzhou and Xiamen but all terminated in Guangzhou. However, north China was not forgotten.

His Central system consisted of twenty-four railways linking eastern metropolises like Tianjin, Shanghai and Nanjing with central and western cities like Luoyang, Hankou, Xi'an and Lanzhou.[31] He downgraded, if not totally eliminated, Beijing, the nucleus of existing railways from his strategy. Even more baffling is the Northeast system.[32] Despite the elaborate rail network already in existence and future lines planned by the Japanese in Manchuria, Sun envisioned an entirely new system for the region. To differentiate, he created a railway hub centred on a fictional city called Dongzhen or East Town whose geographical location was a no man's land. It runs in parallel, hence in conflict, with the Barrier railway, the SMR and the CER. Why duplicate and compete with existing lines?

Throughout his book, Sun never discussed how his government would gather the funds needed for the construction of these railways, nor did he divulge in the expertise and materials required for such a plan. Having spent much of his life fundraising in overseas Chinese communities for his revolution, Sun was something of an expert on the subject. Like Li Hongzhang, he knew how to 'borrow hens to lay eggs'. As Director of the National Rail, he explained that the ROC should attract foreign capital:

> Due to the fact that our government does not have the funds to build railways, we have to borrow foreign capital. But if we borrow in the name of the government, then this would involve international dealings. In my opinion, it is better to do so in the name of private individuals – they would deal with capitalists of foreign countries which have nothing to do with our government. Neither could foreign governments intervene.[33]

Ironically, Dr Sun was following the same policy that had given birth to the Railway Rights Recovery Movement and brought him into power. He proved himself ignorant of railway's history – Jardine and Decauville, among others, were private companies, yet the British and French governments had intervened on their behalf. Worse still, he pushed the responsibilities of China's territorial integrity and financial security onto the shoulders of 'private individuals' – inconceivable for the Qing and irresponsible to suggest as a national leader. Foreign powers once again hungrily leered at the opportunities presented by Sun's railway vision.[34] The first to be built was the Guangzhou to Chongqing line. Sun oversaw the negotiations with the British firm, Pauling and Company. A nineteen-article draft contract was agreed in March 1913:

Article 1: The contractor agrees to lend a large sum of funding to the ROC regime, at an annual interest rate of 0.5 per cent, for the exclusive purpose of building the Guangzhou to Chongqing railway.

Article 9: The contractor has the right to build a branch line to Lanzhou in Gansu province. If the Chinese agree, they could extend the same line to other places.

Article 10: All the field and land next to the railway, after examination, can be used for affiliated territory, stations, repair shops and depots.

Article 15: To encourage Chinese industry, if the value and quality of Chinese material are appropriate, they should be used. Compared with goods and material from other countries, if they are made with the same materials and have the same value, British made goods would enjoy privilege of priority.[35]

These articles reveal a remarkable willingness to rely on foreign loans, considering the ROC was born out of a revolution that opposed borrowing from foreign powers to build railways. An annual interest rate of 0.5 per cent was equivalent to that of the *Final Agreement for the Hukuang Imperial Government Railways*, which led to the Nationalist Revolution. Article 9 echoed the kinds of terms demanded by other foreign powers. Article 10 would undoubtedly allow colonies to grow, while Article 15 offered favourable terms to the British. It seems that Sun was not concerned with borrowing at the cost of China's sovereignty. Wilbur calls Sun's foreign loans 'compromising'.[36] Nowhere did he indicate any concern for the destruction of existing socio-economic life and the environment – unlike late Qing officials who had resisted railways for these matters. His strategy would have opened another phase in railway imperialism. Fortunately, he was soon distracted by political events that threatened his own existence.[37]

Labelled the 'father of the nation', Dr Sun has been credited with the Nationalist Revolution and continues to be commemorated. But his patriotic credentials and leadership qualities often fell short of the man presented by his admirers and scholars. Some have challenged his work and legacy. Wilbur concludes that Sun was 'experienced in almost nothing but fomenting revolution'.[38] He was unrealistic, a dreamer and his vision for railway testifies to that. Charismatic and flexible, his skill lay in delivering inspiring speeches that flattered various audiences. David Strand believed that 'Sun's role in the Chinese Revolution hinged on the fact that he talked so much and so well to so many that his words added up to a particular kind of public deed.'[39] During his visit to Qingdao in 1912, he applauded the 'amazing' speed of progress instigated by the Germans as a 'model for the whole of China'.[40] German sources even suggest that he went on to say: 'in three thousand years China has not achieved what Germany did in Kiaochow in 15 years'.[41] Like Li Hongzhang, Sun appeared to admire Germany, but this did not prevent him from taking Soviet money to found the Huangpu Military Academy. He looked to the British with preference and worked with the Japanese to battle Yuan Shikai.[42] Although the railway revolution brought him to power, he made no difference to its development. Dr Sun learnt from the foreign powers and saw railways as an instrument of nation building. So did his arch enemies – the warlords.

Strategy for warlords

Railway imperialism informed generations of Chinese leaders, both national and regional. Chinese historians have written about warlords and their region building, but few have paid attention to the intimacy between warlord conflicts and railways.[43] William G. Thomas's excellent book analyses the dynamics between the American Civil War and the railroads, arguing that they were the 'twin engines in the development of modern America, operating with independent causes and effects but simultaneously and in relation to one another'.[44] The same could be said of the early ROC era where armed conflicts among warlords and railways played a pivotal role in shaping the warlord era. Railways marked out boundaries and became battlefields; the tracks laid in the preceding decades predetermined the contours of the divided era, compounding existing problems and turning China into the 'land of famine'.[45]

The ROC, founded in 1912, existed only in name. While Dr Sun and his United League-transformed Nationalist Party (Guomindang/GMD) based themselves in Guangzhou, the north was in the hands of the Beiyang government, controlled by four successive warlord regimes: the Yuan Shikai-led Zhili clique (1912–16), the Anhui clique (1916–20), the Zhili clique once more (1920–24) and the Fengtian clique, led by Zhang Zuolin (1924–28). These warlord cliques fought each other for control of the Beiyang government for nearly two decades. Despite the frequent change of leadership, the various Beiyang regimes succeeded in standardising all railway gauges controlled by China.[46] They upgraded the system and improved service; they also implemented a system of national management by introducing regulations and uniforms.[47]

Labelled the 'uncrowned king of Manchuria', Zhang Zuolin deserves attention not just because his career was tied to railways but also because his endeavours reveal geopolitics of railways in the warlord era.[48] Zhang had risen quickly in the ranks of the Fengtian military force in the 1900s.[49] Shrewd and flexible, he allied with whoever suited his interests. For instance, he initially aided the Russians during the Russo-Japanese War, before switching to the Japanese once the latter gained the upper hand. After 1911, he courted the Beiyang government's blessing but resisted their direct control. Through collaboration and intimidation, he managed to legitimise his position as Fengtian Military General and then Inspecting General of the Three Eastern Provinces by 1918.[50] He thrived at a time of constant political change; his main preoccupation was his control of Manchuria, and railways were key to the movement of his troops and the region's economy.

What makes him significant is that he fashioned a three-pronged strategy for railway development in Manchuria: (1) to squeeze out the Russians,

(2) to collaborate with the Japanese in order to constrain their expansion and (3) to build his own lines to undermine foreign powers. The Russian Revolution led to confusion about the CER's future. Zhang assembled a Railway Protection Battalion and stationed them along the CER, ready to stake his claim.[51] But he was not the only one who was keen on the CER. Britain, the United States and France realised that Japan was poised to benefit from the confusion – 'whatever China took back from Russia might eventually ended up in Japan's hands'.[52] To check Japan's ambition, they put together an Inter-Allied Technical Board in March 1919, signed an agreement with Japan and installed John F. Stevens, an American, as president to oversee the CER in transition.[53]

The newly founded USSR published the Karakhan Manifesto, a statement of their policy towards China, in July 1919. It promised: '(a) unconditional return of the CER, (b) renunciation of the Boxer indemnity, (c) rendition of concessions territories and (d) abrogation of extraterritoriality'.[54] This proved to be little more than rhetoric, what Zhou Yuefeng called the 'Lenin moment'.[55] The promise to return the CER was hyperbole and propaganda. As Bruce Elleman correctly identifies, the Soviets were secretly negotiating with the Beiyang regime to continue its control.[56] That may explain why the phrase 'unconditional return of the CER' all but disappeared in the Second Karakhan Manifesto of September 1920.[57] Undeterred by this, Zhang put pressure on the CER leadership, which led to a preliminary agreement between the Beiyang government and the Russo-Chinese Bank in October 1920.[58] The document began:

> In view of the situation created by the complete political disorganization in Russia, rendering temporarily impossible for the said company the maintenance of regular operation; having regard, moreover, to its rights of sovereignty ... By reason of the obligations stated above, the Chinese Government, has, under date of October 2, 1920, notified to the Bank its decision to assume provisionally, pending such arrangement concerning the railway as the Government may reach with the Russian Government that may be recognized by China, the supreme control exercised over the said railway.[59]

But the problem was that the Russo-Chinese Bank did not represent the new Russian government. *Shen Bao* reported in 1923 that the issues of ownership and control had not been resolved.[60] After the Soviet regime finally stabilised itself in 1924, it simply replaced old treaties signed by the Tsarist regime with new ones.[61] The Soviets were no different from their imperial predecessors – they were Bolshevik imperialists. The new treaties simply recycled old terms and conditions. The Soviets agreed to co-manage the CER, but this was no more than empty promise.[62] Zhang kept up the pressure after another agreement was signed in September 1924, and this time

Zhang himself was the signatory.[63] It reduced the CER's concession from eighty to sixty years, a small victory on paper:

> The concession period of eighty years mentioned in Article 12 of the Contract of August 27th (September 8th) 1896, for the Construction and Operation of the Chinese Eastern Railway, shall be reduced to sixty years, upon the expiration of which the line with all its appurtenances will pass free of charge to the Chinese government.[64]

The ownership, management and future of the CER was chaotic in the post-Russian Revolution decade, and historians have disagreed on who held control. Writing in 1936, Tao Shing Chang specified that the CER saw 'dual management'.[65] Sow-Theng Leong argued in 1976 that Manchuria saw 'decolonisation' and that the Chinese 'assumed supreme administration of the CER provisionally'.[66] Bruce Elleman believes that the Soviet Union 'regained majority control' despite the interruption of the revolution.[67] Blaine Chiasson has articulated that the Chinese asserted its control while accommodating the Russian presence, whereas Yiwei Cheng has concluded that there were 'parallel authorities'.[68] The CER demands more and collaborative research.

Zhang also took on the Japanese, but in a very different manner. He managed to assert himself while constraining the Japanese through collaboration. Jun Uchida has written about the 'shifting pattern of alliance and conflict' on Korean railways under Japanese rule: 'each round of negotiation and contestation produced a new matrix of contestation, generating convenient and often precarious alliance among diverse interest group'.[69] The situation in southern Manchuria was similar. Zhang enticed the Japanese, asking them to abandon their 1913-proposed Kaiyuan–Hailong line, in exchange for one from Taonan to Qiqiha'er (see Figure 6.2).[70] This was an attractive offer because Qiqiha'er was located on the horizontal segment of the CER, deep in Russian-controlled Manchuria. The same can be said of several other lines, including one from Changchun to Fuyuan, a town across the border from Khabarovsk, then the largest city in the Russian Far East.[71] These lines were extremely alluring to the Japanese as they offered opportunities to expand north into the Russian sphere.[72] Indeed, Matzusaka referred to the Taonan-Qiqiha'er [洮齐] railway, as an 'invisible Russo-Japanese War'.[73] Zhang reached out to the Japanese because he needed their goodwill in his battles with other warlords.[74] For the Japanese, working with Zhang came with three benefits: it allowed them to march north, it offered Zhang as a client and it would mean avoiding the intervention of the US-led banking consortium, as these were Chinese-led railway initiatives.[75]

Zhang also built his own railways.[76] During his rule as the head of the Beiyang government, Zhang's army expropriated '300 locomotives' and

'6,000 freight and passenger coaches' from north China.[77] With them, he built a number of lines, including the Fengtian to Hailon [奉海], the Kirin to Hailong [吉海], the Hulan to Hailun [呼海] and the Dahushan to Tongliao [大通], among others.[78] Branching out from the SMR, the Fenghai line (236 km) travelled north-eastwards to Hailong.[79] Like the Jihai line, it formed a network to the east of the SMR whereas the Datong did so to the west, as it fanned out from the Barrier railway at Jinzhou (Dahushan), passing Tongliao and terminating in Ang'angxi on the horizontal CER (see Figure 6.2). These lines enabled him to transport goods and troops to and from north China without using the SMR. Avoiding the SMR had the effect of diverting lucrative freight traffic from the Japanese system to the 'Chinese-owned Peking-Mukden line', as Clarence B. Davies postulated.[80] *Shen Bao* opined that Zhang had outsmarted the Japanese.[81] When the Japanese pressed him again in late May 1928 about their multi-phased 'Manchuria–Mongolia rail network', Zhang put one word on the document: *yue* [阅], which means 'read' in the past tense. The Guandong Army decided to eliminate him. Zhang shuttled between Beijing and Fengtian in his own train.[82] When it left the Barrier line and entered the SMR on 4 June 1928, it exploded, killing him at the zenith of his career as the head of the Beiyang government and of Manchuria.[83]

However, his death did not destroy the warlord regime as the Japanese had hoped. His son, Zhang Xueliang, succeeded him and continued his father's three-pronged railway scheme.[84] He did so by pledging his allegiance to the newly founded Nanjing regime in December 1928, something his father had refused to do. This seemed a smart move as it earned him the support of the ROC government now headed by Jiang Jieshi. The 'young marshal', as he was affectionately called, was eager to recover the CER.[85] There was nationwide support for this. Of the 1,030 people interviewed by *Shen Bao*, 810 expressed a willingness to join the struggle against Russia.[86] The Soviets had been using the railway to spread communism in Mongolia and China, which the Nanjing regime had been trying to nip in the bud. In the summer of 1929, with verbal encouragement from Jiang, Zhang began by arresting Soviet diplomats and expelling CER staff.[87] This soon led to armed conflict, where Zhang's army suffered heavy casualties.[88] Zhang asked Nanjing for help, but help never came. Tao Shing Chang called it the 'undeclared war' whereas Michael Walker labelled it the 'war nobody knew'.[89]

The United States tried to intervene again in its effort to prevent the Soviet Union from regaining exclusive control of the CER, using the Kellogg–Briand Peace Pact as a pretext. But only Britain and France supported the American initiative.[90] The US had tried to neutralise railways in Manchuria through the Open Door Policy and the 'Dollar Diplomacy' in 1900s. It grabbed the opportunity offered by the Russian Revolution by installing the Inter-Allied

Board in 1919. The US watched events in Manchuria unfold and weighed up its options, as did Britain. The CER had become the 'barometer of war' in the words of Greg Kennedy; like a 'litmus test', it could detect signs of change in regional and international politics.[91] The British concluded that Russo-Japanese tension in Manchuria would 'consume valuable economic and military resources that otherwise might be marshalled again British interests'.[92] It seems Britain was finally losing its appetite for Manchuria.

A disaster for the ROC, the conflict ended in December 1929 with the Khabarovsk Protocol.[93] The Soviets regained control of the CER and procured the island of Heixiazi (Bolshoi Ussriysky), situated at the confluence of the Amur and Ussuri rivers.[94] Zhang's action might have failed, but the Soviets would not enjoy the CER for long, as the Japanese would invade north Manchuria after the Mukden Incident in September 1931. The Soviets tried to sell the CER. After four years of negotiation, the CER was sold to the Japanese-controlled puppet Manchukuo at 140 million *yen*, not the 625 million *yen* the Russian had asked.[95] No wonder George Lensen called the CER a 'damned inheritance'.[96]

Zhang Xueliang also took on the Japanese, but he was less tactical than his father. Instead of collaborating with them to restrain their expansion, he built new lines on both sides of the SMR, stifling the railway and its feeder lines. The Japanese decided to eliminate the 'young marshal' as they had done his father.[97] The Guandong Army blew up a small segment of the SMR on 18 September 1931 and then claimed that the Chinese had been the perpetrators.[98] To 'retaliate', they launched a full-scale assault on Zhang's army barracks in Mukden. The Mukden or September Eighteenth Incident heralded Japan's full conquest of north Manchuria. The Japanese quickly moved north along the CER, defeating the Russians by February 1932.[99] They even demanded the aforementioned Heixiazi island.[100] Zhang's troops retreated to south of the Mountain–Sea Barrier; he would never see his homeland again. Using railways to oust the Russians and constrain the Japanese had cost the life of the father and the career of the son.

The two Zhang men were not the only warlords to construct railways. Yan Xishan was another. A similar character who navigated difficult waters at a time of constant change, the 'king of Shanxi' ruled the northwest province from 1912 to 1949. Calls for a railway in Shanxi had surfaced in 1905, but the project had been delayed by the same problems that plagued the Chuanhan.[101] Construction finally began by 1932 and the Tongpu [同蒲] railway was completed in 1937.[102] The vertical line travelled the entire province, lacing up key cities including the capital Taiyuan, which had been linked with the Luhan railway at Zhengding, via the Zhengtai line.[103] The Zhengtai gave Yan Xishan access to north and central China via the Luhan, Manchuria via the Barrier railway, and

Shanghai via the Jinpu and Nanjing-Shanghai lines. But Shanxi itself was difficult for these lines to access because the Tongpu used the 1.000-mm narrow gauge rather than the standard 1.435 gauge – Yan was adept at utilising the railway to keep the spoils of Shanxi, China's 'coal capital', for himself.[104] Like Li Hongzhang, Yan was resourceful; he gathered funding, procured expertise and used his soldiers in the construction.[105] His efforts contributed to the regional nature of railway development that was characteristic of the late Qing.[106]

Significant warlords could amass the considerable resources required to build railways. But there were many small warlords who had nothing but rowdy soldiers – they asserted their control over the segments and stations within their 'territory'. Zhang Xun, for example, claimed the Xuzhou segment on the Jinpu line, whereas Ni Sichong exerted his influence on the Bengbu station. They were only two among many. Warlords sought to build and control railways because they were strategic weapons, but they would become 'strategic targets' in conflicts.[107] Spiers elaborated on the intimacy between railways and modern warfare, as it tested an army's 'ability to use, manage and exploit railways'.[108] This can also be seen in the conflicts among warlords. The Zhili clique, led by warlord Wu Peifu, fought three wars with the Fengtian clique, led by Zhang Zuolin, between 1922 and 1925. Zhang used the Barrier railway to move 100,000 troops to south of the Mountain–Sea Barrier in early April 1922. Wu Peifu achieved similar feats with the help of the Luhan line. American journalist Edna Lee Booker reported on the massive congregations of soldiers along the two railways and the chaos they wrought.[109] By the end of April, the two armies engaged in fighting on the two railways. Wu's army managed to push Zhang's forces back to north of the Barrier with the railway as their boundary. The two armies met again in September 1924; this time Zhang briefly triumphed, that is until Wu's return in May 1925.

Such was the scale of armed conflicts among warlords, made more frequent by railways. *Shen Bao* called this 'catastrophe':

> The two wars in the southeast and northeast, and the current conflict in Henan, aside from the damage done to the people, the biggest injury is to the country, in three ways. They used the railway to move soldiers, the trains were held up by them, preventing regular service. They damaged roads and bridges which need repairing. They left the stations in a mess; the locomotives and wagons were so broken that they cannot return to normal operation soon.[110]

The Zhili–Fengtian conflict for control of the Beiyang government was only one among others.[111] Diana Lary counted sixteen warlord conflicts, large or small, in 1928 alone.[112] They wrought havoc to the railways and to the country, leading to the outbreak of famine as Pierre Fuller has

Figure 9.2 Major railways in 1937 (Baidu public domain)

documented.[113] The worst was yet to come. William G. Thomas explains how 'the American Civil War became structured around the railroad network, centred on the boundaries made by junctions and rail lines'.[114] This was exactly the case with the so-called Central Plains War between Jiang Jieshi and three of his warlord opponents: Yan Xishan (Shanxi clique), Feng Yuxiang (Northwest clique) and Li Zongren (Guangxi clique), in 1930 that devastated the country. This conflict saw millions of soldiers battle for months over seven railways – the Barrier, Luhan, Jinpu, Jingsui (extension of the Jingzhang to Baotou in Mongolia), Zhengtai, Longhai and the not-yet completed Yuehan (see Figure 9.2).

Railway traffic in north and central China came to a standstill while tracks, trains and stations became battlegrounds and garrisons. Jiang and his opponents issued commands from their train carriages, where negotiations were also conducted. This all-out war was ended by the 'young marshal' Zhang when he declared his support for Jiang and threatened to march his troops down to the central plains swiftly through the Barrier, the Luhan and Jinpu railways.[115] His declaration ended the warlord era, and, upon its ashes, the Nanjing regime seemed to have finally stabilised. The Jiang regime quickly changed the name of Beijing into Beiping and with it the names of several railways, including: Luhan to Pinghan [平汉], Jinlu to Pingjin [平津], Jingsui to Pingsui [平绥]. Railways predetermined and defined the warlord era. Sadly, these railway wars have received little attention, the warlords' scramble for and development of railways demand more research. Railways defined the conflicts among warlords; they would also determine the course of WWII in China and Asia.

'How wars were won and lost on the railways'[116]

> A train is slowly speeding through the night. Suddenly a man appears, runs alongside and hops onto the train. He clings to the narrow bar, moving closer to the lock of a coach, and opens it. From inside the coach, he throws straw-wrapped goods – rifles – to a team below who collect them from the ground. In a short while, he hops off the train and disappears into the darkness.

This famous scene from the *Railway Guerrilla Squad*, a popular Chinese film from the 1950s, was based on an actual raid that took place on the Linzao line, from Lincheng (site of the Lincheng Outrage) to Zaozhuang, a feeder line of the Jiaoji railway, built by the Germans to access the coal mine. Railway guerrillas were small bands of specialised irregulars whose roles consisted of 'destroying tracks, attacking Japanese supply trains, and robbing them of goods and useful freight – weapons, explosives and in some cases even the payrolls' during the Anti-Japanese War.[117] Wherever there were railways, there were railway guerrillas. The significant ones included: the Shandong Railway Guerrilla Squad active along the Jinpu line, the Baoding Railway Guerrillas and the Railway Destruction Squad on the Luhan/Pinghan.[118] Railway-centred movies were not a Chinese invention; motion, intrigue and action defined many classics, including *The Great Train Robbery*, *Trans-Euro Express* and *Murder on the Orient Express*, to name just a few.

Railways were the 'new weapon of war'.[119] Keith Neilson and T. G. Otte have argued that the American Civil War was the 'first proper railway war' in which railroads gave the Union army a 'significant logistical edge over the Confederate forces'.[120] Railways were strategic in armed conflicts, from warring among European powers to the Arab Revolt and the Anglo-Russo scramble for Central Asia.[121] The Anti-Japanese War in China from 1937 to 1945 was a conflict fought on, along and over railways. It is not an exaggeration to conclude that whoever controlled railways could control China. Stephen MacKinnon and Hans van de Ven are among the few who exposed the role of railways.[122] After driving the Russians out of north Manchuria in 1932, the GA and the SMRC prepared for the conquest of China south of the Great Wall. They began by building the Chengtong [承通] line, from Chengde to Tongzhou in Beijing – this explains why they demanded the Taonan to Rehe line back in 1913.[123] Completed in the spring of 1937, it was ready for the Japanese to march into Beijing on 7 July.

The Nanjing government also prepared to resist the Japanese advance by building its own railways. The regime managed to complete the Yuehan railway in September 1936, just before the Japanese marched down to north China.[124] It extended the Longhai line further west and attempted to re-start the Chuanhan railway.[125] Realising that western China, protected by a chain of impenetrable mountains, would become its wartime base, the regime awarded a contract to the French to build the Chengdu to Chongqing railroad. It also hired the Swedish geographer Sven Hedin, who conducted field survey and proposed a railway in the northwest to connect with Central Asia and Europe, but the quick arrival of war and the lack of resources halted this project.[126]

Railways spearheaded the Japanese advance. The GA depended on railways to move, and their supply line stretched from Korea and Manchuria to south China.[127] After they secured Beijing in July 1937, the Japanese moved on along these railways: the Jinpu to Shandong and Jiangsu provinces, the Pinghan to Hubei and Henan, and the Pingsui to Shanxi and Mongolia.[128] Hans van de Ven pointed out that their westward campaign was successful, taking control of the Pingsui line by October.[129] They managed to capture the northern segment of the Jinpu line, while bombing stations in north and east–central China.[130] To take Shanghai, they pacified Jiading first due to its close proximity to rail lines coming out of the city.[131] Although railways made armies mobile, they were also at the mercy of railways, especially in alien and hostile territories. Railways had to be guarded, which meant committing 'vast resources to defending and deterring attacks on railway property'.[132] The battle lines were thus drawn by railroads, and the most effective way to combat the Japanese was to disrupt their supply line by destroying railways.[133]

Chinese historians have compared railways to line and stations to 'dots' on a line, and anti-Japanese campaigns often centred on both the 'dots' and the 'line'.[134] The Xuzhou campaign in the spring of 1938 was, in fact, a battle for control of Xuzhou station and access to the railways it serviced. Xuzhou sat on the crossroad of the Jinpu and Longhai lines, which offered access to the Jiaoji and the Pinghan lines.[135] Hans Van de Ven has detailed the battle for Xuzhou in which the Jinpu and Longhai became battlegrounds.[136] Led by Seishirō Itagaki, former Chief of Staff for the GA, the Japanese expedition forces converged on Xuzhou, flanking the Chinese army, led by General Li Zongren.[137] They attacked Chinese units guarding the railways first. Stephen MacKinnon detailed the battle at Tangxian station and in Tai'erzhuang.[138] Although the Chinese managed to slow the Japanese advance, it did not prevent the latter from ultimately taking Xuzhou, which by then had become a 'ravaged place'.[139] Having procured the Jinpu and Longhai railways, the Japanese progressed onto the Pinghan, that is north–central China.[140]

Another major Chinese campaign focused on the destruction of the Zhengtai railway, a lifeline for the Japanese. Branching out from Zhengding (Shijiazhuang) on the Pinghan, the Russo-Chinese bank-financed and French-built line snaked horizontally west to Taiyuan on the Tongpu line in Shanxi. To quickly transport coal, the Japanese decided to connect Zhengtai with the Jinpu. This led to the horizontal Shide railway, from Shimen on the Pinghan to Dezhou on the Jinpu.[141] The Zhengtai–Shide line is significant because it laced up major lines in north central and east China, allowing Shanxi's coal access to ports in Tianjin, Qingdao and Shanghai. These lines made north–central China look like a grid (see Figure 9.2). The Japanese tactic was to use 'railways as pillars, roads as chains and forts as locks', this was their 'prisoner's cage policy'; cutting Chinese resistance off from each other could isolate and destroy them in the 'cages'.[142]

The Zhengtai railway was the key to the grid. To protect it, the Japanese built a multitude of forts along the tracks and patrolled in armoured vehicles. The Communists and the Nanjing regime had forged an anti-Japanese alliance in 1936, after the Xi'an Incident, instigated by the 'young marshal' Zhang. The CCP's Red Army was re-grouped into the Eighth Route Army (ERA) led by Marshal Zhu De.[143] The ERA launched the Zhengtai railway blitz on 20 August 1940.[144] With the young Deng Xiaoping as one of the leaders, the ERA dismantled both the Zhengtai and the Tongpu railways by uprooting tracks, blowing up bridges and tunnels, and attacking stations. This campaign was examined by Agnes Smedley in her interview with Zhu De.[145] They destroyed the Jingxing coal mine, a key source of fuel for the Japanese.[146] They also interrupted the Pinghan railway.[147] Michael Lindsay's rare photographs bear witness to the ERA's campaigns (see Figures 9.3 and 9.4).

According to the *Herald*, securing railways and guarding them was Japan's main achievement by January 1940:

> The Japanese call Shenxi and Hopei occupied territory, that is to say, territory of which they are in undisputed possession and where their writ runs supreme. In actual fact, this claim is a very considerable exaggeration. For the territory they exercise effective control is limited to a narrow strip of country about ten miles wide running on either side of the somewhat scanty network of railways in north China. Outside that area, they have nothing whatever to say; one enters the territory of so-called still independent China.[148]

Figure 9.3 Eighth Route Army reconnaissance during the Anti-Japanese War (Image courtesy of the Lindsay family and University of Bristol library, HPC, https://hpcbristol.net/visual/ML02-d001, Identifier ML02-d001)

Figure 9.4 Jingxing coal mine and railway guarded by Japanese pillboxes (Image courtesy of the Lindsay family and University of Bristol library, HPC, https://hpcbristol.net/visual/ML04-d040, Identifier ML04-d040)

This report is significant because it reveals that, although railways helped the Japanese advance, their occupation was constrained to the swath of land on both sides of railways that formed the backdrop for the campaigns.[149] The war on railways continued: the *Herald* reported on railway guerrilla activities along the Nanjing-Shanghai-Hangzhou line and so, too, did the *Times* and *Shen Bao* on lines in southern China.[150] It intensified in May 1944, when the *Times* reported that 'Chinese troops have closed in on Pingshek, 165 miles north of Canton, in continuing attacks designed to break Japanese hold on the Canton-Hankow railroad'.[151]

The Nanjing government retreated to Sichuan with Chongqing as its wartime capital. The regime tried to extend existing lines and build new ones in southern and western China, facilitating the movements of troops, weapons, factories, government agencies and refugees.[152] They included the Hunan to Guangxi [湘桂] line, the Guizhou to Guangxi [黔桂] line, the Lanzhou to Xinjiang [陇新] line (westward extension of the Longhai) and the Baotou to Ningxia [包宁] line (southward extension of the Jingzhang-Pingsui line). We know from the Second Historical Archive in Nanjing that the regime used the Boxer Remittance Fund to repair and construct railways.[153] The southwest and the northwest became the rear and backbone of wartime China.[154] As the Japanese pushed down to southern China, the French-built Haiphong to Kunming railway, which was bombed many times, became a lifeline for international aid.[155] The Japanese captured Haiphong in September 1940, forcing the Jiang regime to make a difficult decision. They dismantled the railway bridge that connects Vietnam with China in Hekou and the tracks from Hekou to Mengzi, to prevent Japanese advance.[156] Despite this, the railway led to diplomatic rows as France, Vietnam and China all claimed ownership during and after the war, as Yang Weizhen has revealed.[157]

Wartime China's only line of communication with the Allies was thus reduced to the 1,146 km long Burma Road. Allied aid travelled from Yangon by train to Lashio in the upper Shan State, where it would take weeks for hundreds of trucks to transport them to Kunming, and then Chongqing. This was slow, costly and dangerous as it zigzagged through hazardous mountains and turbulent rivers. A railway from Lashio to Kunming would speed up the process. But such a line presented technical challenges that had deterred the British. The project required funds, materials and tens of thousands of labourers, all in short supply at this point of the war.[158] As Archibald Little had commented earlier, no private company could afford such an endeavour. Only governments could undertake a task of this magnitude under extreme circumstances. The Allies' plan was for Britain and China to build their separate segments to meet on the Sino-Burmese border. In addition, the Jiang regime would need to extend the line from Kunming to Chongqing, another challenge to the war-torn country.

Construction of the Chinese segment of the Yunnan–Burma or Dianmian [滇缅] railway suffered from shortages of money, equipment, material, manpower and food. This was compounded further by the outbreak of epidemics, desertion and indigenous resistance which had plagued the French. Many newspapers reported on these problems.[159] Director-general Du Zhenyuan went to the United States to seek funding.[160] A graduate of the Li Hongzhang-founded Tangshan Railway College who later studied at Cornell University, Du managed to secure a loan.[161] The Yunnan government did its best to supply 300,000 labourers, many of them women and children.[162] By January 1942, the Chinese segment was near completion.[163] But the Japanese advanced faster than the Allies could build railways. December 1941 saw Japan attack not just Pearl Harbour, but also Hong Kong, Malaya, Singapore and Thailand. Their battle lines now stretched across the tropical terrains of Southeast Asia, making supplies, and hence railways, even more important. To ensure the success of the Burma campaign, the Japanese began to build the Thai–Burma railway from Bangkok to Rangoon in early 1942.[164]

Preliminary research suggests that 62,000 European prisoners of war captured in Malaya, Singapore and Indonesia, as well as 182,496 civilians, were forced to work on this line.[165] They suffered from extreme heat, disease, starvation and lack of medical care, inhumane treatment above all at the hands of the Japanese, immortalised in the film *The Bridge on the River Kwai*. Labelled the 'death railway', it is estimated that a total of 54,613 individuals, 12,399 prisoners and 42,214 civilians, lost their lives.[166] The Japanese moved quickly to take Rangoon in March and advanced to Lashio by May 1942. This forced the Jiang regime to make a painful decision about the nearly completed railway. After several years of extremely hard work with tunnels dug, bridges built and lives lost, the railway had to be destroyed in order to prevent it from falling into Japanese hands. Railways were vital to the Japanese, and so they were targeted by Allied campaigns, including Operation Longcloth led by General Wingate.[167] Like the Anti-Japanese War in China, WWII in Southeast Asia was also a war on, along and over railways.[168]

Railways had become tools in region and nation building, with which warlord and political regimes asserted themselves and maintained control, as foreign powers had done, during the ROC era. Such a situation is comparable to that of Central Africa, where 'the railways proved crucial not only in the creation of the [British] empire and efforts to maintain it, but also in shaping the successor states that replaced it'.[169] Railways were also an effective weapon for war, like airplanes and aircraft carriers – modern warfare required modern technology. They were indispensable in war and war reinforced their strategic importance. The Anti-Japanese War was a conflict that was battled out along, on and over railways. How the Japanese Army conquered China railway by railway demands more research. There can be no doubt that the railway was the most strategic weapon for invaders, but it was also an effective theatre of resistance, a lesson the Communists quickly learnt.

Notes

1 John B. Powell, *My twenty-five years in China* (New York: MacMillan, 1945), 93–94; Lucy Truman Aldrich, 'A week-end with Chinese bandits', *Atlantic Monthly*, vol 132 (July–December 1923): 672–86; Paul French, *Carl Crow a tough old China hand: the life, times and adventure of an American in Shanghai* (Hong Kong: Hong Kong University Press, 2008), 113–32; Caroline Reeves, 'Holding hostages in China, holding China hostage: sovereignty, philanthropy, and the 1923 'Lincheng Outrage'', *Twentieth Century China*, vol 27, 1 (2001): 39–69; James M. Zimmerman, *The Peking Express: the Bandits who stole a train, stunned the West, and broke the Republic of China* (New York: World Affairs, 2023).

2 'The Lincheng outrage', *Herald*, 12 May 1923: 394; 'Jinpulu dajie'an', *Shen Bao*, 15 May 1923: 6–7.

3 'Tientsin-Pukou express wrecked by bandits', *Herald*, 12 May 1923: 380 and 'Lincheng and other outrages', 23 June 1923: 827.
4 'Jinpulu zhongduan', *Shen Bao*, 21 April 1922: 6.
5 Sun Zhongshan, *Jianguo fanglue* (Beijing: Zhongguo chang'an chubanshe, 2011).
6 'Dr. Sun Yat-sen: the president's oath', *Herald*, 6 January 1912: 28.
7 'Zhuandian', *Shen Bao*, 16 January 1912: 2 and 'The abdication', *Herald*, 17 February 1912: 418 and 440.
8 She Jiangdong, 'Lun Zeng Kunhua zaoqi tielu guanli sixiang', *Jindaishi Yanjiu*, vol 4 (1989): 96–105.
9 John Foord, 'Planning a great railway system for the Chinese empire', *Times*, 6 October 1912: SM4.
10 Sun Zhongshan, *Sun Zhongshan quanji* (3 vols, Beijing: Zhonghua shuju, 1982), vol 2, 383.
11 Richard White, *Railroads: the Transcontinentals and the making of modern America* (New York: W. W. Norton, 2011) and William G. Thomas, *The Iron Way: railroads, the Civil War, and the making of modern America* (New Haven: Yale University Press, 2011), 10.
12 Thomas, *Iron way*, 10.
13 Sun Zhongshan, *The international development of China* (Taipei: China Cultural Service, 1953).
14 C. Martin Wilbur, *Sun Yatsen: the frustrated patriot* (New York: Columbia University Press, 1976), 23–26.
15 Sun, *Quanji*, 391, 421, 455 and 463.
16 George E. Anderson, *Railway situation in China* (Washington: Government Printing Office, 1911), 31.
17 Sun, *Fanglue*, 92–96.
18 Murat Ozyuksel, *The Hejaz railway and the Ottoman empire: modernity, industrialization and Ottoman decline* (London: I. B. Tauris, 2014).
19 Spiers, *Engines*, 2–21; Hugo Silveira, 'History, technology, and politics: the political use of railways in Portugal 1850–1900', *Historia Contemporanea*, vol 69 (2022–06): 507–43; Anne Hardgrove, 'Roads, railways and the language of politics', *The Journal of Caribbean History*, vol 52, 1 (2018): 30–67.
20 Sun, *Fanglue*, 93.
21 Sun, *Fanglue*, 186–94.
22 Sun, *Fanglue*, 95.
23 Jia, *Naxieren*, 110.
24 Sun, *Fanglue*, 194–201.
25 Wilbur, *Frustrated patriot*, 23.
26 Wilbur, *Frustrated patriot*, 172–76.
27 Wilbur, *Frustrated patriot*, 172–73.
28 'China's resources', *Times*, 7 October 1923: E6.
29 Sun, *Fanglue*, 145–51 (146).
30 Li En'han, *WanQingde shouhui kuangquan yundong* (Taibei: Zhongyanyuan jinshisuo, 1963).

31 Sun, *Fanglue*, 162–71.
32 Sun, *Fanglue*, 177–85.
33 Sun, *Quanji*, vol 2, 457.
34 'America will help China build railroads', *Times*, 17 December 1916: SM7.
35 Sun, *Fanglue*, appendix, 223–27.
36 Wilbur, *Frustrated patriot*, 75.
37 'Minister of reconstruction wants capital', *Times*, 1 August 1928: 4.
38 Wilbur, *Frustrated patriot*, 75.
39 David Strand, *An unfinished republic: leading by word and deed in modern China* (Berkeley: University of California Press, 2011), 240.
40 Zhu Jianjun, *Dezhan Qingdao 1897–1914* (Beijing: Renmin chubanshe, 2010), 268–74 (271).
41 Mühlhahn, 'Negotiating the nation', 37–56 (53).
42 Marius B. Jansen, *The Japanese and Sun Yat-sen* (Cambridge: Harvard University Press, 1954).
43 Xavier Paules, 'Warlords at work: four crucial realms and four dynamics of state building in Republican China, 1916–1937', *Twentieth Century China*, vol 47, 1 (January 2020): 40–49.
44 Thomas, *Iron way*, 5.
45 Walter H. Mallory, *China: land of famine* (New York: American Geographical Society, 1926).
46 Chang Jui-Te, 'Technology transfer in modern China: the case of railway enterprise (1876–1937)', *Modern Asian Studies*, vol 27, 2 (May 1993): 281–96.
47 Gong Jianlin *et al.*, *Tushuo zhongguo tielu shihua* (Beijing: Tiedao chubanshe, 2011), 28–29.
48 'Chang Tso-Lin the uncrowned king of Manchuria', *Herald*, 18 October 1919: 143.
49 'Liaoyang mazei jinzhuang', *Shen Bao*, 15 July 1906: 3 and 'Dongsheng zoubao', 4 February 1910: 5.
50 'Zhang Zuolin ren xunyueshi zhiyoulai', *Shen Bao*, 12 September 1918: 6.
51 'Zhujun, Zhonghua minguo shiji: Zhongdong tielu hulujun', Shuzi fangzhi, Shengzhi, Junshizhi, Heilongjiang shizhiwang, at www.hljszw.org.cn/bookinfo/201708230037.html, accessed 4 November 2021.
52 Leong, *Diplomatic relations*, 113–15.
53 'The President of the Technical Board (Stevens) to the Secretary of States, Washington, March 15, 1923', No. 861.77/3045, Office of the Historian, https://history.state.gov/historicaldocuments/frus1923v01/d607, accessed 4 November 2021.
54 Leong, *Diplomatic relations*, 131; Davies, *Dragon*, 111–12 and 131; Ma Jiangyun, 'Guojia liyi bianqi yu Su'e zaizhongdong tielu wentishang taidude bianhua', *Zhonggong Dangshi*, 6 (2013): 49–59. 'Su'E zhengfu 1919nian duihua xuanyan', Beiyang zhengfu waijiao dang'an: Zhong'E guanxi: number 03-32-559-02-042, Institute of Modern History Archive (Taipei), at https://archivesonline.mh.sinica.edu.tw/byfonds/detail/volume/?&level=03-32-559-02, accessed 8 July 2023.

55 Zhou Yuefeng, 'Liening shike', *Qinghua Daxue Xuebao*, 5 (2017): 113–28.
56 Elleman, 'Secret diplomacy'.
57 Leong, *Diplomatic relations*, 147–48.
58 MacMurray, *Treaties (1919–29)*, 29–31.
59 MacMurray, *Treaties (1919–29)*, 30.
60 'Zhong'E huiyi geqianzhi donglu wenti', *Shen Bao*, 7 December 1923: 4.
61 MacMurray, *Treaties (1919–29)*, 133–40.
62 'Agreement for the provisional management of the CER, with declaration, May 31, 1924', in MacMurray, *Treaties*, 141–44 (142).
63 'Agreement for the provisional management of the CER, with declaration, May 31, 1924', in MacMurray, *Treaties*, 149.
64 'Agreement for the provisional management of the CER, with declaration, May 31, 1924', in MacMurray, *Treaties*, 150.
65 Chang, *International controversies*, 110–34.
66 Leong, *Diplomatic relations*, 93 and 111.
67 Bruce A. Elleman, 'Sino-Soviet tensions and Soviet administrative control over the Chinese Eastern Railway, 1917–25', in *Manchurian railways*, 59–76; Vul, 'He, who has sown the wind: Karakhan'.
68 Blaine Chiasson, *Administering the colonizer*; Yiwei Cheng, 'Coping with parallel authorities: the early diplomatic negotiations of Soviet Russia and China on the Chinese Eastern Railway, 1917–1925', *Modern Chinese History*, vol 9, 2 (July 2015): 223–43.
69 Jun Uchida, '"A scramble for freight": the politics of collaboration along and across the railways tracks of Korea under Japanese rule', *Comparative Studies in Society and History*, vol 51, 1 (2009–01): 117–50.
70 Matsusaka, *Japanese Manchuria*, 273; 'Jialahan fandui Taoqilu', *Shen Bao*, 14 May 1925: 9 and 'Riren nuli fushe Taoqilu', 26 July 1925: 6.
71 'Lianluo Jitongxian yu Zhongdongxian', *Shen Bao*, 24 October 1929: 9.
72 Matsusake, *Japanese Manchuria*, 273–75.
73 Matsusake, *Japanese Manchuria*, 294.
74 McCormack, *Chang Tso-lin*, 45–52 and 95–99.
75 Matsusake, *Japanese Manchuria*, 289–91.
76 '"Time-worn" treaties no bar to Chinese', *Times*, 15 January 1928: 34.
77 Hallett Abend, 'Chinese railways stripped of cars', *Times*, 8 August 1928: 40.
78 Yi Binglan, *Fengxi yu dongbei tielu* (Beijing: Shehui kexue chubanshe, 2018), 107–33; Paules, 'Warlords at work'.
79 Yi Binglan, 'Dongbei tielu zizhuhuade kaiduan – Fenghai tielu', *Dongbei Shidi*, 6 (2012): 70–76.
80 Clarence B. Davis, 'Railway imperialism in China, 1895–1939', in *Railway imperialism*, 155–73 (167); 'Rifeng guanxi riqu xian'e', *Shen Bao*, 13 April 1928: 9; 'Sino-Japanese bone of contention', *Herald*, 1 Oct 1927: 8.
81 'Fengzhangzhi tielu zhengce', *Shen Bao*, 19 May 1925: 4.
82 'Youchuanbu zoujieshou Xinfeng tieluzhe', *Shen Bao*, 22 November 1907: 10.
83 Coox, 'Kwantung army', in *Informal empire*, 404–7; 'FengZhang jinchen yuzha zhongshang', *Shen Bao*, 5 June 1928: 4 and 9; Paul S. Dull, 'The

assassination of Chang Tso–lin', *The Journal of Asian Studies*, vol 11, 4 (August 1952): 453–63; Hiroaki Kuromiya, *Stalin, Japan and the struggle for supremacy over China, 1894 to 1945* (Abingdon: Routledge Open History, 2023), 117–24.

84 Guo Junsheng, *Zhongdonglu yu zhongdonglu shijian* (Shenyang: Liaoning renmin chubanshe, 2010), 114.

85 'Habu gejie zhenglu yulie', *Shen Bao*, 14 November 1928: 10.

86 'Gejie dui'E wendian', *Shen Bao*, 21 July 1929: 13 and 'Fan'E minyi ceyan', 28 July 1929: 13.

87 'Manchuria seizes 3 Soviet officials', *Times*, 5 June 1929: 6 and 'New China and red Russia come to grips', 21 July 1929: 113.

88 S. C. M. Paine, *The wars for Asia, 1911–1949* (Cambridge: Cambridge University Press, 2012), 85–87; 'Donglu shijian xingshi yanzhong', *Shen Bao*, 16 July 1929: 4–7.

89 Chang, *International controversies*, 144 and Walker, *The war nobody knew*.

90 Chang, *International controversies*, 144–64.

91 Greg Kennedy, 'Barometer of war: British views of imperial security in the Far East, Russo-Japanese relations and the Chinese Eastern Railway, 1929–1935', in *Paths of empire*, 192–216 (202).

92 Kennedy, 'Barometer of war', 210.

93 'Mukden prepared to recognise Habarovsk Protocol', *Herald*, 4 February 1930: 171, and 'Sino-Russian parley', 18 November 1930: 222.

94 The western half the island, an area of about 171 km^2, was returned to China in 2008.

95 Paine, *Wars for Asia*, 89–90.

96 George A. Lensen, *The damned inheritance: the Soviet Union and the Manchurians crises 1924–1935* (Tallahassee: The Diplomatic Press, 1975).

97 Matsusaka, *Japanese Manchuria*, 368.

98 Takeo, *Life along*, 135; O'Dwyer, *Significant soil*, 277–78.

99 'Xinmin shixian', *Shen Bao*, 9 December 1931: 8, 'Jinzhou shixian', 7 January 1932: 17 and 'Rijun qiangzhan Beining tielu', 13 January 1932: 8; 'Japanese occupy Manchuria', *Herald*, 22 September 1931: 398 and 'The fighting in Manchuria', 24 November 1931: 261 and 1 December 1931: 296.

100 'Er'wei shuilu huiyi shuangfang zhengduo hexiazidao', *Shen Bao*, 11 January 1935: 8.

101 'Jinshen qingxiu tielu', *Shen Bao*, 30 October 1905: 4 and 'Jianzhu Tongpu tieluzhi husheng', 14 June 1929: 8.

102 Donald G. Gillin, *Warlord Yen Hsi-shan*; 181–85; 'Tongpulu jiangkaiche', *Shen Bao*, 24 October 1933: 7.

103 'The Chen-tai railway', *Herald*, 12 May 1903: 492.

104 'A journey to Shenxi', *Herald*, 28 December 1932: 490.

105 'Yan Xishan di Jiexiu shicha', *Shen Bao*, 1 December 1933: 4; Jing Zhankui, *Yan Xishan yu Tongput tielu* (Taiyuan: Shanxi renmin chubanshe, 2003), 36–40.

106 Tawney, *Land and labour*, 120–22.

107 Köll, *Transformation*, 69–71.

108 Spiers, *Engines*, 115–43.
109 Edna L. Booker, *News is my job: a correspondent in wartorn China* (New York: Macmillan, 1940), 58–60.
110 'Huochezhi jieyun', *Shen Bao*, 26 March 1925: 3.
111 'War's heavy hand on Chihli', *Herald*, 28 November 1925: 372.
112 Diana Lary, *Warlord soldiers: Chinese common soldiers 1911–1937* (Cambridge: Cambridge University Press, 1985), 5.
113 Pierre Fuller, *Famine relief in warlord China* (Boston: Brill, 2019), 241–53.
114 Thomas, *Iron way*, 105.
115 'Manchurians forces at gates of Peking', *Times*, 22 September 1930: 1.
116 Wolmar, *Engines of war.*
117 Köll, *Transformation*, 199.
118 'Fang Pinghan tielu puohuaidui laoduiyuan Zhou Yusong', *Minzhu Yu Fazhi Shibao*, 23 January 2006; 'Guerillas active round Peiping', *Herald*, 1 February 1939: 184 and 'Guerillas wreck train on Nanking railway', 13 December 1939: 445.
119 Wolmar, *Engines of war*, 132.
120 Neilson and Otte, 'Railpolitick', 10.
121 Showalter, 'The Prussian army', in *Paths of empire*, 21–44; Thomas, *Iron way*; Neil Faulkner, *Lawrence of Arabia's war: the Arabs, the British, and the remaking of the Middle East in WWII* (New Haven: Yale University Press, 2016), 302–22; Wyatt, *Afghanistan*, 62–64, 80, 105–9, 116 and 162.
122 MacKinnon, *Wuhan*, 31–43; Van de Ven, *China at war*, 66–70 and 81–85.
123 Coox, 'Kwantung army', in *Informal empire*, 414–17.
124 'Heavy fighting in north China', *Herald*, 14 July 1937: 52 and 'The Loukouchiao incident', 28 July 1937: 165.
125 Thomas Spain and Oliver Betts, 'Developing China's 'international' railway: the Canton-Hankow line, 1898–1937', *Journal of Transportation History*, vol 40, 3 (December 2019): 322–40.
126 'ZhongRui taizu xibei kexue kaochatuan', *Shen Bao*, 10 May 1927: 9; 'Three routes link China with world', *Times*, 20 August 1939: E5.
127 'Chuankou shituan', *Shen Bao*, 15 July 1937: 4.
128 'Japanese control Hsuchow', *Herald*, 25 May 1938: 312; 'Chengchow falls', *Times*, 10 June 1938: 10 and 'Japanese drive for Hankow', 27 July 1938: 134; Xie Zhonghou *et al.*, *Huabei kangri zhanzhengshi* (Baoding: Hebei renmin chubanshe, 2012), 254–55.
129 Van de Ven, *China at war*, 66.
130 Van de Ven, *China at war*, 69–70.
131 Timothy Brook, 'The pacification of Jiading', in *The scars of war: the impact of warfare on modern China*, eds Diana Lary and Stephen MacKinnon (Vancouver: UBC Press, 2001), 50–74.
132 Spiers, *Engines*, 135.
133 'Mobile arsenals in Shanxi', *Herald*, 19 April 1939: 90.
134 Zhongguo shekeyuan jinshisuo, *Riben qinhua qishinian shi* (Beijing: Sheke chubanshe, 1992), 486.
135 'Hsuchow bombed by Japanese', *Herald*, 29 September 1937: 475; 'Jinpu zhengmian ke hanzhuang', *Shen Bao* (Hankou edition), 19 April 1938: 1.

136 Van de Ven, *China at war*, 81–83.
137 'Guandongjun canmouzhang Banyuan', *Shen Bao*, 25 February 1937: 7.
138 MacKinnon, *Wuhan*, 31–43; Köll, *Transformation*, 196.
139 Diana Lary, 'A ravaged place: the devastation of the Xuzhou region 1938', in *Scars of war*, 98–116.
140 'Huabei youjidui huoyue gelutiegui', *Shen Bao*, 1 March 1939: 3; 'Guerillas blow up train at Hsuchow', *Herald*, 3 January 1940: 2; 'China still holds rails', *Times*, 23 July 1942: 7 and 'New rail centers seized by Chinese', 23 August 1942: 9.
141 'Shide tieluzhi jihua', *Shen Bao*, 7 February 1908: 12 and 'Tan Huabei xianzhuang', 12 August 1942: 5.
142 Shekeyuan jinshisuo, *Qishinian*, 490–94; Kangzhanshi bianxiezu, *Zhongguo kangri zhanzhengshi* (Beijing: Renmin chubanshe, 2011), 417–21.
143 'Chinese Reds made 8th Route Army', *Herald*, 15 September 1937: 394.
144 'Chinese forces attack railway', *Herald*, 28 August 1940: 319 and 'Guerillas active in north China', 9 October 1940: 43.
145 Agnes Smedley, *The great road: the life and times of Chu Teh* (New York: Monthly Review Press, 1956), 362–63.
146 'Eighth Route Army in action', *Herald*, 18 September 1940: 436.
147 'Pinghanlu jiaotong zhongduan', *Shen Bao*, 24 August 1940: 4.
148 'Guerilla warfare in north', *Herald*, 10 January 1940: 38.
149 'Japanese machine attacks railway line', *Herald*, 3 November 1937: 179 and 'Concerted attack on railways', 8 March 1939: 405.
150 'Yue youjidui weixie Rijun', *Shen Bao*, 20 January 1939: 8; 'Guerillas mining local railways', *Herald*, 21 May 1941: 283; 'Japanese closer in Kiangsi drives', *Times*, 17 June 1942: 1.
151 'Two-way push opens on China railway', *Times*, 4 May 1944: 9, 'Japanese gain in China', 25 January 1945: 12, 'China cut in two by railway loss', 1 February 1945: 11, and 'Chinese attack on railway', 8 March 1945: 13.
152 'Chinese built new rial link', *Times*, 17 February 1942: 7 and 'Xibu tielu jianshe', *Shen Bao* (Hong Kong edition), 29 March 1939: 3.
153 'Jiechu gengkuan hezhang qingdan', Jiaotongbu, file number 20.2.1653: 17, 24, 37, 118, 122, The Second History Archive, Nanjing, China.
154 Li Ziming, *Huoche shangde minguo* (Beijing: Tiedao chubanshe, 2014), vol 1, 84–91.
155 'Japan deliver secret ultimatum to French', *Times*, 11 January 1940: 1, Hallett Abend, 'Vital munitions kept from China', 20 March 1940: 17 and R. Tillman Durdin, 'Japan may attack French Indo-China', 16 June 1940: 36; 'Bombing of Mengzi', *South China Morning Post*, 15 June 1941: 14.
156 Zhuang Xingcheng *et al.*, *Dianyue tielu shiliao huibian* (Kunming: Yunnan renmin chubanshe, 2014); Wang Yuzhi, *Dianyue tielushi yanjiu* (Beijing: Renmin chubanshe, 2020).
157 Yang Weizhen, 'Postwar Sino-French negotiation about Vietnam, 1945–1946', in *Negotiating China's destiny in WWII*, eds Hans van de Ven *et al.* (Stanford: Stanford University Press, 2014), 205–19.
158 'Fazhan Xinan jiaotong xingzhu Chuandian tielu', *Shen Bao*, 25 October 1938: 10 and 'Jiaotongbu donggong xingzhu liangda tielu', 7 November

1938: 10; Li Nengyan, 'Kangzhan shiqi Xinan jiaotong jianshe', *Yuxi Shifanxueyuan Xuebao*, vol 35, 1 (2019): 30–37; Emma Reisz, 'Projecting the road: topological photograph on the Yunnan-Burma frontier', *The Chinese Historical Review*, vol 25, 2 (July 2018): 143–62.

159 'Work began on Yunnan Burma railway', *Herald*, 15 February 1939: 274, 'Work on Yunnan-Burma line', 15 November 1939: 274 and 'China, Burma railway to be speeded', 15 January 1941: 86.

160 'The Ambassador in China (Johnson) to the Secretary of State, Chungking, March 26, 1941', No. 845C.7793/10: telegram, Office of the Historian, at https://history.state.gov/historicaldocuments/frus1941v05/d535, accessed 22 July 2021.

161 'U.S. loan for Burma railway', *Herald*, 26 March 1941: 474.

162 Hellmut De Terra, 'Chinese rush road to Burma railway', *Times*, March 16, 1938: 12; Cui Gang, 'Dianmian tielu yu kangzhan jingshen', *Xinan Jiaotong Daxue Xuebao Shehui Kexueban*, vol 16, 5 (2015): 8–15.

163 'Dianmian tielu gongcheng datiwancheng', *Shen Bao*, 9 January 1942: 2.

164 Peter Davies, *Across the Three Pagodas Pass: the story of the Thai-Burman railway* (Amsterdam: Amsterdam University Press, 2022).

165 Paul Kratoska, 'Labor in the Malay peninsula and Singapore under Japanese occupation', in *Asian labour*, 237–48 (242).

166 Kratoska, 'Labor in the Malay peninsula'. See also 'Burma-Thailand railway and the Hellfire Pass, 1942 to 1943', Australian Government Department of Veterans' Affairs, at https://anzacportal.dva.gov.au/wars-and-missions/burma-thailand-railway-and-hellfire-pass-1942-1943, accessed 2 December 2023.

167 'British in 3–month Burma foray', *Times*, 21 May 1943: 1 and 'Chindits in Burma', 3 May 1944: 1.

168 'Foe near Imphal', *Times*, 8 April 1944: 1 and 'Burma Road terminus', 8 March 1945: 13; Michael A. Charney, *Imperial military transportation in British Asia: Burma 1941–1942* (London: Bloomsbury Academic, 2019).

169 Hanes III, 'Railway Politics', in *Railway imperialism*, 41–69.

10

'High speed empire': China dreams[1]

On 6 December 1949, fifty-three years after Li Hongzhang's journey to Russia for the coronation of Tsar Nicholas II, Mao Zedong boarded a train in Beijing.[2] He was travelling to the Soviet Union for Joseph Stalin's 70th birthday. Mao's willingness to make the long journey indicates that Sino-Russian relations had not changed despite the ideological alignment. His train travelled along the Barrier railway, then on the SMR and CER, changing at Harbin before joining the Trans-Siberian line and arriving at Moscow on 16 December. Mao fell in love with railway travel. From then on, he would journey around China on a train that bore his name, where he was often depicted reading, writing, smoking and relaxing, even in his pyjamas. Mao had the last laugh as the People's Republic of China (PRC) inherited all railways on Chinese soil, regardless of who built them. Examining the CCP's endeavour will expose where its predecessors failed. The ability to envision, invest and coordinate at the national level, to create research/education facilities and command a disciplined labour force, made an enormous difference. The CCP used railways to project itself onto the international stage during the Cold War and dominate the world economy today. The current mania for highspeed rail opens a unique window onto the CCP's global ambition.

Back to the 'cradle of conflict'

Historians have acknowledged that the communist revolution began with labour activism.[3] It actually began on railways and this would change the course of the revolution. In the spring of 1920, the Beijing Marxism Study Group was formed with Li Dazhao, a librarian at Peking University, as its leader. Two members, Deng Zhongxia and Zhang Guotao, were sent to Changxindian, the first station on the Pinghan line and home to railway workers. Hans van de Ven propounded this when discussing communist cells in the early 1920s.[4] They opened a school and taught the workers how to read and write, but this was a cover for spreading Marxism and organising them into unions. Jia Benyi called it the 'battle for railway workers'.[5] This course of action was propagated at the CCP's first congress held in July 1921 and endorsed by the Communist International. At the latter's Far Eastern meeting in January 1922, Lenin received three Chinese delegates; one of them was Deng Pei, a railway worker. This railway underground work soon expanded to other stations, particularly at Zhengzhou, where the horizontal Longhai crosses the vertical Pinghan, making it the ideal battleground for agitation. They led a series of strikes along these railways, all put down by the Beiyang regime, culminating in the '7 February Massacre' in 1923.[6] Hundreds lost their jobs, their unions demolished and leaders killed.[7]

The CCP's railway-anchored revolution had failed and with it the urban revolution. The young CCP practised Marxism by the book; they weaponised the railway and used it to rouse the class consciousness of the proletarian. The party saw railway and mining workers as the main political force in its quest for power. These intellectual revolutionaries failed to comprehend that although many of them supported the struggle for better pay and working conditions, few would commit at the cost of their jobs. Forming a kind of aristocracy among the emerging working class, railway workers had less to gain with strikes but much more to lose. However, this early failure allowed a different voice to emerge – Mao's idea of rural revolution, where peasants would form the backbone. This rural revolution began with the Nanchang uprising in August 1927 – the CCP chose Nanchang because of the Nanxun railway as it helped them prepare in secret and transport soldiers.[8] When their rural revolution failed in 1934, the CCP embarked on the Long March; they marched into areas where there were no rail networks, hence difficult for the Nanjing regime to reach them. Moscow became increasingly important in the CCP's struggle for survival; the CCP's 6th Congress was held there.[9] More than 140 delegates journeyed to and from the USSR via the CER. When Zhang Zuolin and Zhang Xueliang did their best to reclaim the CER and expel the Russians, the CCP allied with the 'wolf', to invoke the earlier comparison. This takes us back to Manchuria where it all began.

On the evening of 8 August 1945, a few hours before the US dropped the second atomic bomb on Nagasaki, nearly 150,000 Soviet troops mounted a huge offensive on Japanese-occupied Manchuria.[10] The Russians were fulfilling the Yalta Protocol – 'in two or three months after Germany has surrendered and the war in Europe is terminated, the Soviet Union shall enter war with Japan'.[11] Their aim was to regain their old possessions – the CER and Port Arthur.[12] They quickly defeated the Japanese and hoisted Soviet flags in Harbin, along the stations, and in Port Arthur. Forty years since the Russo-Japanese War in 1905, and thirteen years after Japan captured

Harbin in 1932, the Russians had taken back what they thought belonged to them. They immediately changed the gauge to Russian standard, making it difficult for the Barrier railway to access Manchuria. Japanese defeat did not liberate Manchuria, nor did it see the return of the CER–SMR to China, as Article Two of the Yalta Protocol stipulates: 'The former rights of Russia violated by the treacherous attack of Japan in 1904 shall be restored.' Two of the three terms dealt with Dalian–Port Arthur and the railways:

(b) The commercial port of Dairen shall be internationalised, the pre-eminent interests of the Soviet Union in this port being safeguarded and the lease of Port Arthur as a naval base of the U.S.S.R. restored.
(c) The Chinese-Eastern Railroad and the South Manchurian Railroad, which provide an outlet to Dairen, shall be jointly operated by the establishment of a joint Soviet-Chinese company, it being understood that the pre-eminent interests of the Soviet Union shall be safeguarded and that China shall retain sovereignty in Manchuria.[13]

Despite siding with the Allies and making huge sacrifices during WWII, China found itself once again in a situation similar to that with Shandong and Jiaoji railways after WWI. Wu Sufeng has discussed Jiang Jieshi's frustration with this.[14] A week after the Soviet attack, the USSR and the Jiang regime signed the *Sino-Soviet Treaty of Friendship and Alliance*.[15] The Treaty specified that they would joint-manage the railways and Port Arthur for thirty years. Once again, 'friendship' and 'alliance' had come at the cost of China's sovereignty as it was clear that the railways were 'largely in Soviet hands'.[16] The Protocol stipulated that Soviet troops were to withdraw from Manchuria three months after the Japanese surrender. However, Russian withdrawal had always been a tricky business. The only difference this time was that Anglo-Russo rivalry had been replaced by Soviet–US competition. Under the pretext of withdrawal, Soviet troops began to remove whatever they could lay their hands on. Their plunder was investigated and documented by Edwin Pauley in 1946, and the results are staggering, particularly in terms of railways:

1. Railway shop capacity was removed by an estimated 50 percent.
2. The Chinese National Government is unable to account for 90 percent of the rolling stock which was in Manchuria at the time of Japanese surrender.
3. Reports from Chinese Nationalist state that 15 percent of the trackage in Communist-held territory was torn up and removed by the Soviets.
4. The estimated total loss was $221,340,000.[17]

The Pauley Mission reported that 'Railway communication[s] are at a low ebb. Principal causes are: Soviet removal of rolling stock and of machine tools from repair shops; the destruction by Chinese Communists of many bridges and culverts.'[18] The trail of Russian destruction stretched far beyond the railways to other industrial assets, such as electrical infrastructure, coal, iron, steel, metals, chemicals, cement, lumber and liquid fuel; they also took $3,000,000 gold bullion, 591,030,000 Manchurian *yuan*, as well as textiles and food stuff.[19] This inadvertently exposed how much the Japanese had managed to amass. The land had enriched the Japanese, and now the Soviets, who were delaying their withdrawal, still stationed along the railways and in the concession zones. Historians have traced the origin of the Cold War in Asia to the Korean War.[20] It really began with the Soviet presence in Manchuria. Just as Russia's refusal to withdraw in the aftermath of the Boxer Rebellion had led to the Russo-Japanese War, the slow Soviet exit unsettled the United States and complicated the Nationalist-Communist battle for China.

The GMD and CCP were ready to settle their scores after the Japanese defeat. Their contest for supremacy began in Manchuria, the most industrialised region with the most developed railway network. On the eve of Japanese surrender, GMD forces were scattered in southern China and Southeast Asia fighting with the Allies. The CCP armies based in the north and northwest were geographically closer to Manchuria. Victor Cheng has detailed how a small army led by Commander Zeng Kelin captured the Barrier railway and rode into southern Manchuria by August 1945.[21] Zeng's unit was followed by others, including the Fourth Field Army led by General Lin Biao in September. To catch up with the CCP, Jiang Jieshi asked the United States to ship and airlift GMD troops to Manchuria.[22] The US could undertake this easily as the Marine Corps had sailed into Qingdao in September 1945, occupying the city and other locations until June 1949. This manoeuvre violated the Yalta Protocol, which stated that 'the pre-eminent interest of the Soviet Union shall be safeguarded'.[23] It agitated the Soviets who soon strengthened their forces along the railways. *Shen Bao* called this the 'darkest moment' and believed another world war was brewing.[24] It also led to CCP protest in March 1946.[25] The contest between the two parties began on the Barrier railway in April 1946, while George Marshall, the US special envoy, was still in China trying to negotiate a coalition government:[26]

> Chinese Communist troops have begun an 'all-out offensive' against the Peiping-Mukden railway, in north-eastern China, in a bid to prevent Chinese government troops from taking over in Manchuria ... Thousands of Communist soldiers are attacking Suichung, thirty-seven mile northeast of Shanhaiguan (about halfway between Peiping and Mukden where the railway runs along the coast of the Gulf of Liaotung) ... Fengtai, China's most important railway junction just south of Peiping, was also being attacked ... More

than 15,000 Communists, it is reported, have already entered the vital central Manchurian railway junction of Harbin.[27]

The old Barrier railway, with its construction interrupted by the First Sino-Japanese War, the suppression of the Boxers and the Russo-Japanese War, continued to be the battleground for conflicts as it had been during the warlord era. Knowing the crucial role railways would play in the battle to come, the CCP quickly assembled a special unit called 'Railway Guards'. Engineers and conductors who had worked for the CER and SMR were absorbed into the unit, and they quickly trained soldiers for work on railways as Steven Levine pointed out.[28] To retaliate against the US for its involvement in Manchuria, Soviet troops gave CCP forces not just access to railways but also weapons, leading to accusations from the GMD and the Americans.[29] With Soviet backing, the Fourth Field Army secured north Manchuria with Harbin as its headquarters when the GMD forces arrived in south Manchuria.[30] The two forces engaged in battles on and along the SMR-CER networks. Harold Tanner is one of the few who has written about the role of railways in this conflict, focusing on the battle for Siping in 1946, where CCP forces sabotaged railways to prevent GMD troops from moving northwards.[31] When GMD forces repaired the Changchun to Mukden line in June 1947, the CCP's Railway Guards destroyed it in October.[32]

The situation was precarious. South Manchuria was in the hands of GMD forces, with Americans standing by, whereas the north was occupied by CCP troops, with the help of Soviet Union.[33] Railways had turned Manchuria into a 'bridgehead' in the 1890s, and the region remained a 'cradle of conflict' in the 1940s. By the summer of 1948, the CCP forces were transformed into the People's Liberation Army (PLA), and they were ready to launch offensives against the GMD in south Manchuria. The Railway Guards grew into the Railway Column. Tanner highlights the importance of railways in the showdown: 'Railways shaped the field of battle in Manchuria, determining how both Communist and Guomindang commanders deployed their forces, the routes along which they chose to advance, and the strategic points that they fought to control.'[34] Working at night, the Railway Column cut deep into GMD-occupied south Manchuria, destroying GMD supply lines and restoring damaged lines to enable the PLA to surprise the unprepared GMD forces the next morning.[35] Their work was instrumental in the battle for Jinzhou, a keystone in the Liaoshen campaign, in which the PLA secured southern Manchuria by November 1948.

The CCP had won the battle for Manchuria which Mao called the 'first son of the People's Republic'. Chen Lixu has claimed that the CCP actually considered making Harbin its capital in 1948.[36] The CCP brought in leaders of other political parties and held the preparatory meeting for the First

Political Consultative Conference there in 1948. Steven Levine has characterised the CCP victory in Manchuria as an 'anvil'.[37] The region's human resource, industrial power and railways would support their southward march. With Manchuria under its belt, the PLA launched the Pingjin and the Huaihai campaigns. The former was conducted along the Barrier railway in the hope of taking Beijing and Tianjin, whereas the latter was fought along the Jinpu, with a focus on Xuzhou where the GMD had battled against the Japanese in 1938.[38] The Railway Column was fundamental to PLA victories. Their slogan exemplified this: 'Wherever the field army marches to, there we construct railways.'[39] No wonder Marshal Zhu De called them the 'People's Iron Army'.[40] Mao remarked that the 'Railway Column is what the PLA could not do without'.[41] It was imperialists-built railways that carried the PLA to a quick and decisive victory in Manchuria and north China.

Following their victory, Mao Zedong and the CCP's top leaders boarded a train at the Gaopiedian station of the Western Tombs line on 25 March 1949. The Empress Dowager's private railway had now come to serve the CCP leadership. They rode on the Gaoyi and old Luhan railways to Beijing.[42] Soon after the establishment of the PRC, Mao was eager to visit Stalin, bringing us back to his trip to Moscow. After gifts were presented and the other guests had departed, Mao stayed behind.[43] He did not take his leave until the *Sino-Soviet Treaty of Friendship, Alliance and Mutual Assistance* was signed on 14 February 1950.[44] Mao negotiated fiercely with his hosts.[45] The new treaty would see the return of the CER–SMR and Port Arthur, as published in the 'Communique on Railway':

> In accordance with the relations of friendship and collaboration which developed between the Soviet Union and the Chinese People's Republic, there was signed in Moscow, 14 Feb 1950, an agreement on the Chinese Changchun railway on the strength of which the Soviet government handed over without compensation to the government of the Chinese People's Republic for its complete disposal all its right concerning joint-administration of the Chinese Changchun railway with all properties appertaining to the railway. [46]

Li Hongzhang had his revenge five decades after his death. The old CER was returned to China on 31 December 1952 in a ceremony led by Premier Zhou Enlai. It was renamed the Chinese Changchun railway [中长], Changchun being the point of breakage after the Russo-Japanese War. Mao Zedong had once compared the PRC to a blank piece of paper upon which one might write the most beautiful words and draw the most delightful pictures.[47] But the PRC was certainly not a blank slate regarding railways (see Figure 10.1).

The various lines built by the late Qing, the foreign powers, the warlords, the Beiyang and Nanjing regimes had given the CCP a good start. Writing in

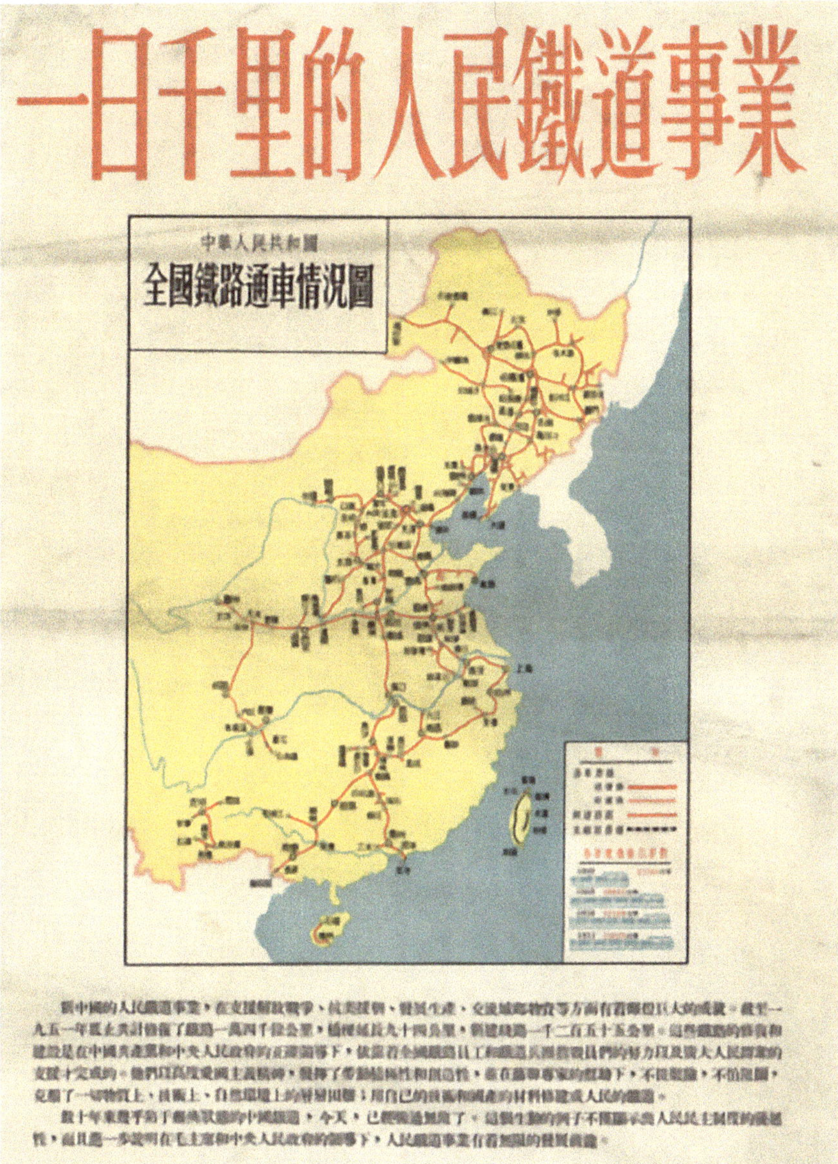

Figure 10.1 Map of operating railways in 1952 (Stefan R. Landsberger collection, BG e37/111, at https://chineseposters.net/posters/e37-111)

1955, E-tu Zen Sun believed that railway development was an 'unfinished task'.[48] Her assessment was proved correct as the CCP embarked on massive construction. However, the Soviet Union was frustrated with the loss of the CER and Port Arthur, the warm water harbour which only the Korean Peninsula could supply. Such considerations led to Stalin's endorsement of Kim Il Sung's invasion of South Korea and the Korean War.[49] The origins of both the Korean War and the Cold War are intimately tied to railways and Manchuria.

Where its predecessors failed

Clarence Davis claimed that 'The Sino-Japanese War of 1894 brought a revolution in railway construction to China'.[50] However, rather than a 'revolution', what 1894 ushered in was the scramble for railway concessions. The real revolution took place after 1949. Railways had helped the CCP initiate its revolution and win the battle for China, now they would aid them in nation building. The Railway Column was transformed into the Railway Corps (RC) in May 1949.[51] In contrast to its counterparts in other parts of the world, the RC was put to civilian use and delivered on its mandate with military speed and efficiency. Sadly, few scholars have researched the RC.[52] Teng Daiyuan, a veteran who had joined the CCP in 1924, became the first Minister of Railway. He laid out the work ahead:

> An advanced socialist country like the Soviet Union has 150,000km of railroads. They have 170,000,000 people but those who are employed in railway work are more than 3,000,000 strong. We have 470,000,000 people but only 26,000km of railways with only 300,000 staff. This is far from what we desire. Nationally we only have 4000 or so engineers. If we want to increase our railways by 15,000km, we need 40,000 or so engineers.[53]

The PRC looked to the Soviet Union for inspiration and aid; forty-two engineers and scientists arrived in August 1949 to help with railway construction.[54] C. K. Leung is one of the few scholars who has written about railways in the PRC era. He articulates that Mao laid out the government's vision in a work titled 'On the Ten Great Relationships' in 1956, where Mao outlined the need to balance the coastal half of the country with the interior.[55] Called the 'Arm of the Army' by Elisabeth Köll, the RC was key to deliver Mao's vision.[56] At its peak in 1974, the RC boasted fifteen divisions and more than 500,000 troops, with an engineering college that trained specialists, conducted research and published its own newspaper. The RC had its own anthem and performance ensemble that travelled around the country to entertain soldiers on construction sites.[57] The RC would build the most difficult railways, from Kashgar on the Kazakhstan–Kyrgyzstan–Tajikistan border, to Xiamen on the coast of the Taiwan Strait and the hazardous Yunnan-Guizhou plateau.[58] If Dr Sun's *Strategy to Build a Nation* was a dream, the Mao regime was transforming that dream into a reality.

The railways the PRC inherited were distributed unevenly. Extensive rail networks existed in Manchuria, north and east–central China, while south China was connected by key lines that could be extended. In contrast, western China was a blank slate, and thus the CCP regime prioritised this part of the country in order to integrate the region under the banner of 'Socialist Construction'. The first line to be built by the RC was the Chengyu [成渝] line, from Chengdu to Chongqing. Deng Xiaoping, then Party Secretary of the Southwest Bureau, led the project. This was not the first railway he built. Using materials captured from enemies, he actually put together a narrow-gauge line from Handan on the Luhan to Changzhi in Shanxi during the Civil War.[59] This time, he had more resources at his disposal. Deng mobilised more than 100,000 Sichuanese workers, peasants, students and townsfolk.[60] Rather than 'drawing' shares, as was the case for the Chuanhan, the CCP 'drew' labour, a more productive way to tap into local resources. Beginning in June 1950, the 505 km railroad was completed by June 1952.[61] This meant that the western segment of the never-built Chuanhan was completed. Zhang Lan, a participant of the Chuanhan railway protection movement, marvelled at this:

> After the Nationalist Revolution, the Beiyang and ROC regimes held power for four decades, politics remained chaotic and people stayed poor, Sichuan remained hard to access, nothing changed. But the southwest has only been liberated for two and half years, and the thousand *li* Chengyu railway has been completed.[62]

State investment, central coordination, Soviet expertise and a motivated labour force were essential for producing results. This was only the beginning of Sichuan's change in fortune. Soon the RC moved to construct the Baocheng [宝成] line, from Baoji in Shaanxi province to Chengdu, linking Chengdu with northwest China. The choice of Baoji was significant as it sits toward the western end of the horizontal Longhai railway. Commenced in 1905 and continued in fits and starts throughout the ROC era, the Longhai was finally completed in 1953. The 668 km Baocheng line was completed by 1957, it meant that Sichuan now had access to northwest and north central China.[63] In the same year, the old Luhan, now called the Jinghan [京汉], was linked with the Yuehan in Hankou by a bridge, turning it into the Beijing to Guangzhou, or Jingguang [京广] railway. By the end of the First Five-Year Plan in 1957, the regime had built thirty-three new railways, including the

Yingxia [鹰厦], from Yingtan in Jiangxi province to Xiamen, in preparation for the attack on Taiwan.

The RC was soon dispatched to construct the Chengkun [成昆] railway in 1958.[64] Stretching southward, it links Chengdu with Kunming, the terminus of the Haiphong–Yunnan line. The Yunnan Plateau had challenged the French and discouraged the British. However, the RC succeeded in laying the 1,100 km line, which included 991 bridges and 427 tunnels. Of the 300,000 RC soldiers who laboured on it, more than 2,000 died (see Figure 10.2).[65] There are twenty-two memorial sites along the railroad dedicated to those who lost their lives.[66] The Great Leap Forward and the Cultural Revolution interrupted construction; the line was finally completed in March 1974.[67]

As the Cold War intensified in the 1960s, so too did railway construction. China was physically and diplomatically caught between the USSR and the USA. In response to this threat, the CCP decided to move its military and heavy industries to western China in 1964, in what was termed the Third Front (TF) project. They divided the country into three 'fronts'. The coastal regions made up the First Front, what lay between this and the Jingguang railway constituted the Second, with the vast region to its west as the Third.[68] Barry Naughton highlights its 'deleterious effect on national development,' arguing that the regime should have focused on the more developed coastal region.[69] Meyskens disagrees, articulating that the TF rail construction 'integrated large portions of inland China into the national railroad system' and accelerated rail development.[70] Hindsight seems to lend support to Meyskens's assessment as this integration laid foundation for economic development in the post-Mao era to come.

Moving military and heavy industries to western China meant more railways. Guizhou province, for example, saw the construction of the Hunan to Guizhou [湘黔], Sichuan to Guizhou [川黔] and Yunnan to Guizhou [滇黔] lines. The mountainous Guizhou benefited from the strategic project, so too did Sichuan. In 1968, the RC began to build the Xiangyu [襄渝] railway.[71] Starting in Xiangyang, Hubei province, the Xiangyu travelled westward to Chongqing. It took ten years to complete the 916 km railway, an indication of the challenges involved.[72] Sichuan was finally connected with Hubei – the dream of Chuanhan shareholders. Sichuan would benefit further from programs such as 'Alleviate Poverty' when post-Mao economic reform extended to underdeveloped areas. Many were poverty-stricken because they lived in mountainous regions with no access to railways; western Hubei was a prime example of this.

The Yiwan [宜万] railway was born, or reborn to be more precise, as this was the eastern segment of the old Chuanhan, from Yichang in western Hubei to Wanxian in Chongqing. The Yiwan project was part of the 10th Five-Year Plan and construction began in 2002. The technical challenge was

Figure 10.2 The Railway Corps constructing the Chengkun line (Tiedaobing xinan zhihuibu, *Tiedaobing zhandouzai chengkunxian* (No place of publication, 1970), no page number, pamphlet held at the Guo Tingyi library, Institute of Modern History, Taipei)

enormous as the line needed to traverse five mountain chains, with jagged, sometimes vertical, columns that separated central China from Sichuan (see Figure 8.3). The challenge lay not just in the highlands but also in rivers and underground tributaries. They often flooded tunnels, swallowing machines and killing workers.[73] Since pumping or diverting the flow did not work, the Railway Ministry brought academics/engineers from around the country to the site where they experimented and delivered solutions, on a government deadline – completion before the centennial of the Nationalist Revolution.[74] The technical difficulties are reflected in the numerous manmade structures that pepper the line. Of the 377 km long railway, 278 km was made up of tunnels and bridges. The line snakes through 159 tunnels, some of which took five or six years to complete; 54 of them are double tracked and 5 are over 10km long.[75] It crosses 253 bridges, of which 21 are enormous and double track. The most difficult and most expensive railway built thus far, the Yiwan was completed in December 2010 in time for the centenary of the Nationalist Revolution, by which time the post-Mao regime had spent more than 22.5 billion *yuan*, a staggering 66 million *yuan* per kilometre.[76] Hindsight seems to suggest that the Chuanhan could never have been built as the late Qing did not possess the capital, technology, material and labour force needed.

Yet, given that the RC was dissolved in 1984, who built the Yiwan? In theory the 380,000 strong RC were demobilised as a result of economic reform, which Meyskens calls the 'demilitarisation of Chinese socialism'.[77] In reality, they were absorbed by state-owned enterprises, such as the China Railway Construction Company (CRCC). The Yiwan was constructed by the 16th Bureau of the CRCC, or the 51st Regiment of the 11th Division of the old RC.[78] Economic reform transformed a former military unit into a profit-seeking machine, where military-style efficiency reached new heights. 'Serving the people', the slogan that dominated the Mao era, was replaced by the post-Mao ethos of 'time is money and efficiency is life'. Railways provide a distinct window into the formula for success behind post-Mao economic miracle.

The Chengyu, Baocheng, Chengkun, Xiangyu and Yiwan placed Sichuan on the map of China's rail network. Sichuan would continue to benefit from the regime's highspeed scheme, and the BRI was announced in 2013. In that year, construction began on the Chenglan [成兰] railway, which linked Chengdu with Lanzhou, terminus of the old Longhai. The year 2014 saw more projects initiated, including the Chuangzang [川藏] line between Chengdu and Tibet, and the Xicheng [西成] line connecting Xining, the capital of Qinghai province, with Chengdu. A century after the Chuanhan-inspired revolution, Chengdu now sits in the heart of a rail network that fans out in all directions. What sweet revenge for the people of Sichuan! The post-Mao regime has marched further west. The Qinghai to Tibet [青藏] and Yunnan to Tibet [滇藏] lines were completed by 2023, whereas the Gansu to Tibet [甘藏] is under construction, with the Xinjiang to Tibet [新藏] line under survey. Starting from Urumqi, the capital of the Uyghur Autonomous Region, this last line travels along the border with Kazakhstan, Kyrgyzstan, Tajikistan, Afghanistan, Pakistan, Kashmir, India, Nepal and Bhutan. The post-Mao regime is realising Dr Sun's dream for a Plateau System. Assuming the CCP is capable of such a feat, the question remains: who will use this railway as it snakes through uninhabitable terrains and what will be the cost to the environment?

All the lines mentioned above are highspeed. In 2007 China rolled out its first generation of highspeed rail: the *Harmony* series travelling at 250 km/h, and consisting of 'four vertical and four horizontal' lines. The four verticals are: (1) Beijing to Harbin [京哈], which travels parallel to the old Barrier line, the SMR and CER; (2) Beijing to Shanghai [京沪] that journeys alongside the old Jinpu; (3) Beijing to Hong Kong [京港] that runs alongside the old Luhan, Yuehan and Canton to Kowloon railway, reducing travel time from about 36 hours to 8½ hours. Three of the four verticals followed the old routes. The fourth, from Hangzhou to Shenzhen [沪深] follows Dr Sun's plan for a link between the Yangzi and the Pearl River deltas.[79] The four horizontals are: (1) Xuzhou to Lanzhou [徐兰], which parallels the old Longhai; (2) Qingdao to Taiyuan [青太], which links the old Jiaoji with the Shide and Zhengtai; both (3) Shanghai to Kunming [沪昆] and (4) Shanghai to Chengdu [沪蓉] are based on Dr Sun's plans. Dr Sun's strategy and lines built by Li Hongzhang, Zhang Zhidong, Yuan Shikai, foreign powers, warlords and the ROC regimes, have continued to enable the CCP's nation building.

The CCP regime began to introduce an even faster highspeed franchise called the *Renaissance*, which travels at 350 km/h, on 'eight vertical and eight horizontal' lines in 2017. The *Renaissance* is equipped with more monitors and alarms than its predecessor for increased safety. At a height of 4,050 mm, it is larger than its 3,700 mm predecessor and more comfortable, with light and temperature adjustment and reduced discomfort to the ear. Built to last, it also looks sleeker, as the bulky components on top of train carriages – the generator and air condition machine – are integrated into the body of the train. They consist of a few new routes, including a new vertical coastal corridor from Dandong on the Korean frontier to the Beibu Gulf on the Vietnam border lacing up littoral cities like Dalian, Tianjin, Qingdao, Shanghai, Xiamen, Guangzhou, Shenzhen and another from Hohhot in Inner Mongolia to Nanning in Guangxi – all to facilitate easy access to Southeast Asia. The existing Beijing to Hong Kong line would make a detour, crossing the Taiwan Strait to Taipei. It also includes an upgraded horizontal corridor, an extension of the old Longhai to Urumqi in Xinjiang. Sichuan has benefited again from this upgrade. Two new vertical lines, the Beijing to Kunming [京昆] and the Lanzhou to Guangzhou [兰广], pass Chengdu; as do two new horizontal lines, the Xiamen to Chengdu [夏渝] and the Yangzi River corridor from Shanghai to Chengdu – lines the British and the French had dreamt of.

The *Renaissance* is significant not because it is faster and safer but because it was made with and operates on the first iteration of Chinese patented and franchised rail technology – unlike the *Harmony*, whose technology was imported from Canada (Bombardier), Germany (Siemens) and Japan (Kawasaki). The *Renaissance* marked the beginning of a new phase in the race for highspeed rail. Since this watershed moment, China has intensified research and is now in the process of developing trains that can travel as fast as or even faster than airplanes. When the world was gripped by the COVID-19 pandemic in July 2021, China debuted 'the world's fastest train' travelling at speeds of 600 km or 373 miles per hour in Qingdao.[80] China is also experimenting with high-temperature superconducting maglev-evacuated tube transport, which will run at more than 1,500 km/h, reducing the travel time from Beijing to Shanghai from 4–5 hours to 1–1:30 hours. Why is China racing ahead when it is already a 'high speed economy'?[81] It appears that wherever the late Qing and ROC failed, the CCP has succeeded. China now thrives on what it lacked during the late Qing: funding, expertise and material. What makes the CCP regime different from its predecessors? The answer is multifaceted.

The CCP learnt their lessons from the foreign powers – railways are the hardware of an industrial and powerful nation that had enabled them to encroach, divide and exploit China. To catch up with them, China must harness the power of this strategic technology. This determination drove the CCP to invest where the late Qing and the ROC had failed. Unlike its predecessors, the PRC enjoyed relative peace and stability after 1949 (the turmoil of the Cultural Revolution notwithstanding), and the Cold War drove them to prioritise railway construction in the name of national security. Elisabeth Köll astutely concludes that railroads 'needed the support of a stable, although not necessarily strong, state for their successful development as business and administrative institutions'.[82] Not content with merely playing catch up, the post-Mao regime has a strong desire to outshine its competitors and lead technological advance. This can be seen from the name of the *Renaissance* franchise, indicative of the 'great renaissance of the Chinese nation', a phrase that has often been evoked by Xi Jinping. It can also be seen from such international initiatives as the BRI.

In summing up the Qing's failure to secure railway autonomy, En-han Lee believed that they lacked not just funding, but also 'engineering and technological talent' and 'managerial and organizational entrepreneurship'.[83] The CCP's regime tackled the issue head on and in a sustained manner. In the 1950s, they sought help from the Soviets, who responded by sending aid and experts, although they were later withdrawn as a result of the Sino-Soviet split. This act of communist solidarity trained a generation of Chinese scientists and engineers, while many more studied in the Soviet Union. The CCP made massive investments in railway education and research. The Chinese Academy of Railway Science, founded in 1950, is home to thousands of scientists scattered in numerous labs. Universities devoted to the development of railway technology were established. Some, such as Shanghai Jiaotong and Xi'an Jiaotong, have now grown to be world-class institutions that lead cutting-edge research. Every railway has its own occupational school, devoted to training engineers and technicians.[84] This dual education system has churned out professionals who have taken China's railway development to greater heights. The aforementioned July 2021 debut, for example, mobilised more than thirty universities, research institutions and enterprises.

Covell Meyskens is exact in arguing that 'labour, technology and the state' were the three ingredients of success in railway construction; he believed that at least 5.5 million people were put to work, and this militarised labour force included rural militia and regional rail workforce, in addition to the RC.[85] Aside from that, regional leaders, Deng Xiaoping for example, often mobilised local workers, peasants, students and the ordinary. If a lack of responsibility and coordination was the hallmark of failure during the late Qing, the CCP regime seems to have embraced the ethos that delivery is key.[86] The two-pronged approach of central planning, management and a disciplined labour force that delivered with military speed and proficiency has been an effective one. The Empress Dowager used eunuchs to drive her toy railway. The United States, Canada, Russia, Germany, Japan and France used Chinese labour to build their railways. Now, the CCP uses them to power its economic development.

The success of railways has enabled China to catch up and rise above. But at what cost? Shen Baozhen predicted in 1865 that 'erasing mountains and rivers would create enormous disaster for the future'. He foreshadowed what would be the concern of today's environmentalists. Railway construction in the name of economic development has wrought damage to the environment; they look like chains on the Chinese landscape (see Figure 10.3). The highspeed network operates on a separate system. China, in other words, has two different rail systems. It is saturated by railways, wasting energy and resources as they demand continued investment, the highspeed franchises in particular. Landslides, for example, have threatened the Chengkun railway while sandstorms have challenged construction in Xinjiang, burying tracks and building sites.[87] Many lines cross rivers, wetlands, mountains

Figure 10.3 Railway map of China in the early 2020s (Baidu public domain)

and tunnels that are prone to changes in weather pattern, thereby demanding constant maintenance. Old railways and trains are becoming obsolete. The regime is trying to salvage them. Some have been turned into museums and tourist attractions.[88] Some have been transformed into 'market trains', offering extremely low fares to enable people in remote areas to travel-sell their produce.[89] Others have become 'silver trains' as the regime tries to galvanise domestic consumption by encouraging seniors to travel.[90] The post-Mao economy is dependent on railways; this could have consequences the regime may not be prepared for. Chinese people might enjoy rail travel, but the railways also dictate how they move. Railways have actually become an instrument of control as the CCP has extended its hands to every corner of the country, enabling the regime to 'nip rebellions in the bud', to evoke the phrase by Li Hongzhang. But the CCP have also exported its railway vision and revolution.

'When China rules the world'[91]

> Western countries have speculated about how much political significance the undertaking may have. But there can be no doubt that the project – by far Communist China's most ambitious and costliest foreign-aid effort – is expected to gain Peking considerable prestige and influence in Africa.[92]

This 1971 *New York Times* headline announced the construction of the Tanzania – Zambia railway. Why was China building a railroad in Africa when the country itself was cash-strapped, underdeveloped and in the chaos wrought by the Cultural Revolution, while facing real threat from the Soviet Union and the United States? Answering this question takes us to decolonisation and the Cold War. Gordon Barrett has authored a superlative book on how the PRC used science as a diplomatic strategy to assert itself.[93] Railway diplomacy opens another frontier.

The young PRC had few friends on the world stage. Premier Zhou Enlai embarked on a charm offensive to win allies. He impressed many at the Bandung conference in 1955 and throughout the Third World. Though poor and small in size, these countries all had seats in the UN. The CCP was eyeing the ROC's seat in the UN and needed their votes. Decolonisation had left political vacuums in the departure of western powers.[94] Suffering at the hands of the same imperialists enabled China to speak a language understood by many in the Global South. Mounting Sino-Soviet tension made it even more urgent for China to reach out. Zhou visited ten African countries in 1963–64, announcing that China would provide foreign aid to newly decolonised countries in Asia, Africa and Latin America.[95] Richard Hall and Hugh Peyman have termed this 'red flare from the East'.[96] The PRC promised 'interest-free or low-interest loans' and made it clear that 'the Chinese Government strictly respects the sovereignty of the recipient countries, and never attaches any conditions or asks for any privileges'.[97] The Chinese offer appeared preferable to the treaties imposed by European powers.

President Julius Kambarage Nyerere of the newly independent Tanzania visited China in February 1965.[98] Part of German East Africa since 1885, Tanzania was awarded to Britain at the 1919 Paris Peace Conference, a fate that Shandong shared. This had angered Belgium and Portugal; both managed to claim their share after the conference. Like China, Tanzania was sliced up like a melon, offering the CCP valuable ammunition in alliance-making. Nyerere had been looking for help to build a railway that would free his country from dependence on white-ruled apartheid South Africa. The project required funding and expertise which they, like the late Qing, did not possess. He had attempted to raise the funds from the World Bank, but the project was deemed 'unfeasible' and 'later efforts to interest other governments and private groups in the undertaking fell through'.[99] No one saw any political nor economic incentive in helping Tanzania.

'Kaunda waves the little red book' heralded President Kenneth Kaunda's visit to the PRC in June 1967.[100] Zambia had also suffered at the hands of European powers. A land-locked country, it faced serious challenges after gaining independence from Britain.[101] Rich in copper, Zambia was a 'wealthy pauper' as rail transportation through Rhodesia (Zimbabwe) and South Africa were very expensive.[102] A railway through Tanzania would be ideal as it provided access to the sea.[103] But the country had neither the capital nor the expertise needed to tackle one of the most challenging terrains on earth: the East African Rift. Kaunda had approached Britain, France, West Germany and the United States for help but to no avail.[104] China was the only option left on the table and the three countries reached an agreement in September 1967.[105] China would build the Tanzania–Zambia, the Tanzam [坦赞] railway with an interest-free loan of nearly 1 billion RMB (9.88) without any conditions attached – the PRC's largest foreign aid program. Route surveys began in May 1968, when China announced their second project: the Mali–Guinea line (see Figure 10.4).[106]

Figure 10.4 Tanzania–Zambia and Mali–Guinea railways (Baidu public domain map)

The 1,860 km or 1,065 mile railway travels from Kapiri Mposhi in the heart of Zambia to Dar es Salaam, the largest city on the coast of Tanzania. Construction of the first segment, from the coast to Mlimba, began in October 1970.[107] The rainy season turned the land into an endless swamp that few modern machineries could handle. The environment was harsh; some Chinese workers were killed by wild animals and diseases. When construction approached Mlimba, good news came from the UN on 25 October 1971. A resolution calling for the PRC to replace the ROC, put forward by Albania and Algeria, was passed with 76 votes. Support from countries such as Tanzania and Zambia had been vital. The first segment of the railway was finished as China took its seat at the UN in November.[108] As a token of appreciation, the PRC presented the UN with a state gift: an ivory sculpture of the Chengkun railway. China was proud of its railway achievement.[109] As the Tanzam carved deeper into the hinterland for the next segment to Makambako, the land began to climb and so, too, would the railway, as high as 1670 metres. This was terrain as difficult as that of southwest China, and the Railway Corps were called in to help.[110]

The next challenge was the Rift valley, which required the construction of tunnels and bridges in some of the most hazardous places on earth. Overcoming huge obstacles, the RC completed the Tanzania section by August 1973. Then the rivers, swamps and floodplains of Lake Malawi greeted them with more challenges.[111] Working with their 'African brothers', the RC completed the railway, and it became operational by July 1976.[112] By then, China had sent more than 100 tons of materials, some purchased from Europe with foreign currency. Hall and Peyman suggest that at least 25,000 Chinese and 50,000 Africans laboured on the project.[113] They dug 22 tunnels, and built 320 bridges and 93 stations, but more than 60 Chinese workers lost their lives.[114] Some scholars have sung China's praises for its 'proletarian internationalism'.[115] But others have been critical:

> Despite the building by the Chinese of the Tanzam Railway connecting Zambia (Northern Rhodesia) with Tanzania (Tanganyika and Zanzibar), and efforts to use the Benguela Railway to the Angolan coast, these alternate routes could never handle enough traffic to free the Zambian economy from its need for the southern connection. Eventually, with little fanfare, traffic between the mining industry north of the Zambesi and the white-dominated regimes south of the river revived.[116]

Hall and Peyman believe that China's reward was 'primarily goodwill'.[117] But this was a political project: the PRC needed a political theatre to assert itself in its contest with the Soviet Union and the United States. Africa provided the optimal stage. Mao's claims to be the leader of the Third World came with responsibilities. The regime starved its own people in order to establish itself on the international stage. The railway did benefit Tanzania and Zambia by reducing their dependence on Zimbabwe, South Africa and facilitating the movement of people and goods. But China gained more by replacing the ROC at the UN and in the Security Council.

The Tanzam was built during the Cultural Revolution, when Mao's political crusade to destroy 'capitalist roaders' ruined China's economy and devastated its people. Yet his message to Africa was drastically different: 'Africa's current task is to struggle against imperialism, not against capitalism.'[118] The regime was thus undertaking abroad what seemed inconceivable inside China, namely, helping Africa develop capitalism. The African project and the Cultural Revolution seemed to have goals at odds with each other, yet Mao justified it. Although the CCP dogmatically insisted on following communism, it was also capable of applying capitalist practices whenever and wherever necessary. This apparent paradox in the CCP's approach reveals its flexibility and core value – whatever serves its political ends, even when they contradict with what they preach. Thus, capitalism thrived behind the scenes during the anti-capitalist Cultural Revolution. This double-standard defined Mao-era diplomacy, and it has thrived in the post-Mao era where capitalism now flourishes in a communist country. The political biography of railways provides a unique window into the CCP's double act, and it exposes the contradiction between its ideology and practice in China and beyond, both then and now.

Building the Tanzam helped the CCP to fashion a type of diplomacy which would enhance its global ambition in the post-Mao era. After its completion in 1976, Chinese engineers stayed behind to maintain the railway and train local staff.[119] They were soon joined by state-owned railway enterprises like the China Civil Engineering Construction Corporation (CCECC). Furthermore, the post-Mao regime has since injected US$42 million to upgrade the old Tanzam in 2012.[120] They also completed the Benguela railway, which runs through Congo and Angola in 2019, connecting with the Tanzam to form a transcontinental line from the Atlantic to the Indian Ocean (see Figure 10.4).[121] China has continued to build railways on the continent: from Addis Ababa to Djibouti, from Mombasa to Nairobi, from Lagos to Ibadan, and the new capital line for Egypt.[122] China's railway affair with Africa has endured with renewed energy and purpose.[123] Tim Zajontz has labelled this 'China's new railway imperialism in Africa'.[124] The post-Mao regime is constructing not just railways but also roads, airports, stadiums, schools, hospitals and public facilities. It is extracting natural resources and exerting influence in Africa, and it all began with a railway.

Chinese railways have also invaded Southeast Asia which Tom Miller has called 'advancing down the Mekong'.[125] China is realising the British dream to connect Burma with southwest China – a line from Chongqing to the Indian Ocean via Myanmar.[126] The most difficult segment within China, from Kunming to the Burmese border, was completed in 2021; then it journeys to Lashio and Mandalay before terminating at Yangon. China seems to have succeeded where Britain failed. This is a strategic line vital to China's economic development and global ambition; it cuts the cost of foreign trade through the South China Sea and the Strait of Malacca, giving China the shortest and most direct access to the Indian Ocean. China is also building a network of lines in the old Indochina and the Malay Peninsula, aside from completing the Jakarta to Bandung highspeed in 2023.[127]

T. G. Otte and Keith Neilson have argued that railways were an 'instrument in the tool kit of the Great Powers'.[128] Now, the 'instrument' is enabling the post-Mao regime. Chinese railways have marched west along the old Silk Road under the umbrella of the BRI which Miller styled 'empire building along the New Silk Road'.[129] The BRI links China directly with Europe, Nile Green has called it 'from the silk road to railroad'.[130] One travels from Lianyungang, origin of the Longhai line on the east coast, to Rotterdam in Holland. Another zigzags from Yiwu, the 'world's largest small commodity market', to Madrid. Dr Sun had wanted to 'connect with railways in India and Europe (those going to Baghdad, Damascus, and Cairo)'. Chinese railway construction has indeed made progress towards Bagdad and Damascus.[131] Could they reach Cairo, or perhaps even further to the Cape, as Cecil Rhodes had envisioned? China constructed the Ankara–Istanbul highspeed rail and, in an ironic twist, the CRCC is poised to build the planned highspeed line from Moscow to Kazan which will terminate in Beijing.[132] Chinese-made high speed wheels were exported to Germany.[133] China has now technologically surpassed the imperial powers that had invented railways, and whose attempts at developing highspeed franchise, such as HS2 in the UK, have encountered the same problems that plagued the Chuanhan a century ago.[134]

The post-Mao regime is building a railway empire; this has disturbed many.[135] Some accuse China of 'debt trap diplomacy' and exporting its illiberal values. Former US Vice-President Mike Pence declared that China wanted to dominate the world 'economically, militarily and technologically'.[136] From victim of railway imperialism, China has now become a railway expansionist. Li Hongzhang and Dr Sun might have been pleased. Although China has neither appropriated land nor stationed forces, it has certainly used railways to extract resources and spread Chinese influence or 'soft power'. As can be seen from the *Renaissance* franchise, China is eager to promote Chinese standard. The desire to outshine the once-powerful West arguably exposes the CCP's insecurity, but to actively outstrip the Western standard also reveals something more significant – China's ambition to lead, some might say dominate, the world. China's technological progress and economic success have astounded the world; it has also emboldened the regime, for if China could dominate the world as Pence has asserted, it can possibly also 'rule the world', to borrow from Martin Jacques, through its railways.[137]

Railways have enabled the CCP to develop the country and dominate the global economy. But behind the seeming success, there are hidden dangers of unknown quantity. Their economic success and dominance are hinged on railways. Is the post-Mao regime aware that it would take enormous resource, constant effort and long-term commitment to maintain such a vast domestic, regional and global network? They have invested millions to maintain the old Tanzam railway. What about other projects in Africa, Central Asia and Southeast Asia where railways are at the mercy of not just natural disasters but also sabotage and armed conflicts? They could easily destroy the lines and disrupt service. Quetta in Pakistan is a case in point as its BRI line has been attacked by insurgents.[138] The Chongqing to Yangon line is another example as it travels through war-prone territories controlled by warlords and challenging topographies that Myanmar has no resources or expertise to maintain. Strategic and symbolic investments in underdeveloped countries come with real risks the CCP may not be ready for. Who indeed could have predicted that China, who resisted the railways and suffered a great deal at the hands of railway imperialism, would now surpass those foreign powers who had invented and introduced them? Railways helped the CCP to consolidate its rule at home and make friends on the international stage. The post-Mao era saw a Great Leap Forward, and the railway has shown us the extent of China's rise and its global ambition. Despite the growing success in railway construction and technology, the post-Mao reform remains as apolitical as late Qing reform. It was the lack of political reform that led to the collapse of the Qing, but whether history will repeat itself remains to be seen.

Notes

1 Doig, *High speed empire.*

2 'Mao zhuxi fangsu', *Renmin Ribao* (*Remin* hereafter), 18 December 1949: 1; C. L. Sulzberger, 'Soviet said to plan treaty with Mao', *Times*, 22 October 1949: 6 and 'Mao's schedule', 29 January 1950: 115.

3 Elisabeth Perry, *Shanghai on strike: the politics of Chinese labor* (Standford: Stanford University Press, 1993).

4 Hans van de Ven, *From friend to comrade: the founding of the Chinese communist party 1920–1927* (Berkeley: University of California Center for Chinese Studies, 1991), 70.

5 Jia, *Naxieren*, 162.

6 'Jinghanlu bagong zhongzhi canju', *Shen Bao*, 10 February 1923: 7 and 'Jinghan luchaohou xingshi yanzhong', 23 February 1923: 13.
7 Perry, *Shanghai*, 76.
8 Li Ziming, *Minguo*, vol 1, 92–96.
9 Chong-Sik Lee, *Revolutionary struggle in Manchuria: Chinese communism and Soviet interest 1922–1945* (Berkeley: University of California Press, 1983).
10 'Red Army strikes', *Times*, 9 August 1945: 1.
11 'The Yalta conference', Yale Law Library, at https://avalon.law.yale.edu/wwii/yalta.asp, and 'Yalta conference agreement', Woodrow Wilson Center, at https://digitalarchive.wilsoncenter.org/document/116176, accessed 31 July 2021.
12 'Invasion of Manchuria', Harry S. Truman Library, at www.trumanlibrary.gov/education/presidential-inquiries/invasion-manchuria, accessed 31 July 2021 and 'Official record of the Yalta conference', *Times*, 17 March 1955: 47.
13 'The Yalta conference', Yale Law Library.
14 Wu Sufeng, 'The Nationalist government's attitude toward post-war Japan', in *Negotiating China's destiny*, 193–219 (200).
15 'Zhongsu youhao tongmeng', *Shen Bao*, 27 August 1945: 1; Hao Jiangdong, 'Zhanhou chuqi Sidalin duihua 'shuanggui' zhengcede yanjiu', *E'luosi Yanjiu*, 2 (2015): 95–122 and Elleman, *International competition*, 147–51.
16 Tillman Durdin, 'Russians operate Manchurian line', *Times*, 3 February 1948: 3.
17 Pauley, *Report*, 40.
18 Pauley, *Report*, 52.
19 Pauley, *Report*, 38–45.
20 Niu Jun, 'The birth of the People's Republic of China and the road to the Korean war', in *The Cambridge history of the Cold War*, eds Melvyn P. Leffler and Odd A. Westad (Cambridge: Cambridge University Press, 2010), 221–41.
21 Victor Cheng, *China between peace and war: Mao, Chiang and the Americans, 1945–1947* (Canberra: ANU Press, 2023), 80.
22 'Reds say civil war is on in Manchuria', *Times*, 17 February 1946: 1; Odd Arne Westad, *Cold War and revolution*, 109–11.
23 'The Yalta conference', Yale Law Library.
24 'Dongbei jushi', *Shen Bao*, 9 February 1946: 1 and 'Dongbei chongtu', 26 February 1946: 1.
25 'Zhou En-lai to Lieutenant General Alvan C. Gillem, Jr, Chungking, March 15, 1946', Marshall mission files, Lot 54–D270, Office of the Historian, at https://history.state.gov/historicaldocuments/frus1946v09/d347, accessed 8 August 2021.
26 'Civil war spreads', *Times*, 29 October 1945: 1.
27 'Communists open north China drive', *Times*, 11 April 1946: 1; 'Harbin shinei wuzi', *Shen Bao*, 21 September 1946: 1.
28 Steven Levine, *Anvil of victory: the communist revolution in Manchuria, 1945–1949* (New York: Columbia University Press, 1987), 168.
29 'China army chief accuses Russians', *Times*, 25 June 1947: 1 and 'State Department allegations of Soviet violations', 30 May 1948: 2.
30 'China Reds encircle Manchurian capital', *Times*, 16 November 1945: 1.

31 Tanner, *Battle for Manchuria*, 106–34; Cheng, *Peace and war*, 85–86; Yang Xingming, 'Dongbei jiefang zhanzhengzhong danshengde tiedaobing budui', *Heilongjiang Shengzhi*, 8 (2013): 23–24.
32 'Major victories listed by Nanking', *Times*, 10 June 1947: 24 and 'Manchurian drive is pressed by the Reds', 4 October 1947: 6.
33 Zach Fredman, *The tormented alliance: American servicemen and the occupation of China 1941–49* (Chapel Hill: University of North Carolina Press, 2022), 163–95.
34 Tanner, 'Railways in communist strategy', in *Manchurian railways*, 149–70.
35 Tielushi bianji yanjiu zhongxin, *Zhongguo tielu dashiji* (Beijing: Tiedao chubanshe, 1996), 168.
36 Chen Lixu, 'Zhonggong Zhongyang cengshexiang dingdu Ha'erbin', *Dangshi Zonglan*, 9 (2018): 19–21.
37 Levine, *Anvil of victory*.
38 'Tiedao zongdui ganxiu Jinpulu', *Renmin*, 13 March 1949: 1, Chen Yongjin, 'Jiefangqu renmin zhiyuan dajun nanxia', 15 April 1949: 1, 'Jinpu beining longhai tielu', 12 May 1949: 1 and 'Tiedao bingtuan', 23 January 1950: 2.
39 Li Guangsheng, 'Yezhanjun dadaonali, tielujiu xiudaonali', *Renmin*, 13 September 1974: 2.
40 Zhongguo renmin jiefangjun tiedaobing, *Renmin tiejun* (Beijing: Tiedao bingtuan zhengzhibu, 1950), cover.
41 Liu Weiyang, *Yongyuande tiedaobing* (documentary), episode 1 (27:44), China Central Television, at www.crcc.cn/col29504/art/2018/art_29504_2424289.html, accessed 2 November 2021.
42 Tielushi, *Dashiji*, 173.
43 'Mao to discuss Sino-Soviet pact', *Times*, 2 January 1950: 1.
44 'Chinese Reds sign 30–year alliance', *Times*, 15 February 1950: 1; 'ZhongSu youhao tongmeng huzhu tiaoyu', *Renmin*, 14 February 1951: 1.
45 'Record of conversation between I. V. Stalin and Chairman Mao Zedong on 16 December 1949', Woodrow Wilson Center, at https://digitalarchive.wilsoncenter.org/document/87836/download, accessed 31 July 2021.
46 'Text of Soviet-Chinese announcement', *Times*, 16 September 1952: 4 and Masafumi Asada, 'The return of the Chinese Changchun railway to China and Sino-Soviet relations, 1949–1952', *Asian Studies*, vol 60, 1 (May 2015): 5–23.
47 'Renmin tiedao zhazhi shiyue chuangkan', *Renmin*, 1 September 1949: 2.
48 Zen Sun, 'Pattern of railway development'.
49 Shen Zhihua (translated by Neil Silver), *Mao, Stalin and the Korean war: trilateral communist relations in the 1950s* (London: Routledge, 2012), 102–19.
50 Clarence B. Davis, 'Railway imperialism in China', 155–73 (159).
51 'Tiedaobu chengli tiedaobingtuan', *Renmin*, 31 May 1949: 1.
52 Michael Y. M. Kau provided a four-page narrative on the RC, see *Revival: the People's Liberation Army and China's nation building* (London: Taylor and Francis, 2017), chapter 13.
53 'Teng Daiyuan buzhang jianghua', *Renmin*, 12 June 1949: 1.

54 Tielushi, *Dashiji*, 179; 'Fazhan guomin jingjide', *Renmin*, 13 August 1955: 2; Renmin Tiedaobao, *Renmin tiedaobao tingkan jiniance* (Beijing: Renmin tiedaobao, 1958), 29; M. Gardner Clark, *Development of China's steel industry and Soviet technical aid* (Ithaca: Corner University, 1973).
55 Mao Zedong, 'Lun shida guanxi', *Renmin*, 26 December 1976: 1; Leung, *Railway patterns*, 150–55.
56 Köll, *Transformation*, 240.
57 Zhang, 'Socialist builders'.
58 Köll, *Transformation*, 235.
59 Gong, *Tushuo*, 97.
60 'Chengyu tielu', *Renmin*, 18 October 1951: 2 and Cheng Zaihua, 'Sulian zhuangjia he Chengyu tielu', 14 February 1952: 2.
61 'Chongqing Chengdu relie qingzhu', *Renmin*, 2 July 1952: 1 and 'Chengyu tielu', Wenhua zongheng at Sichuan shengqing wang, www.scsqw.cn/whzh/jdbn/content_53575, accessed 10 November 2021.
62 Zhang Lan, 'Qingzhu Chengyu tielu tongche', *Renmin*, 2 July 1952: 2 and 'Chengyu tielu', 3 July 1954: 3.
63 Ye Xin, 'Baocheng tielu', *Renmin*, 11 July 1955: 2 and Xiong Yuzhong, 'Women xiutongle jianxiande Baocheng tielu', 2 January 1958: 3.
64 'Chengkun tielu', *Renmin*, 19 August 1953: 1.
65 'Xinlao Chengkunxian jianzheng shiji bianqian', Renmingwang, at http://finance.people.com.cn/n1/2021/0620/c1004-32135087.html, accessed 9 August 2021.
66 'Liaobuqide Chengkun tielu! Yizhan yibei jixianlie', Zhongqing zaixian, https://m.cyol.com/gb/articles/2021-05/29/content_Xb9RlSp2z.html, accessed 9 August 2021.
67 'Chengkun tielu shengli jiancheng tongche', *Renmin*, 23 March 1974: 1.
68 Covell Meyskens, 'The makings of China's Cold War motor city', in *Material contradictions in Mao China*, eds Jennifer Altehenger and Denise Ho (Seattle: Washington University Press, 2022), 200–18.
69 Naughton, 'The Third Front'.
70 Meyskens, 'Third Front railroads'.
71 Chen Yuanmo, 'Xiangyu tielu chuangjianji', *Renmin*, 15 June 1978: 3.
72 'Xiangyu tielu', *Renmin*, 15 June 1978: 1.
73 Zhang Zhifeng, 'Yiwan tielu', *Renmin*, 10 August 2007: 5 and Yang Xiwei, 'Yiwan tielu', 24 November 2007: 2.
74 Wang Liwu, 'Zhongguo tiejian shiliuju jituan', *Renmin*, 10 November 2008: 15.
75 Zhang Zhifeng, 'Chuanyue zhulu jinqu', *Renmin*, 19 August 2010: 9.
76 Lu Yanan, 'Yiwan tielu kaitong', *Renmin*, 23 December 2010: 4.
77 Covell Meyskens, *Mao's Third Front: the militarization of Cold War China* (Cambridge: Cambridge University Press, 2020), 227–36.
78 Tan Bing, 'Nanjiang tiedaobing', *Wenshi Yuekan*, 4 (2008): 24–26.
79 Sun, *Fanglue*, 173.
80 Chao Zheng and Deng Wangqiang, 'Shisu liubaigongli gaosu cifu jiaotong xitong xiaxian', *Beijing Ribao*, 21 July 2021.

81 Keith Bradsher, 'A high speed economy', *Times*, 13 February 2011: B1.
82 Köll, *Transformation*, 297.
83 Lee, *Quest*, 134 and 136.
84 Philip Wing Keung Chan, *Public education reform and network governance: lessons from Chinese state-owned enterprise schools* (London: Routledge, 2019).
85 Meyskens, 'Third Front railroads'.
86 Xiao Ma, *Localised bargaining: the political economy of China's highspeed railway program* (New York: Oxford University Press, 2022).
87 Qian Zheng *et al.*, 'Investigation of landslides that occurred in August on the Chengdu-Kunming railway, Sichuan, China', *Geosciences (Basel)*, vol 9, 12 (November 2019): 497; Cheng Jian-jun *et al.*, 'Characteristics of the disastrous wind-sand environment along railways in the Gobi area of Xinjiang, China', *Atmospheric Environment (1994)*, vol 102 (2015–02): 344–54.
88 'Qianmen station: witness to countless historical events' [前门火车站无数历史事件的见证], 6 May 2019, Beijing Municipal Government, at www.beijing.gov.cn/renwen/zt/2019gjbwgr/zgtdbwgzymg/tbzymzgtp/201905/t20190506_1877115.html, accessed 5 March 2025.
89 Luo Wangshu, 'Slow trains, fast dreams: a story of China's railway impact', *China Daily*, 20 February 2025, at www.chinadaily.com.cn/a/202502/20/WS67b7038ca310c240449d65e4_1.html, accessed 5 March 2025.
90 Chris Lau, 'China hopes new fleet of 'silver trains' will encourage its elderly to travel', 12 February 2025, at https://edition.cnn.com/2025/02/12/travel/china-trains-elderly-economy-intl-hnk, accessed 5 March 2025.
91 Martin Jacques, *When China rules the world* (London: Penguin, 2012).
92 'Tanzania-Zambia railway', *Times*, 29 January 1971: 65.
93 Gordon Barrett, *China's Cold War science diplomacy* (Cambridge: Cambridge University Press, 2022).
94 Donovan C. Chau, *Exploiting Africa: the influence of Maoist China in Algeria, Ghana and Tanzania* (Annapolis: Naval Institute Press, 2014); Ian Taylor, 'Mao Zedong's China and Africa', *Twentieth Century Communism*, vol 15 (September 2018): 47–72.
95 Jay Walz, 'Aid to Africans offered by Chou', *Times*, 17 December 1963: 1.
96 'Chou opens visit to African lands', *Times*, 15 December 1963: 27. Hall and Peyman, *Uhuru railway*, 40.
97 'The Chinese government's eight principles for economic aid and technical assistance', Woodrow Wilson Center, at https://digitalarchive.wilsoncenter.org/document/chinese-governments-eight-principles-economic-aid-and-technical-assistance-other-countries, accessed 31 July 2021.
98 Zhou Nan, 'Woguo yuanjian Tanzan tielude juece Beijing', *Shijie Zhishi*, 19 (2015): 61–63.
99 'Tanzania defends Chinese rail aid', *Times*, 23 November 1969: 23 and 'World Bank rejection', 29 January 1971: 65.
100 Hall and Peyman, *Uhuru railway*, 87.
101 Lawrence Fellows, 'Rhodesia moving toward showdown with Zambia', *Times*, 30 September 1964: SU2-1 and 'Rhodesia and Zambia split up railway', 10 July 1967: 43.

102 William A. Payne, 'Zambia's paradox: a wealthy pauper', *Times*, 25 Jan 1965: 47 and 'Rail shipment of Zambia copper through Rhodesia', 27 May 1966: 61.
103 'Zambia hopes for a new rail line to the sea', *Times*, 25 January 1965: 79; Hall and Peyman, *Uhuru railway*, 45.
104 Lawrence Fellows, 'China to build a Zambia-Tanzania railroad', *Times*, 15 January 1968: 13.
105 'Zhong, Tan, Zan sanguo zhengfu qianding', *Renmin*, 6 September 1967: 4; 'Peking to aid rail project for Zambia and Tanzania', *Times*, 10 September 1967: 8.
106 'Chinese to help build Mali-to-Guinea railroad', *Times*, 30 May 1968: 3; 'Tansangniya duan chuce gongzuo', *Renmin*, 21 October 1968: 5; R. W. Apple Jr, 'Chinese pressing Zambia rail line', *Times*, 28 September 1969: 11.
107 'Yuanjian Tanzan tielu gongcheng jishu renyuan', *Renmin*, 5 Jan 1970: 6 and Charles Mohr, 'Kaunda lead Tanzam rail ceremonies', *Times*, 27 October 1970: 2.
108 'First stretch is completed in Tanzania-Zambia line', *Times*, 19 November 1971: 12 and 'China joins U.N. Council', 24 November 1971: 4.
109 'Woguo songgei Lianheguo zongbude lipin', *Renmin*, 16 October 1974: 6.
110 Zhang Jixue, 'Wode 'Tanzantielu' qinghuai 1 & 2', *Guoji Yuanzhu*, 2 (2015): 64–68 and 5 (2015): 26–33; Sun Yongfu, 'Canyu yuanjian tanzantielu de jinliyu ganshou', *Shijie Zhishi*, 20 (2015): 67–69.
111 'Tanzan tielu zai jianshezhong', *Renmin*, 26 December 1973: 5; Lu Datong, 'Tanzan tielu yuanjian gongcheng qinliji', *Shiji Zhishi*, 6 (2016): 41–50; Guo Jieru, 'Li Shufen de gushi', *Guoji Yuanzhu*, 4 (2016): 107–11.
112 'Tanzhong liangguo gongren he jishu renyuan', *Renmin*, 1 May 1973: 5; 'China turns over Tan-Zam railroad', *Times* 15 July 1976: 5 and 'Zambia-Tanzania railway is monument to Chinese builders' prestige', 24 April 1977: 55.
113 Hall and Peyman, *Uhuru railway*, introduction.
114 'Tuanjie zhandou zhuyouyi', *Renmin*, 25 March 1974: 6.
115 Hall and Peyman, *Uhuru railway*, introduction; Martin Bailey, *Freedom railway: China and the Tanzania-Zambia railway link* (London: Collings, 1976).
116 Hanes III, 'Railway politics', in *Railway imperialism*, 41–69 (64).
117 Hall and Peyman, *Uhuru railway*, 190.
118 Mao Zedong, 'Africa's current task is to struggle against imperialism, not against capitalism', Woodrow Wilson Centre, at https://digitalarchive.wilsoncenter.org/document/mao-zedong-africas-task-struggle-against-imperialism, accessed 31 July 2021.
119 'Qiongbu diyi fuzongtong shicha Tanzan tielu', *Renmin*, 29 March 1975: 5.
120 'Tazara receives US$42 million railway funding from China', *African Review of Business and Technology*, Monday 1 October 2012.
121 Liu Lei *et al.*, 'Tanzan tielu xiyan gongcheng hongguan zouxiang fang'an yanjiu', *Tieluyunshu yu Jinji*, vol 34, 10 (2012): 87–92; 'Angola's 1300km, Chinese built railway is finished', *Global Construction Review*, 15 August 2014.
122 Noah Fowler, 'China says it built a railway in Africa out of altruism', *Los Angeles Times*, August 4, 2017; Nicholas Muller, 'The Chinese railways

remolding East Africa', *The Diplomat*, 25 January 2019; Duncan Miriri, 'Kenya opens $1.5 billion Chinese built railway linking Rift Valley town with Nairobi', *Reuters*, 16 October 2019.

123 Julia C. Strauss, 'The past in the present: historical and rhetorical lineages in China's relations with Africa', *The China Quarterly*, vol 199 (2009): 777–95.

124 Tim Zajontz, 'Railway imperialism in East Africa', in *Oxford handbook*, 515–34.

125 Tom Miller, *China's Asian dream: empire building along the New Silk Road* (London: Zed Books, 2017), 95.

126 'China plans railway from Chungking to sea', *Herald*, 25 December 1940: 486 and 'China opens rail line with access to Indian Ocean via Myanmar', *The Irrawaddy*, 30 August 2021.

127 Dragan Pavlicevic and Agatha Kratz, 'Testing the China threat paradigm: China's high-speed railway diplomacy in Southeast Asia', *Pacific Review*, vol 31, 2 (4 March 2018): 151–68; Wu and Chong, 'Developmental railpolitics'.

128 Neilson and Otte, 'Railpolitick', 7.

129 Miller, *Asian dream*, book title.

130 Nile Green: 'From the Silk Road to the railroad (and Back): the means and meaning of the Iranian encounter with China', *Iranian Studies* 48, 2 (2015): 165–92.

131 Sinan Mahmoud, 'Iraq woos China for transport project linking Asia and Europe', *The National News*, 2 June 2023.

132 'Moscow-Kazan high-speed rail line', *Railway Technology*, 14 November 2021.

133 'Chinese high-speed train wheels export to Germany', *China Daily*, 23 July 2018.

134 Dustin Gardiner, 'California high-speed rail's latest threat', *San Francisco Chronicle*, 16 June 2021.

135 Tim Zajontz *et al.* eds, *Africa's railway renaissance: the role and impact of China* (New York: Routledge, 2024); Trin Aiyara, 'The long and winding railway: domestic politics and the realization of China-initiated high-speed railway projects in Thailand', *Chinese Political Science Review*, vol 4, 3 (2019–09): 327–48; Slavomir Horak, 'Turkmenistan in Eurasian railway geopolitics', *Central Asian Survey*, vol 42, 1 (2023–01): 171–90.

136 Mike Pence, 'The administration's policy towards China', speech delivered at the *Hudson Institute* on 4 October 2018, at www.hudson.org/events/1610-vice-president-mike-pence-s-remarks-on-the-administration-s-policy-towards-china102018, accessed 11 August 2021.

137 Jacques, *When China rules*.

138 Charlie Campbell, 'The Quetta bombing shows that China's growing global presence comes with risk', *Time*, 21 April 2021, at https://time.com/5957182/quetta-bombing-pakistan-china-belt-road-bri/, and Saleen Ahmed, 'Pakistan insurgents take dozens of hostages in attack on train', *Reuters*, 12 March 2025, at www.reuters.com/world/asia-pacific/firing-passenger-train-pakistan-tv-channels-report-2025-03-11/, accessed 14 March 2025.

Conclusion

Railway Imperialism in China: a Political Biography has taught us new lessons through a unique and revealing lens. First, it has shed new light on how late Qing reform and railway development more specifically, were made more complex by Western and Japanese encroachment. It has challenged the conventional view that the late Qing court and officials were opposed to the introduction of railways, as epitomised by the destruction of the Wusong. This volume has argued that their opposition was based on sound reasons. Many officials understood the need to reform if China was to defend itself against foreign intrusion and even competed with each other for control over reform projects. They were aware that modern technologies had given their enemies the upper hand and encouraged their aggression. However, they were also clear that reform, although necessary, could not come at the cost of the Qing's sovereignty. Though they appreciated that railways would strengthen the country, these officials insisted that the demand for it had to come from the Chinese themselves, rather than the foreign powers who only sought their own gains. They were gravely concerned with the destruction of the existing socio-economic order and the environment. This political biography exposes fundamental flaws with late Qing reform: reactionary, ad hoc and regional where control rested in the hands of individual officials and foreign powers. This stands in sharp contract with post-Mao reform and railway development: visionary, systematic and national, even global.

With the ascent of railway imperialism in the 1890s, the fears of the late Qing officials finally came to pass. The history and politics of railway development allows us to precisely map out the speed and scope of Western expansion. Railways had become the new 'engines for empire'. European powers, the United States and Japan tried to outmanoeuvre each other in order to secure railway loans and contracts, using them to carve out their sphere of influence and control. The scramble for China transformed from seeking 'settlements' to the hunt for railway concessions as they offered privileges that were previously insurmountable; that is, to acquire territories beyond Treaty-ports and far into the interior, to demand military presence, to take advantage of China's natural and human resources, and to rule the railway zone as colonies. A railway-driven empire was less costly than war and much more profitable to maintain. Previous generations of historians have been concerned with whether these railways benefited China. Although foreign powers did not set out to aid China, the railways they left behind did propel the country onto the path of industrialisation, with the CCP as the ultimate beneficiary.

Focusing on railway imperialism has provided new perspectives on the collapse of the Qing, as provincial officials, local gentry/merchants and the ordinary people were galvanised into action by the 1900s. They used railways to prevent foreign encroachment and to protect local interests. The Railway Rights Recovery Movement fuelled the fire of anti-Manchu sentiment, but in Sichuan it was manipulated by constitutional monarchists and then hijacked by the United League, with the aid of organised crime, in their common goal to overthrow the Qing. The genuine grievances of Chuanhan railway shareholders have been reduced to a footnote in history. Imitating the foreign powers, the Beiyang, the various warlord and Nanjing regimes, all embarked on constructions of their own. Like the foreign powers, they used railways to carve out their spheres and serve their regional, rather than national, goals. It is evident that the Communists have been the best students of railway imperialism. Railways enabled them to jumpstart their revolution, to sabotage the Japanese, to win the battle for China and to project themselves on the international stage. The CCP inherited not just lines built by imperial powers but also their technology and an unfinished vision, which the post-Mao regime has taken further, using railways to conquer the world through the BRI.

The political biography of railways reads like a biography of Li Hongzhang, the self-styled imperial 'plasterer' whom foreigners called the 'ablest statesman in China'.[1] He had a hand in nearly every railway commissioned and constructed by the time he died in 1901. Tenacious and pragmatic, he was daring, even devious, in realising his aims, ingratiating himself with the Empress Dowager, competing with his colleagues, pitting foreigners against one another, and using their money and expertise to fund and build railways. His efforts, long criticised, ultimately paid off as the CCP inherited a solid infrastructure in Manchuria, north and east–central China. Does this legacy redeem him? He was one of many individuals who mattered a great deal: Zhang Zhidong, Yuan Shikai, Sergei Witte, Ferdinand von Richthofen, Paul Doumer and Zhang Zuolin, among others. These figures left their individual imprint on railways and politicised their development. They shaped the accidental, regional and personal nature that characterised railway development until the Communist era.

It is not an exaggeration to say that the political biography of railways is also a biography of Manchuria, a region whose history permeates many chapters. The region's rise was tied in with railways and with it British, Russian, German, Japanese and American ambitions for and expansion into China. As Canton was to the opium trade, Manchuria was to railways. Railways turned Manchuria into a 'bridgehead', and thus it would remain the 'cradle of conflict' for over half a century. The fate of Manchuria was repeated in Shandong, where the Germans, the British and the Japanese contested over the Jiaoji-Jinpu railways; in the northwest with the Jingzhang line, as the British and the Russians challenged each other's claims; in the Yangzi River valley where Britain, France, Belgium and the United States wrestled over the Chuanhan and Yuehan lines; and in the southwest where Britain and France jostled for lines into Yunnan and Sichuan. The pattern is clear – where there are railways, there are also competition and conflict. Although the model advanced in this book helps us better comprehend railway-driven empires, they deserve more research. Britain's ubiquity deserves more examination than it has received here. More comparative studies would illuminate the scramble for railways that defined the Age of Empire.

This volume has considered an aspect of Chinese history that has been largely overlooked: that all the wars that took place on Chinese soil were closely associated with railways. The railway had truly become not just the 'engines for empire' but also the 'new weapon of war'. Like weapons themselves, the railway was both strategically and logistically important to modern warfare. Wars were indeed 'won and lost' on, along and over railways: the First Sino-Japanese War, the suppression of the Boxer Rebellion, the Russo-Japanese War, WWI in Shandong, the conflicts among warlords, WWII in China and Southeast Asia, and the Civil War between the GMD and CCP. But the study of wars has overshadowed the research of railways. Like opium, which has traditionally defined the history of modern China, railways also defined the history of China, especially at times of war. My hope is that this book will inspire more research into the history and politics of railways, especially during war as well as peacetime too. This latter point is important, as the rise of post-Mao China, as was also true of the various empires, can be measured by the growth and expansion of their railways.

'Of the future it is wisest not to prophesy', Percy Kent philosophised in 1907.[2] The political biography of railways in China is far from finished. Railways built by foreign powers, warlord, ROC and Mao regimes, are dated. The cost to maintain and subsidise them is high; even more so are the highspeed franchises. Railways might have been 'engines for empire' but empires have fallen. Such a fate might well befall the BRI, as it is subjected to many variables: from terrorism and natural disaster to armed conflicts. The

CCP's grand scheme has not been profitable. This explains why state-owned companies were forced to undertake unprofitable projects. It also explains the gradual shift 'from a focus on public infrastructure, such as transport projects, toward revenue-backed projects'.[3] Regardless, 'China has met one of its biggest goals – extending its influence'.[4] It is worth the cost for now but in the long run, the BRI might 'follow the path of its ancient predecessor, with fragmentation contributing to decline and eventual collapse'.[5] History and politics of railways in and beyond China clearly deserve more research than this humble volume. They will teach us new lessons about the past and old lessons about the future.

Notes

1 'The death of Li Hung-Chang', *Herald*, 13 November 1901: 913.
2 Kent, *Enterprise*, 1.
3 S. A. Kader, 'China's BRI remains strategic and agile amid global uncertainty – Dr. Christoph Nedopil', *Zwaya Project*, 7 April 2025, at www.zawya.com, accessed 15 April 2025.
4 Tessa Wong, 'Belt and Road Initiative: is China's trillion-dollar gamble worth it?' *BBC News*, 17 October 2023, at www.bbc.co.uk/news/world-asia-china-67120726, accessed 5 March 2025.
5 Nadia Clark, 'The rise and fall of the BRI', *Council on Foreign Relations*, 6 April 2023, at www.cfr.org/blog/rise-and-fall-bri, accessed 5 March 2025.

Selected bibliography

Primary sources

Archival

The First Historical Archive (Beijing).
Hansard Parliamentary Debates.
The Institute of Modern History Archive (Taipei).
Jardine Matheson Archive.
The National Archive.
Robert Hart Project at Queen's University Belfast.
Rothschild Archive.
Royal Engines Museum.
The Second Historical Archive (Nanjing).
Sheng Xuanhuai Archive (Hong Kong).

Newspapers

African Review of Business and Technology.
Beijing Ribao.
Bradford Observer.
Bulletin du Comite de L'Asia française.
China Daily.
The Chinese Students' Monthly.
La Dépêche Coloniale.
The Diplomat.
Exeter and Plymouth Gazette.
Global Construction Review.
The Guardian and the Observer.
The Irrawaddy.
The London and China Telegraph.
Los Angeles Times.
La Mayenne.
The National News.
The National Observer.
New York Times.
North China Herald.

Pacific Review.
Railway Technology.
Renmin Ribao.
Reuters.
Sacramento Daily Union.
San Francisco Chronicle.
South China Morning Post.
The Standard.
The Times (London).

Municipal and provincial archives

Beijing Municipal Government, at www.beijing.gov.cn
Heilongjiang shizhiwang, at www.hljszw.org.cn/
Jilin difangzhi, at http://dfz.jl.gov.cn/
Liaoningsheng dang'anguan, at www.lnsdag.org.cn
Qingdao dang'an xinxiwang, at www.qdda.gov.cn
Renminwang, at www.people.com.cn/
Sichuan shengqing wang, at www.scsqw.cn/
Yunnan dang'an guan, at www.ynda.yn.gov.cn/
Zhongqing zaixian, at www.cyol.net/

Websites

Australian Government Department of Veterans' Affairs, at https://anzacportal.dva.gov.au
BBC News, at https://bbc.com
Beiträge zur Geschichte Tsingtaus–1897 bis 1953, at https://tsingtau.org/
China Central Television, at https://tv.cctv.com/
Chinese Posters, at https://chineseposters.net
CNN Travel, at https://edition.cnn.com
Council on Foreign Relations, at www.cfr.org
Guangming wang, at www.gmw.cn
Historical Photographs of China, Bristol University www.hpcbristol.net
Library of Congress, at https://loc.org
Harry S. Truman Library, at https://trumenlibrary.gov
Heilongjiang Shenhui kexueyuan, at hlass.org.cn
Hudson Institute, at www.hudson.org/
The Industry History of Hong Kong Group, at https://industrialhistoryhk.org
Office of the Historian (USA), at https://history.state.gov/
Stop HS2, at http://stophs2.org/
The Support and Communication Centers for People Returning from China, at www.sien-center.or.jp
Woodrow Wilson Center, at https://digitalarchive.wilsoncenter.org
Yale Law Library, at https://law.yale.edu
Zawya Project, at www.zawya.com
Zhongguowang, at www.china.org.cn/english

Printed sources (both original and secondary)

Adasa, Masafumi, 'The return of the Chinese Changchun railway to China and Sino-Soviet relations, 1949–1952', *Asian Studies*, vol 60, 1 (May 2015): 5–23.

Adasa, Masafumi, 'Making a Vancouver in the Far East: 'the trinity transportation system' of the Chinese Eastern railway 1896–1917', in *Russia and its northeast Asian neighbors: China, Japan and Korea 1858 to 1945*, ed. Kimitaka Matsuzato (Lanham: Lexington Books, 2017), 65–88.

Afinogenov, Gregory, *Spies and scholars: Chinese secrets and imperial Russia's quest for world power* (Cambridge: Belknap Press of Harvard University Press, 2020).

Akiko, Yosano, *Travels in Manchuria and Mongolia: a feminist poet from Japan encounters pre-war China,* translated by Joshua A. Fogel (New York: Columbia University Press, 2001).

Altan, Selda, 'Politics of life and labor: French colonialism in China and Chinese coolie labor during the construction of the Yunnan-Indochina railway', *International Labor and Working Class History*, vol 101 (2022): 77–99.

Altan, Selda, *Chinese workers of the world: colonialism, Chinese labour and the Yunnan Indo-China railway* (Stanford: Stanford University Press, 2024).

Altree, Wayne, 'A half-century of the administration of the state railways of China', *Papers on China*, vol 3 (1949): 83–121.

Anderson, George E., *Railway situation in China* (Washington: Government Printing Office, 1911).

Anonymous, 'Coal and railways at the north', in *Papers on China* (Hong Kong: no publisher or date of publication), 35–37.

Appadurai, Arjun, 'Introduction, commodities and the politics of value', in *The social life of things: commodities in cultural perspective*, ed. Anjun Appadurai (Cambridge: Cambridge University Press, 1986), 1–62.

Armstrong, Alex, *Shantung: a general outline of the geography and history of the province* (Shanghai: Mercury, 1891).

Austin, Ian, *Ulyssess S. Grant and Meiji Japan 1869–1885* (London: Routledge, 2019).

Bakich, Olga, 'Origins of the Russian community on the Chinese Eastern Railway', *Canadian Slavonic Papers*, vol 27, 1 (March 1985): 1–14.

Bakich, Olga, 'Russian city in China: Harbin before 1917', *Canadian Slavonic Papers*, vol 28, 2 (June 1986): 129–48.

Ballantyne, Tony and Antoinette Burton, 'Empires and the reach of the global', in *A world connecting, 1875–1945*, ed. Emily S. Rosenberg (Cambridge: Harvard University Press, 2012), 285–434.

Barak, On, *Powering empire: how coal made the Middle East and sparked global carbonisation* (Oakland: University of California Press, 2020).

Barry, Arthur J., 'Railway development in China', *Journal of Royal Society of Arts*, vol LVII, 2948 (21 May 1909): 541–58.

Becker, Julius, '"To Grab, When the Grabbing begins" German foreign and colonial policy during the Sino-Japanese war of 1894/95 and the Triple Intervention of 1895', *The International History Review*, vol 44, 1 (2022): 1–20.

Berghoff, Hartmut, Frank Biess and Ulrike Strasser, eds, *Exploration and entanglement: Germans in the Pacific world* (Oxford: Berghahn Books, 2018).

Biggerstaff, Knight, 'The secret correspondence of 1867–68: view of leading Chinese statesmen regarding further opening of China to Western influence', *Journal of Modern History*, vol 22, 2 (June 1950): 122–36.

Biggerstaff, Knight, *The earliest modern government schools in China* (Ithaca: Cornell University Press, 1961).

Bigham, Clive, *A year in China, 1899–1900, with some account of Admiral Sir E. Seymour's expedition* (London: Macmillan, 1901).

Bonker, Dirk, 'Global politics and Germany's destiny from an 'East Asian perspective': Alfred von Tirpitz and the making of Wilhelmine navalism', *Central European History*, vol 46, 1 (March 2013): 61–96.

Booker, Edna, *News is my job: a correspondent in wartorn China* (New York: Macmillan, 1940).

Braisted, William, 'The United States and the American China development company', *The Far Eastern Quarterly*, vol 11, 2 (1952): 147–65.

Bretelle-Establet, Florence, 'Resistance and receptivity: French colonial medicine in southwest China, 1898–1930', *Modern China*, vol 25, 2 (April 1999): 171–203.

Bright, Charles, *Imperial telegraphic communication* (London: P. S. King & Son, 1911).

Brisson, Ulrike, 'Elizabeth von Heyking: China through the eyes of a female aristocrat', in *Sino-German encounters and entanglements: transnational politics and culture 1890 to 1950*, ed. Joanne Miyang Cho (Cham: Palgrave Macmillan, 2021), 29–53.

Brook, Timothy, 'The pacification of Jiading', in *The scars of war: the impact of warfare on modern China*, eds Diana Lary and Stephen MacKinnon (Vancouver: UBC Press, 2001), 50–74.

Brown, Tristan, *Laws of the land: Fengshui and the state in Qing China* (Princeton: Princeton University Press, 2023).

Bulow, Bernhard von, *Imperial Germany* (New York: Dodd and Mead, 1914).

Burton, Wilbur, 'French imperialism in China', *Current History*, vol 39, 4 (1 January 1934): 428–31.

Cady, John, *The roots of French imperialism in Eastern Asia* (Ithaca: Cornell University Press, 1954).

Cai Er'kang, *Li fuxiang lipin ou'meiji* (Taibei: Wenhai chubanshe, 1972).

Cai Er'kang, *Li Hongzhang lipin oumeiiji* (Changsha: Yuelu chubanshe, 1986).

Cao, Yin, 'The Yunnan-Burma railway, 1860s–1940s: imagining, planning and rejecting a railway that was never built', *Journal of Southeast Asian Studies*, vol 54, 2 (June 2023): 298–315.

Carl, Katherine, *With the Empress Dowager* (New York: The Century Co., 1905).

Carlson, Ellsworth, *The Kaiping mines 1877–1912* (Cambridge: Harvard University Press, 1971).

Carnegie Endowment, *The Consortium: the official text of the Four Power Agreement* (Washington, DC: The Endowment, 1921).

Catalogue of Treaties 1814 to 1918 (Washington: Government Printing Office, 1919).

Cater, James Hugh, *Creating a Chinese Harbin: nationalism in an international city 1916–32* (Ithaca: Cornell University Press, 2002).

Chang, Jui-Te, 'Technology transfer in modern China: the case of railway enterprise (1876–1937)', *Modern Asian Studies*, vol 27, 2 (May 1993): 281–96.

Chang, Sung-Hwan, 'Russian designs on the Far East', in *Russian imperialism from Ivan the great to the revolution*, ed. Taras Hunczak (New Brunswick: Rutgers University Press, 1974), 299–321.

Che, Ling, 'Jiaohui, tielu, gonglu yu jindai Yunnan xiyide fazhan', *Qujing Shifan Xueyuan Xuebao*, vol 32, 1 (2013): 5–9.

Chen, Lixu, 'Zhonggong Zhongyang cengshexiang dingdu Ha'erbin', *Dangshi Zonglan*, 9 (2018): 19–21.
Chen, Pingyuan and Xia Xiaohong, *Tuxiang wanqing* (Tianjin: Baihua wenyi chubanshe, 2001).
Cheng, Victor, *China between peace and war: Mao, Chiang and the Americans, 1945–1947* (Canberra: ANU Press, 2023).
Cheng, Weirong, *Jindai dongbei tielu fushudi* (Shanghai: Shehui kexue chubanshe, 2008).
Cheng, Yiwei, 'Coping with parallel authorities: the early diplomatic negotiations of Soviet Russia and China on the Chinese Eastern Railway, 1917–1925', *Modern Chinese History*, vol 9, 2 (July 2015): 223–43.
Chere, Lewis, *The diplomacy of the Sino-French war (1883–1885): global complications of an undeclared war* (Notre Dame: Cross Culture Publications, 1998).
Chi, Madeleine, 'Shanghai-Hangchow-Ningbo railway loan: a case study of the rights recovery movement', *Modern Asian Studies*, vol 7, 1 (January 1973): 85–106.
Chiasson, Blaine, *Administering the colonizer: the Chinese administrative takeover of the Chinese Eastern Railway concession, 1918–29* (Vancouver: UBC Press, 2014).
Christensen, Peter, *Germany and the Ottoman railways: art, empire, and infrastructure* (New Haven: Yale University Press, 2017).
Chung, Sue Fawn, 'The much-maligned empress dowager: a revisionist study of the Empress Dowager Tz'u-his (1835–1908)', *Modern Asian Studies*, vol 13, 2 (1979): 177–96.
Colquhoun, A. R., *Across Chryse: being the narrative of a journey of exploration through the south China border lines from Canton to Mandalay* (London: Sampson Low, 1883).
Colquhoun, A. R., *The mastery of the Pacific* (New York: Macmillan, 1902).
Conger, Sarah, *Letter from China with particular reference to the Empress Dowager and the women of China* (Chicago: A. C. McClurg, 1910).
Conrad, Sebastian, *German imperialism: a short history* (Cambridge: Cambridge University Press, 2014).
Cooper, John, *The unexpected story of the Nathanial Rothschild* (London: Bloomsbury, 2015).
Coox, Alvin, 'The Kwantung Army Dimension', in *The Japanese informal empire in China, 1805–1937*, eds Peter Duus and Ramon H. Myers (Princeton: Princeton University Press, 1991), 395–428.
Dai, Wu-san, 'Modern railway technology transfer to China: the case of Kiaotsi railway, 1898–1914', *East Asian Science, Technology and Society*, vol 10 (February 2016): 211–14.
Dai, Zhili, Sichuan baolu yundong shiliao huizuan (Taibei: Zhongyanyuan Jinshisuo, 1994).
Darwent, Charles, *Shanghai: a handbook for travellers and residents* (Shanghai: Kelly and Walsh, 1920).
Das, Pallavi, *Colonialism, development, and the environment: railways and deforestation in British India 1860–1884* (New York: Palgrave Macmillan, 2015).
Daughton, J. P., *In the forest of no joy: the Congo-Ocean railroad and the tragedy of French colonialism* (New York: W. W. Norton, 2021).
Davenport, Arthur, *Report on the trading capabilities of the country traversed by the Yunnan mission* (London: Harrison and Sons, 1877).

Davies, H. R., *Report on the part of Yunnan between the Bhamo frontier and the Salween* (Rangoon: Government Printing, 1894).
Davies, H. R., *Yun-nan: the link between India and the Yangtze* (Cambridge: Cambridge University Press, 1909).
Davies, John Paton, *Dragon by the tail: American, British, Japanese and Russian encounter with China and one another* (New York: Norton, 1972).
Davis, Bradley Camp, *Imperial bandits: outlaws and rebels in the China-Vietnam borderlines* (Seattle: University of Washington Press, 2017).
Davis, Clarence, 'Railway imperialism in China, 1895–1939', in *Railway imperialism*, eds Clarence Davis and Kenneth Wilburn (Westport: Greenwood Press, 1991), 155–73.
Dennett, Tyler, *Americans in eastern Asia: a critical study of the United States' policy in the Far East in the nineteenth century* (New York: Barnes and Noble, 1922).
Der Ling, *Two years in the Forbidden City* (New York: Moffat, Yard and Company, 1911).
Der Ling, *Old Buddha* (New York: Dodd, Mead, 1928).
Deschanel, Paul, *Le question du Tonkin* (Paris: Berger-Levrault, 1883).
Dezong Jinghuangdi shilu (Beijing: Zhonghua shuju, 1986).
Dianshizhai huabao (Guangzhou: Guangdong renmin chubanshe, 1983).
Ding, Richang, *Fuwu gongdu* (Taibei: Huawen shuju, 1968).
Diogo, M. P. and B. J. Navarro, 'Re-designing Africa: railways and globalization in the era of the new imperialism', in *Technology and globalization: networks of experts in world history*, eds D. Pretel and L. Camprubi (Cham: Palgrave Macmillan, 2018), 105–28.
Dirlik, Arif and Roxann Prazniak, 'The 1911 revolution: an end and a beginning', *China Information*, vol 25, 3 (November 2011): 213–31.
Doig, Will, *High speed empire: Chinese expansion and the future of Southeast Asia* (New York: Columbia University Global Reports, 2018).
Dukes, Paul, *Russia in Manchuria: a problem of empire* (Abingdon: Routledge, 2022).
Duus, Peter and Ramon H. Myers, eds, *The Japanese informal empire in China, 1805–1937* (Princeton: Princeton University Press, 1991).
Dyer, Colin, 'Singapore in 1897 as seen by Paul Doumer', *Journal of the Malaysian Branch of the Royal Asiatic Society*, vol 90, 313 (2, 2017): 113–16.
Elliot, Mark, 'The limits of Tartary: Manchuria in imperial and national geographies', *The Journal of Asian Studies*, vol 59, 3 (August 2000): 603–46.
Eskildsen, Robert, '1911: a review', *Modern China*, vol 2, 2 (April 1976): 141–84.
Eskildsen, Robert, 'Reconsidering 1911: lessons from a sudden revolution', *Journal of Modern Chinese History*, vol 6, 1 (2012): 1–14.
Eskildsen, Robert, *Transforming empire in Japan and East Asia: the Taiwan expedition and the birth of Japanese imperialism* (Basingstoke: Palgrave Macmillan, 2019).
Elleman, Bruce, 'The Soviet Union's secret diplomacy concerning the Chinese Eastern Railway 1924–1925', *The Journal of Asian Studies*, vol 53, 2 (May 1994): 459–86.
Elleman, Bruce, *Diplomacy and deception: the secret history of Sino-Soviet diplomatic relations 1917–1927* (Armonk: M. E. Sharpe, 1997).
Elleman, Bruce, 'Sino-Soviet tensions and Soviet administrative control over the Chinese Eastern Railway, 1917–25', in *Manchurian railways and the opening of China: an international history*, eds Bruce Elleman and Stephen Kotkin (New York: Routledge, 2010), 59–76.

Elleman, Bruce, *International competition in China, 1899–1991: the rise, fall and restoration of the Open-Door policy* (London: Routledge, 2015).
Elleman, Bruce, *Wilson and China: a revised history of the Shandong question* (London and New York: Routledge, 2015).
Elleman, Bruce and Stephen Kotkin, eds, *Manchurian railways and the opening of China: an international history* (New York: Routledge, 2010).
Ericson, Steven, *The sound of the whistle: railroads and the state in Meiji Japan* (Cambridge: Council on East Asian Studies, Harvard University, 1996).
Esherick, Joseph, 'Harvard on China: the apologetics of imperialism', *Bulletin of Concerned Asian Scholars*, vol 4, 4 (1972): 9–16.
Fawcett, Brian, 'The Chinese Labour Corp in France, 1917 to 1921', *Journal of the Hong Kong Branch of the Royal Asiatic Society*, vol 40, 1 (January 2000): 33–111.
Feuerwerker, Albert, *China's early industrialization: Sheng Hsuan-Huai (1844–1916) and Mandarin enterprise* (Cambridge: Harvard University Press, 1958).
Fichter, James, *British and French colonialism in Africa, Asia and the Middle East* (Cham: Springer, 2019).
Fisher, John, 'Lord Curzon and British strategic railways in Central Asia before, during and after the First World War', in *Railways and international politics: paths of empire, 1848–1945*, eds T. G. Otte and Keith Neilson (London: Routledge, 2006), 137–55.
Fleming, William, 'Profits and visions: British capital and railway construction in Argentina, 1854–1886', in *Railway imperialism*, eds Clarence Davis and Kenneth Wilburn (Westport: Greenwood Press, 1991), 71–84.
Fogel, Joshua, 'The Japanese and the Jews: a comparative analysis of their communities in Harbin, 1898–1930', in *New frontiers: imperialism's new communities in East Asia, 1842 to 1953*, eds Robert Bickers and Christian Henriot (Manchester: Manchester University Press, 2017), 88–108.
Fourniau, Charles, 'La genese et l'evolution de l'affaire du Tonkin', *Revue Historique*, vol 246, 2/500 (October 1971): 377–408.
Fraleigh, Matthew, 'Transplanting the flower of civilization: the 'peony girl' and Japan's 1874 expedition to Taiwan', *International Journal of Asian Studies*, vol 9, 2 (July 2012): 177–209.
French, William, 'In the path of progress: railroads and moral reform in Porfirian Mexico', in *Railway imperialism*, eds Clarence Davis and Kenneth Wilburn (Westport: Greenwood Press, 1991), 85–102.
Fuller, Pierre, *Famine relief in warlord China* (Boston: Brill, 2019).
Ganz, A. Harding, 'The German navy in the Far East and the Pacific: the seizure of Kiautschou and after', in *Germany in the Pacific and Far East 1870–1914*, eds John A. Moses and Paul M. Kennedy (St Lucia: Queensland University Press, 1977), 115–36.
Gao, Lecai, *Riben Manzhou yimin yanjiu* (Beijing: Renmin chubanshe, 2000).
Gao Zhizhong, *MingQing gongci yu gongting wenhua yanjiu* (Beijing: Fangzhi chubanshe, 2014).
Garnier, Francis, *The French in Indochina: with a narrative of Garnier's explorations in Cochin-China, Annam and Tonquin* (London: T. Nelson, 1884).
Gillin, Donald, *Warlord Yen Hsi-shan in Shanxi province 1911–49* (Princeton: Princeton University Press, 1967).
Goddard, Stacie, *When right makes might: rising powers and world order* (Ithaca: Cornell University Press, 2018).
Godshall, Wilson, *Tsingtau under three flags* (Shanghai: Commercial Press, 1929).

Golitsyn, B. B. (translated by Li Shuxiao *et al.*), *Zhongdong tielu huludui canjia 1900nian manzhoushijian jilue* (Beijing: Shangwu yinshuguan, 1984).

Gordon, Maria M. Ogilvie, 'Ferdinand von Richthofen's Diaries from China', *Scottish Geographical Magazine*, vol 24, 6 (1908): 310–17.

Gottschall, Terrell, *By Order of the Kaiser: Otto von Diederichs and the rise of the imperial German Navy, 1865–1902* (Annapolis: Naval Institute Press, 2003).

Gower, Adam, *Jacob Schiff and the art of risk: America's financing of Japan's war with Russia (1904–05)* (New York: Springer International, 2018).

Green, Nile, 'From the Silk Road to the railroad (and Back): the means and meaning of the Iranian encounter with China', *Iranian Studies*, vol 48, 2 (2015): 165–92.

Groeneveld, Sabina, 'Far away at home in Qingdao (1897–1914)', *German Studies Review*, vol 39, 1 (February 2016): 65–79.

Gundry, R. S., *China and her neighbours: France in Indo-China, Russia and China, India and Thibet* (London: Chapman and Hall, 1893).

Guojia tushuguan, *Qingdai baokan tuhua jicheng* (Beijing: Guojia tushuguan, 2001).

Gwyn, David, *The coming of railways: a new global history 1750–1850* (New Haven: Yale University Press, 2023).

Hall, Richard and Hugh Peyman, *The great Uhuru railway: China's showpiece in Africa* (London: Gollancz, 1976).

Hanes, W. Travis, 'Railway politics and imperialism in Central Africa, 1889–1953', in *Railway imperialism*, eds Clarence Davis and Kenneth Wilburn (Westport: Greenwood Press, 1991), 41–69.

Harcave, Sidney (translator), *The memoir of Count Witte* (Armonk: M. E. Sharpe, 1990).

Hardgrove, Anne, 'Roads, railways and the language of politics', *The Journal of Caribbean History*, vol 52, 1 (2018): 30–67.

He, Hanwei, *Jinghan tielu chuqi shilue* (Hong Kong: Chinese University, 1979).

Headland, Isaac, *Court life in China: the capital, its officials and people* (New York: Revell, 1909).

Headrick, Daniel, *The tools of empire: technology and European imperialism in the nineteenth century* (New York: Oxford University Press, 1981).

Headrick, Daniel, *The tentacles of progress: technology transfer in the age of imperialism 1850–1940* (New York: Oxford University Press,1988).

Hong, Rengan, *Hong Rengan xuanji* (Beijing: Zhonghua shuju, 1978).

Hou, Chi-ming, *Foreign investment and economic development in China, 1840–1937* (Cambridge: Harvard University Press, 1965).

Hsu, Mongton Chih, *Railway problems in China* (New York: Columbia University, 1915).

Huenemann, Ralph, *The dragon and the iron horse: the economics of railroads in China, 1876–1937* (Cambridge: Council on East Asian Studies, Harvard University, 1984).

Ibos, Pierre, *Le chemin de fer du fleuve rough et la pénétration française au Yunnan* (Paris: Henri Charles-Lavauzelle, 1906).

Jacques, Martin, *When China rules the world* (London: Penguin, 2012).

Jia, Benyi, *Zhongguo zaoqi tielude naxieren heshi* (Beijing: Tiedao chubanshe, 2014).

Jiaotongbu, *Jiaotongshi luzhengbian* (Shanghai: Jiaotongbu, 1935).

Jing, Zhankui, *Yan Xishan yu Tongput tielu* (Taiyuan: Shanxi renmin chubanshe, 2003).

Joseph, Philip, *Foreign diplomacy in China 1894–1900* (London: G. Allen and Unwin, 1928).

Kangzhanshi bianxiezu, *Zhongguo kangri zhanzhengshi* (Beijing: Renmin chubanshe, 2011).
Karuka, Manu, *Empire's tracks: indigenous nations, Chinese workers, and the Transcontinental railroad* (Oakland: University of California Press, 2019).
Kaske, Elisabeth, 'Sichuan as a pivot: provincial politics and gentry power in late Qing railway projects in southwestern China', in *Southwest China in a regional and global perspective: metals, transport, trade and society*, eds Ulrich Theobald and Jin Cao (Leiden: Brill, 2018), 379–423.
Keiichi, Tsuneishi, 'Unit 731 and the Japanese imperial army's biological warfare program', in *Japan's wartime medical atrocities: comparative inquires in science, history and ethics*, eds Jing-Bao Nie, Nanyan Guo, Mark Selden and Arthur Kleinman (London: Routledge, 2010), 23–31.
Kelly, Patrick, *Tirpitz and the Imperial German Navy* (Bloomington: Indiana University Press, 2011).
Kennedy, Greg, 'Barometer of war: British views of imperial security in the Far East, Russo-Japanese relations and the Chinese Eastern Railway, 1929–1935', in *Railways and international politics: paths of empire, 1848–1945*, eds T. G. Otte and Keith Neilson (London: Routledge, 2006), 192–216.
Kent, Percy, *Railway enterprise in China: an account of its origin and development* (London: Edward Arnold, 1907).
Kinder, Claude, *Railways and collieries of north China* (London: The Institution of Civil Engineers, 1891).
Kinney, Henry, *Modern Manchuria and the south Manchurian railway* (Dairen: Japan Advertiser Press, 1928).
Kirby, William, *Germany and Republican China* (Stanford: Stanford University Press, 1984).
Klein, Maury, *The life and legend of E. H. Harriman* (Chapel Hill: University of North Carolina Press, 2011).
Kneitz, Agnes, 'German water infrastructure in China: colonial Qingdao 1898–1914', *NTM Zeitschrift fur Geschichte der Wissenschaften, Technik und Medizin*, vol 24, 4 (December 2016): 421–50.
Köll, Elisabeth, 'Chinese railroads, local society and foreign presence: the Tianjin-Pukou line in pre-1949 Shandong', in *Manchurian railways and the opening of China: an international history*, eds Bruce Elleman and Stephen Kotkin (New York: Routledge, 2010), 106–23.
Köll, Elisabeth, *Railroads and transformation of China* (Cambridge: Harvard University Press, 2019).
Kopytoff, Igor, 'The cultural biography of things: commoditisation as process', in *The social life of things: commodities in cultural perspective*, ed. Anjun Appadurai (Cambridge: Cambridge University Press, 1986), 64–91.
Kratoska, Paul, 'Labor in the Malay peninsula and Singapore under Japanese occupation', in *Asian labour in the wartime Japanese empire: unknown histories*, ed. Paul Kratoska (London: Routledge, 2015), 237–48.
Lao, Chongmin, ed., *Mantie dang'an ziliao huibian* (Beijing: Shehui kexue wenxian chubanshe, 2011).
Lary, Diana, *Warlord soldiers: Chinese common soldiers 1911–1937* (Cambridge: Cambridge University Press, 1985).
Lary, Diana, 'A ravaged place: the devastation of the Xuzhou region 1938', in *The scars of war: the impact of warfare on modern China*, eds Diana Lary and Stephen MacKinnon (Vancouver: UBC Press, 2001), 98–116.

Lattimore, Owen, *Manchuria: cradle of conflict* (New York: Macmillan, 1935).
Le Fevour, Edward, *Western enterprise in late Ch'ing China: a selective survey of Jardine, Matheson and Company's operations, 1842–1895* (Cambridge: Harvard University Press, 1968).
Lee, En-han, *China's quest for railway autonomy, 1904–1911: a study of the Chinese railway rights recovery movement* (Singapore: Singapore University Press, 1997).
Lee, Robert, *France and the exploitation of China 1885–1901* (Hong Kong: Oxford University Press, 1989).
Legendre, A. F., *Le Far West chinois. Deux Années au Setchouen* (Paris: Plon-Nourrit, 1905).
Lei, Yi, *Li Hongzhang yu wanqing sishinian* (Taiyuan: Shanxi renmin chubanshe, 2013).
Leibo, Steven, *Transferring technology to China: Prosper Giquel and the self-strengthening movement* (Berkeley: Institute of East Asian Studies, University of California, 2005).
Lenoir, Désirée, *Le consul qui en savait trop: les ambitions secrètes de la France en Chine, 1886–1904* (Paris: Nouveau Monde, 2011).
Lenoir, Désirée, *Dianyue tielu yu Yunnan wangshi*, translated by Xu Tao and Zhang Ruizi (Nanjing: Yilin chubanshe, 2019).
Lensen, George, *The damned inheritance: the Soviet Union and the Manchurians crises 1924–1935* (Tallahassee: The Diplomatic Press, 1975).
Leong, Sow-Theng, *Sino-Soviet diplomatic relations, 1917–1926* (Canberra: Australia National University Press, 1976).
Leung, Chi-Keung, *China: railway patterns and national goals* (Chicago: University of Chicago Department of Geography, 1980).
Leung, Yuen-sang, *The Shanghai taotai: the linkage man in a changing society 1843–90* (Singapore: Singapore University Press, 1990).
Leutner, Mechthild, *Musterkolonie Kiautschou: die expansion des Deutschen Reich in China 1897 bis 1995* (Berlin: Verlag, 1997).
Levine, Steven, *Anvil of victory: the communist revolution in Manchuria, 1945–1949* (New York: Columbia University Press, 1987).
Li, Binggang, ed., *Riben zai Dongbei nuyi laogong diaocha yanjiu* (Beijing: Shehui kexue chubanshe, 2009).
Li, Bingheng, *Li Bingheng ji* (Beijing: Zhonghua shuju, 2013).
Li, Cunkuan, 'Guoji waijiaoshi shangde da pianju: lun Guangxu zhong'e miyue', *Shehui Kexue Zhanxian*, vol 2 (1987): 182–91.
Li, Danke, 'Popular culture in the making of the anti-imperialist and nationalist sentiment in Sichuan', *Modern China*, vol 30, 4 (October 2004): 470–505.
Li, Hongzhang, *Li Wenzhonggong zougao* (Shanghai: Guji chubanshe, 1997).
Li, Hongzhang, *Li Hongzhang quanji* (Hefei: Anhui jiaoyu chubanshe, 2008).
Li, Jie, 'Phantasmagoric Manchukuo: documentaries produced by the South Manchuria railway company, 1932–1940', *Positions: East Asian Culture Critique*, vol 22, 2 (2014): 329–69.
Li, Siyi, *Tielu xiandai xin* (Taibei: Shibao wenhua chubanshe, 2020).
Liang, Ernest, *China: railways and agricultural development, 1875 to 1935* (Chicago: University of Chicago Department of Geography, 1982).
Lin, Qingyuan and Wang Daocheng, *Shen Baozhen xinza kaozhu* (Chengdu: Bashu chubanshe, 2013).
Little, Archibald, *The Far East* (Oxford: Clarendon Press, 1905).

Little, Archibald, *Across Yunnan: a journey of surprises* (London: Sampson Low & Marsten, 1910).
Liu, Heung Shing, *China in revolution: the road to 1911* (Hong Kong: Hong Kong University Press, 2011).
Liu, Mingchuan, *Liu Mingchuan wenji* (Heifei: Huangshan shushe, 1997).
Liu, Xuefeng, 'Zaifeng and late Qing railway policy', in *China: how the empire fell*, eds Joseph Esherick and George Wei (London: Routledge, 2011), 222–58.
Liu, Yan'ni and Chen Sheng, *Chuanzheng yu jindai zhongguo haijun* (Fuzhou: Fujian renmin chubanshe, 2016).
Lorin, Amaury, '"La civilisation suit la locomotive": le *credo* ferroviaire de Paul Doumer, gouverneur général de l'Indochine (1897–1902)', *Revue d'histoire des chemins de fer*, vol 35, 1 November (2006): 41–54.
Lu, Yixu, 'Germans and the death-throes of the Qing: Mo Yan's The sandalwood torture', in *German colonialism revisited: African, Asian and Oceanic experiences*, eds Nina Berman, Klaus Mühlhahn and Patrice Nganang (Ann Arbor: University of Michigan Press, 2014), 271–83.
Lu, Yixu, 'Colonial Qingdao through Chinese eyes', in *The cultural legacy of German colonial rule*, ed. Mühlhahn, Klaus (Berlin: Walter de Gruytar, 2017), 127–43.
Luo, Zhiyuan, 'Yike daoding yidixue, yigen zhenmu yitiaoming', *Chuangzao*, vol 2 (2010): 108–9.
McCormack, Gavan, *Chang Tso-lin in northeast China, 1911–1928: China, Japan and the Manchurian idea* (Stanford: Stanford University Press, 1977).
McDonald, Kate, 'Asymmetrical integration: lessons from a railway empire', *Technology and Culture*, vol 56, 1 (January 2015): 115–49.
MacKinnon, Stephen, *Wuhan, 1938: war, refugees, and the making of modern China* (Berkeley: University of California Press, 2008).
McLean, D., 'Chinese railways and the Townley Agreement of 1903', *Modern Asian Studies*, vol 7, 2 (March 1973): 145–64.
McMeekin, Sean, *The Berlin-Baghdad express: the Ottoman empire and Germany's bid for world power* (Cambridge: Harvard University Press, 2021).
MacMurray, John Van Antwerp, *Treaties and agreements with and concerning China 1894–1919* (New York: Oxford University Press, 1921).
MacMurray, John Van Antwerp, *Treaties and agreement with and concerning China 1919–1929* (Washington: Carnegie Endowment, 1929).
Mahltig, Philipp, 'German railways in China: technology as a site of knowledge', *Journal of Transport History*, vol 40, 3 (December 2019): 301–21.
Mallory, Walter, *China: land of famine* (New York: American Geographical Society, 1926).
Marbotte, Pierre, *Un chemin de fer au Yunnan: l'aventure d'une famille francaise en chine*, translated by Xu Tao (Kunming: Yunnan meishu chubanshe, 2010).
Marcus, Lilit, 'China debuts world's fastest train', CNN Travel, 21 July 2021.
Marks, Steven, *Road to power: the Trans-Siberian railroad and the colonization of Asian Russia 1850–1917* (Ithaca: Cornell University Press, 1991).
Martin, Bernd, 'The Prussian expedition to the Far East (1860–1862)', *Newsletter for Modern Chinese History*, 6 (1988): 38–52.
Matsusaka, Yoshihisa Tak, *The making of Japanese Manchuria, 1904–1932* (Cambridge: Harvard University Asia Center, 2003).
Matsuzato, Kimitaka, 'Introduction: Russia as the decisive factor in the modern history of northeast Asia', in *Russia and its northeast Asian neighbors: China, Japan*

and Korea 1858 to 1945, ed. Kimitaka Matsuzato (Lanham: Lexington Books, 2017), vii–xv.
Mayers, William F., *The treaty ports of China and Japan* (London: Trubner & Co, 1867).
Mayers, William F., *Treaties between the empire of China and foreign powers* (Shanghai: North China Herald Office, 1877).
Meister, Sylvia and Wang Jin (translator), *Didang yougu: 1903 zhi 1910 yiwei Sulishi gongchengshi qinlide dianyue tielu* (Kunming: Yunnan renmin chubanshe, 2018).
Meyskens, Covell, 'Third Front railroads and industrial modernity in late Maoist China', *Twentieth-Century China*, vol 40, 3 (October 2015): 238–60.
Meyskens, Covell, *Mao's Third Front: the militarization of Cold War China* (Cambridge: Cambridge University Press, 2020).
Mi, Rucheng, *Zhongguo jindai tielushi ziliao: 1863–1911* (Beijing: Zhonghua shuju, 1963).
Michaud, Jean, *'Incidental' ethnographers: French Catholic missions on the Tokin-Yunnan frontiers, 1880 to 1930* (Leiden: Brill, 2007).
Miller, Chris, *We shall be masters: Russian pivots to East Asia from Peter the Great to Putin* (Cambridge: Harvard University Press, 2021).
Miller, Tom, *China's Asian dream: empire building along the New Silk Road* (London: Zed Books, 2017).
Mitter, Rana, '1911: the unanchored Chinese revolution', *China Quarterly*, vol 208 (December 2011): 1009–20.
Mokros, Emily, 'Reconstructing the imperial retreat: politics, communications, and the Yuanming Yuan under the Tongzhi emperor 1873–4', *Late Imperial China*, vol 33, 2 (2012): 76–118.
Moore, Gregory, *Defining and defending the open door policy: Theodore Roosevelt and China, 1901–1909* (Lanham: Lexington Books, 2015), 111–37.
Moses, John, 'The coolies labour question and German colonial policy in Samoa, 1900–14', in *Germany in the Pacific and Far East 1870–1914*, eds John A. Moses and Paul M. Kennedy (St Lucia: Queensland University Press, 1977), 234–62.
Mühlhahn, Klaus, 'Negotiating the nation: German colonialism and Chinese nationalism in Qingdao: 1897–1914', in *Twentieth century colonialism and China: localities, everyday and the world*, eds Bryna and David Goodman (London: Routledge, 2012), 37–56.
Mühlhahn, Klaus, ed., *The cultural legacy of German colonial rule* (Berlin: Walter de Gruyter, 2017).
Mühlhahn, Klaus, 'A new imperial vision?: the limits of German colonialism in China', in *German colonialism in a global age*, eds Bradley Naranch and Geoff Eley (Durham: Duke University Press, 2020), 129–46.
Myers, Ramon, 'Japanese imperialism in Manchuria: the South Manchuria Railway Company, 1906–1933', in *The Japanese informal empire in China, 1805–1937*, eds Peter Duus and Ramon H. Myers (Princeton: Princeton University Press, 1991), 101–32.
Naranch, Bradley, "Colonized body', 'Oriental machine': debating race, railroads, and the politics of reconstruction in Germany and East Africa, 1906–1910', *Central European History*, vol 33, 3 (2000–01): 299–338.
Naughton, Barry, 'The Third Front: defence industrialisation in the Chinese interior', *The China Quarterly*, vol 115 (September 1988): 351–86.

Neilson, Keith and T. G. Otte, 'Railpolitick', in *Railways and international politics: paths of empire, 1848–1945*, eds T. G. Otte and Keith Neilson (London: Routledge, 2006), 1–15.
Nish, Ian, *Origins of the Russo-Japanese war* (New York: Longman, 1985).
Nish, Ian, *The history of Manchuria, 1840–1948: a Sino-Russo-Japanese triangle* (Folkestone: Renaissance Press, 2016).
O'Dwyer, Emer, *Significant soil: settler colonialism and Japan's urban empire in Manchuria* (Cambridge: Harvard University Asia Center, 2015).
Ogata, Sadako, *Defiance in Manchuria: the making of Japanese foreign policy 1931–32* (Berkeley: University of California Press, 1964).
Orbach, Danny, 'By not stopping: the first Taiwan expedition (1874) and the roots of Japanese military disobedience', *The Journal of Japanese Studies*, vol 42, 1 (2016): 29–55.
Orbach, Danny, *Curse on this country: the rebellious army of imperial Japan* (Ithaca: Cornell University Press, 2017).
Otte, T. G., "The Baghdad railway of the Far East", in *Railways and international politics: paths of empire, 1848–1945*, eds T. G. Otte and Keith Neilson (London: Routledge, 2006), 112–36.
Ouyang, Yuefeng, 'Cong tongchou dikai dao fendao yangbiao', *Wuyi Daxue Xuebao*, vol 10, 3 (2008): 42–45.
Ozyuksel, Murat, *The Hejaz railway and the Ottoman empire: modernity, industrialization and Ottoman decline* (London: I. B. Tauris, 2014).
Paine, S. C. M., *Imperial rivals: China, Russia and their disputed frontier* (Armonk: M. E. Sharpe, 1996).
Paine, S. C. M., *The wars for Asia, 1911–1949* (Cambridge: Cambridge University Press, 2012).
Park-Barjot, Rang-ri, 'Le patronat français des travaux publics et les réseaux ferroviaires dans l'empire français: l'exemple de chemin de fer au Yunnan (1898–1913), *Publications de la Société française d'histoire des outre-mer*, vol 6, 1 (2008): 653–70.
Paules, Xavier, 'Warlords at work: four crucial realms and four dynamics of state building in Republican China, 1916–1937', *Twentieth Century China*, vol 47, 1 (January 2020): 40–49.
Pauley, Edwin, *Report on Japanese assets in Manchuria* (Washington, DC: USGPO, 1946).
Pelcovits, Nathan, *Old China hand and the foreign office* (New York: King's Crown Press, 1948).
Pereira, Hugo Silveira, 'The camera and the railway: framing the Portuguese empire and technological landscapes in Angola and Mozambique, 1880s–1910s', *Technology and Culture*, vol 64, 3 (2023–07): 737–59.
Perry, Elisabeth, *Shanghai on strike: the politics of Chinese labor* (Stanford: Stanford University Press, 1993).
Pholsena, Vatthana, 'Technology and empire: a colonial narrative of the construction of the Tonkin-Yunnan railway', *Critical Asian Studies*, vol 47, 4 (2015–10): 537–57.
Picot, Henry Philip, *Railways in western Asia* (London: Central Asia Society, 1904).
Plens, Claudia, 'Railways: landmarks and scars in the Atlantic Rainforest', in *Archaeologies of the British in Latin America*, ed. Charles E. Orser (Cham, Switzerland: Springer, 2019), 201–18.

Pletcher, David, *The diplomacy of involvement: American expansion across the Pacific 1784 to 1900* (Columbia: University of Missouri Press, 2001).
Pong, David, 'Confucian patriotism and the destruction of the Wusong railway, 1877', *Modern Asian Studies*, vol 7, 4 (1973): 647–76.
Pong, David, *Shen Baozhen and China's modernization in the nineteenth century* (Cambridge: Cambridge University Press, 1994).
Pomeranz, Kenneth, *The great divergence: China, Europe and the making of the modern world economy* (Princeton: Princeton University Press, 2021).
Powell, John, *My twenty-five years in China* (New York: MacMillan, 1945).
Qingdao dang'anju, *Diguo zhuyi yu jiaohaiguan* (Beijing: Dang'an chubanshe, 1986).
Qingdaoshi dang'anguan, *Tieti xiade zui'e: Riben zai Qingdao lueduo laogong shimo* (Beijing: Dang'an chubanshe, 2003).
Qinhuangdao gangwuju, *Qingchao xuwenxian tongkao* (Taibei: Xinxing shuju, 1963).
Qinhuangdao gangwuju, *Qinhuangdaogang tielu yunshu fazhanshi* (Beijing: Haiyang chubanshe, 1993).
Quan, Hansheng, *Zhongguo jinjishi luncong* (Hong Kong: CUHK, 1972).
Rankin, Mary, 'Nationalistic contestation and mobilisation politics: practice and rhetoric of the railway-rights recovery at the end of the Qing', *Modern China*, vol 28, 3 (July 2002): 315–61.
Rapier, Richard, *Remunerative railways for new countries with an account of the first railway in China* (London: E. & F. N. Spon, 1878).
Rapier, Richard, 'Brief account of the Wusong railway', in *Minutes of the proceedings of the Institution of Civil Engineers*, 59 (1880): 274–77.
Ravenstein, Ernst, *The Russians on the Amur: its discovery, conquest and colonization* (London: Trubner, 1861).
Rawlinson, John, *China's struggle for naval development 1839–1895* (Cambridge: Harvard University Press, 1967).
Reichman, Henry, 'The 1905 revolution on the Siberian railroad', *Russian Review*, vol 47, 1 (1988): 25–48.
Reid, Anthony, *Imperial alchemy: nationalism and political identity in Southeast Asia* (Cambridge: Cambridge University Press, 2010).
Reinhardt, Anne, *Navigating semi-colonialism: shipping, sovereignty, and nation-building in China, 1860–1937* (Cambridge: Harvard University Asia Center, 2018).
Richthofen, Ferdinand von, *Schantung und seine eingangspforte Kiautschou* (Berlin: Dietrichs Reimer, 1898).
Richthofen, Ferdinand von, *Tagebucher aus China band I* (Berlin: Dietrich Reimer, 1907).
Robinson, Ronald, 'Introduction: railway imperialism', in *Railway imperialism*, eds Clarence Davis and Kenneth Wilburn (Westport: Greenwood Press, 1991), 1–5.
Rohl, John, *Wilhelm II: the Kaiser's personal monarchy 1888–1900* (Cambridge: Cambridge University Press, 2004).
Rosenbaum, Arthur, 'Railway enterprise and economic development: the case of the imperial railways of north China', *Modern China*, vol 2, 2 (April 1976): 227–72.
Rosenbaum, Arthur, 'The Manchuria bridgehead: Anglo-Russian rivalry and the Imperial Railway of North China, 1897–1902', *Modern Asian Studies*, vol 10, 1 (February 1976): 41–64.
Rousseau, Jean-François, 'An imperial railway failure: the Indochina-Yunnan railway, 1898–1941', *The Journal of Transport History*, vol 35, 1 (2014): 1–17.

Schrecker, John, *Imperialism and Chinese nationalism: Germany in Shantung* (Cambridge: Harvard University Press, 1971).

Schrecker, John, 'Kiautschou and the problem of German colonialism', in *Germany in the Pacific and Far East 1870–1914*, eds John A. Moses and Paul M. Kennedy (St Lucia: Queensland University Press, 1977), 185–208.

Sethia, Tara, 'Railways, Raj, and the Indian states: policy of collaboration and coercion in Hyderabad', in *Railway imperialism*, eds Clarence Davis and Kenneth Wilburn (Westport: Greenwood Press, 1991), 103–20.

Sewell, Bill, *Constructing empire: the Japanese in Changchun, 1905–1945* (Vancouver: UBC Press, 2019).

Shehui Kexueyuan Jinshisuo, *Riben qinhua qishinian shi* (Beijing: Shehui kexue chubanshe, 1992).

Shelford, Fred, *The development of West Africa by railways* (London: Royal Colonial Institute, 1904).

Showalter, Dennis, 'Railroads, the Prussian army, and the German way of war in the nineteenth century', in *Railways and international politics: paths of empire, 1848–1945*, eds T. G. Otte and Keith Neilson (London: Routledge, 2006), 21–44.

Sichuansheng dang'anguan, *Baolu zhendang: Sichuan baolu yundong dang'an tuji* (Chengdu: Sichuan daxue chubanshe, 2011).

Siegelbaum, Lewis, 'Another "Yellow Peril": Chinese migrants in the Russian Far East and the Russian reaction before 1917', *Modern Asian Studies*, vol 12, 2 (April 1978): 307–30.

Silveira, Hugo, 'History, technology, and politics: the political use of railways in Portugal 1850–1900', *Historia Contemporanea*, vol 69 (2022–06): 507–43.

Sinn, Elisabeth, 'The fugitive in paradise: Wang Tao and cultural transformation in late nineteenth-century Hong Kong', *Late Imperial China*, vol 35, 1 (1998): 56–81.

Sladen, E. B., *Official narrative of and papers connected with the expedition to explore the trade routes to China via Bhamo* (Rangoon: British Burma Press, 1869).

Smedley, Agnes, *The great road: the life and times of Chu Teh* (New York: Monthly Review Press, 1956).

Smirnov, Dimitri, 'The sonic occupation of Central Asia', in *Sonic histories of occupation: experiencing sound and empire in a global context*, eds Russell Skelchy and Jeremy Taylor (London: Bloomsbury Academic, 2022), 173–94.

Stephenson, Rowland, *Report upon the feasibility and most effectual means of introducing railway communication into the empire of China* (London: J. E. Adlard, 1864).

South Manchuria Railway Company, *Manchuria: land of opportunities* (New York: South Manchuria Railway Company, 1924).

Spiers, Edward, *Engines for empire: the Victorian army and its use of railways* (Manchester: Manchester University Press, 2015).

Sprye, Richard, *Correspondence between Captain Richard Sprye and William-Ewart Gladstone* (Whitefish: Kissinger Publishing, 2010 reprint).

Squier, Lee Welling, *A lamb to the slaughter: an American girl's experience in the Orient from the China-Japan war to the relief of Peking* (Greenburg: Patriot Publishing, 1901).

Steen, Andreas, 'Germany and the Chinese coolie: labor, resistance and the struggle for equality, 1884–1914', in *German colonialism revisited: African, Asian and Oceanic experiences*, eds Nina Berman, Klaus Mühlhahn, Patrice Nganang (Ann Arbor: University of Michigan Press, 2014), 147–60.

Steinmetz, George, *The devil's handwriting: precoloniality and the German colonial state in Qingdao, Samoa, and Southwest Africa* (Chicago: The University of Chicago Press, 2007).
Strand, David, *An unfinished republic: leading by word and deed in modern China* (Berkeley: University of California Press, 2011).
Strauss, Julia, 'The past in the present: historical and rhetorical lineages in China's relations with Africa', *The China Quarterly*, vol 199 (2009): 777–95.
Sturdevant, Sandra, 'Imperialism, sovereignty, and self-strengthening: a reassessment of the 1870s', in *Reform in nineteenth century China*, eds Paul Cohen and John Schrecker (Cambridge: Harvard East Asia Center, 1976), 63–70.
Sun, E-tu Zen, *Chinese railways and British interests 1898–1911* (New York: King's Crown Press, Columbia University, 1954).
Sun, E-tu Zen, 'The pattern of railway development in China', *The Far Eastern Quarterly* (pre-1986), vol 14, 2 (February 1955): 179–99.
Sun, Zhongshan, *Jianguo fanglue* (Beijing: Zhongguo chang'an chubanshe, 2011).
Sun, Zhongshan, *Sun Zhongshan quanji* (Beijing: Zhonghua shuju, 1982).
Takeo, Ito, *Life along the South Manchurian Railway: the memoirs of Ito Takeo*, translated by Joshua A. Fogel (Armonk: M. E. Sharpe, 1988).
Talbot, Frederick, *The railway conquest of the world* (London: William Heinemann, 1911).
Tan, Bing, 'Nanjiang tiedaobing', *Wenshi Yuekan*, 4 (2008): 24–26.
Tan, Guiluan, *Zhongdong tielude xiuzhu yu jinying: Er'guo zaihua shilide fazhan* (Taipei: Lianjin chubanshe, 2016).
Tan, Issac C. K., 'Science and empire: tracing the imprint of dactylography in Manchuria, 1924–1945', *Japan Forum*, vol 32, 4 (2020): 531–54.
Tanner, Harold, 'Railways in communist strategy and operations in Manchuria, 1945–48', in *Manchurian railways and the opening of China: an international history*, eds Bruce Elleman and Stephen Kotkin (New York: Routledge, 2010), 149–70.
Tanner, Harold, *The battle for Manchuria and the fate of China, Siping 1946* (Bloomington: Indiana University Press, 2013).
Tao, Shing Chang, *International controversies of the Chinese Eastern Railway* (Taipei: Cheng Wen, 1971).
Tawney, Richard, *Land and labour in China* (London: Allen and Unwin, 1932).
Thomas, W. A., 'An intra-empire capital transfer: the Shanghai rubber company boom 1901–1912', *Modern Asian Studies*, vol 32, 3 (1998): 739–60.
Thomas, William, *The Iron Way: railroads, the Civil War, and the making of modern America* (New Haven: Yale University Press, 2011).
Thomason, Conrad, 'Ambiguities of modernisation: the China case', *The China Geographer*, vol 5 (Fall 1976): 13–27.
Thompson, Roger, 'If Shanxi's coal is lost, then Shanxi is lost!: Shanxi's coal and an emerging national movement in provincial China 1898–1908', *Modern Asian Studies*, vol 45, 5 (September 2011): 1261–88.
Thompson, Virginia, *French Indo-China* (London: George Allen and Unwin, 1937).
Thorne, Christopher, *The limits of foreign policy: the West, the League and the Far Eastern crisis of 1931–33* (London: Hamilton, 1972).
Thorner, Daniel, *Investment in empire: British railway and steam shipping enterprise in India 1825–1848* (Philadelphia: University of Pennsylvania Press, 1950).
Tiedaobing xinan zhihuibu, *Tiedaobing zhandouzai chengkunxian* (No place of publication, 1970).

Tirpitz, Alfred von, *My memoir* (New York: Dodd and Mead, 1919).
Trang, Phan T. H., 'Paul Doumer: aux origines d'un grand projet: le chemin de fer indochinois', *Historie, Economie et Société*, vol 30, 3 (September 2011): 115–40.
Tsai, Weipin, 'The Qing empire's last flowering: the expansion of China's post office at the turn of the twentieth century', *Modern Asian Studies*, vol 49, 3 (May 2015): 895–930.
Tucker, David, 'Labor policy and the construction industry in Manchukuo: systems of recruitment, management and control', in *Asian labour in the wartime Japanese empire: unknown histories*, ed. Paul Kratoska (London: Routledge, 2015), 25–57.
Uchida, Jun, '"A scramble for freight": the politics of collaboration along and across the railways tracks of Korea under Japanese rule', *Comparative Studies in Society and History*, vol 51, 1 (2009–01): 117–50.
Van de Ven, Hans, *From friend to comrade: the founding of the Chinese communist party 1920–1927* (Berkeley: University of California Center for Chinese Studies, 1991).
Van de Ven, Hans, *Breaking with the past: the Maritime Customs Service and the global origins of modernity in China* (New York: Columbia University Press, 2014).
Van de Ven, Hans, *China at war: triumph and tragedy in the emergence of the new China* (Cambridge: Harvard University Press, 2018).
Van Dijk, Kees, *Pacific strife: the great powers and their economic and political rivalries in Asia and the Western Pacific 1870–1914* (Amsterdam: Amsterdam University Press, 2015).
Vul, Nikita, 'He, who has sown the wind: Karakhan, the Sino-Soviet conflict over the Chinese Eastern Railway, 1925–26', *Modern Asian Studies*, vol 48, 6 (November 2014): 1670–94.
Vul, Nikita, 'Chinese Eastern Railway and Russian minority groups in Manchuria', *Modern History of Russia*, vol 11, 1 (2021): 89–102.
Walker, Michael, *The 1929 Sino-Soviet War: the war nobody knew* (Lawrence: University Press of Kansas, 2017).
Wang, Di, *Violence and order on the Chengdu plain: the story of a secret brotherhood in rural China, 1939–49* (Stanford: Stanford University Press, 2018).
Wang, Ermin and Wu Lunni, *Sheng Xuanhuai shiye handiangao* (Hong Kong: CUHK, 1993).
Wang, Hsien-chun, 'Merchants, mandarins, and the railway: institutional failure and the Wusong railway, 1874–1877', *International Journal of Asian Studies*, vol 12, 1 (January 2015): 31–53.
Wang, Suzy, 'Medicine-related war crimes trials and post-war politics and ethics: the unresolved case of unit 731, Japan's bio-warfare program', in *Japan's wartime medical atrocities: comparative inquires in science, history and ethics*, eds Jing-Bao Nie, Nanyan Guo, Mark Selden and Arthur Kleinman (London: Routledge, 2010), 32–58.
Wang, Tao, *Taoyuan wenlu waibian* (Shanghai: Guji chubanshe, 2010).
Wang, Xiliang, 'Jindai zhongguo dongbei cenlinde zhiminkaifa yushengtai kongjian bianqian', *Heilongjiang Shenhuikexueyuan Lishiyanjiu*, 1 (2017 online).
Wang, Ying, *Li Hongzhang yu wanqing zhongwai tiaoyuede yanjiu* (Changsha: Hunan renmin chubanshe, 2011).
Waugh, Daniel, 'Richthofen's 'Silk Roads': toward the archaeology of a concept.' *The Silk Road*, vol 5, 1 (Summer 2007): 1–10.

Wcislo, Francis, *Tales of imperial Russia: the life and times of Sergei Witte 1849–1915* (Oxford: Oxford University Press, 2011).
Weiss, Sheila, 'The race hygiene movement in Germany, 1904–1945', in *The well-born science: eugenics in Germany, France, and Russia*, ed. Mark B. Adams (New York: Oxford University Press, 1990), 8–68.
Weng, Fei, *Li Hongzhang yu huaijunde chuangjian* (Hefei: Huangshan shushe, 2012).
Weng, Tonghe, *Weng Tonghe riji* (Shanghai: Zhongxi shuju, 2012).
Wenxiang, Jia Zhen and Bao Jun, *Chouban yiwu shimo: Tongzhichao* (Beijing: Zhonghua Shuju, 1960).
Westwood, J. N., *A history of Russian railways* (London: George Allen and Unwin, 1964).
White, Richard, *Railroads: the Transcontinentals and the making of modern America* (New York: Norton, 2011).
Wilber, C. Martin, *Sun Yatsen: the frustrated patriot* (New York: Columbia University Press, 1976).
Wilbur, Kenneth, 'Engines of empire and independence: railways in South Africa, 1863–1916', in *Railway imperialism*, eds Clarence Davis and Kenneth Wilburn (Westport: Greenwood Press, 1991), 25–40.
Wilhelm II, *The Kaiser's memoir* (New York: Harper Brothers, 1922).
Wilson, Sandra, 'The 'new paradise': Japanese emigration to Manchuria in the 1930s and 1940s', *The International History Review*, vol 17, 2 (May 1995): 249–86.
Wittfogel, Karl, *Oriental despotism: a comparative study of total power* (New Haven: Yale University Press, 1957).
Wright, Mary, *The last stand of Chinese conservatism: the T'ung-chih restoration 1862–1874* (Stanford: Stanford University Press, 1962).
Wright, Winthrop, *British-owned railways in Argentina: their effect on the growth of economic nationalism 1854–1948* (Austin: University of Texas Press, 2021).
Wolff, David, *To the Harbin station: the liberal alternative in Russian Manchuria, 1898–1914* (Stanford: Stanford University Press, 1999).
Wolmar, Christian, *Engines of war: how wars were won and lost on the railways* (New York: Public Affairs, 2010).
Wong, R. Bin, 'Centennial perspective on China's 1911 revolution', *China Information*, vol 25, 3 (November 2011): 275–82.
Woodruff, Smith, *The German colonial empire* (Chapel Hill: University of North Carolina Press, 1978).
Wu, Shang-Su and Alan Chong, 'Developmental railpolitics: the political economy of China's high speed rail projects in Thailand and Indonesia', *Contemporary Southeast Asia*, vol 40, 3 (December 2018): 503–26.
Wu, Shellen, 'The search for coals in the age of empires: Ferdinand von Richthofen's odyssey in China, 1860–1920', *The American Historical Review*, vol 119, 2 (April 2014): 339–63.
Wu, Shellen, *Empires of coal: fuelling China's entry into the modern world, 1860–1920* (Stanford: Stanford University Press, 2019).
Wu, Shellen, 'Engineering empire: German influence on Chinese industrialization, 1880–1925', in *Exploration and entanglement: Germans in the Pacific world*, eds Hartmut Berghoff, Frank Biess and Ulrike Strasser (Oxford: Berghahn Books, 2018), *Exploration and entanglement*, 153–70.
Wu, Yong and Liu Zhixiang, *Gengzi xishou congtan* (Shanghai: Shanghai shuju, 1996).
Wyatt, Christopher, *Afghanistan and the defence of empire: diplomacy and strategy during the great game* (London: I. B. Tauris, 2011).

Xie, Xueshi, *Mantieshi ziliao: meitiepian* (Beijing: Zhonghua shuju, 1987).
Xie, Xueshi, ed., *Mantie dang'an shiliao huibian* (Beijing: Shehui kexue wenxian chubanshe, 2011).
Xu, Tao and Zhang Xinzi, *Lingshi Fang Suya* (Nanjing: Yilin chubanshe, 2019).
Xue, Xiantian, *Zhongdong tielu hulujun yu dongbei bianjiang zhengju* (Beijing: Shehui kexue wenxian chubanshe, 1993).
Yang, Honglin, *Cixi huiluan: 1901 niande yici teshu luxin* (Beijing: Sanlian shudian, 2017).
Yangwu yundong wenxian huibian (8 vols, Taibei: Shijie shuju, 1963).
Yen, Ching-hwang, 'Chang Yu-nan and the Chaochow railway (1904–08)', *Modern Asian Studies*, vol 18, 1 (February 1984): 119–35.
Yokell, Matthew, 'Investing in German 'Hong Kong': the building of a global economic presence in Qingdao 1898–1914', in *Sino-German encounters and entanglements: transnational politics and culture 1890 to 1950*, ed. Joanne Miyang Cho (Cham: Palgrave Macmillan, 2021), 77–106.
Young, C. Walter, *Japanese jurisdiction in the South Manchuria Railway areas* (Baltimore: The Johns Hopkins Press, 1931).
Young, John, *The research activities of the South Manchuria Railway company 1907–45* (New York: Columbia University Press, 1966).
Younghusband, Francis, *Among the celestials: a narrative of travels in Manchuria, across the Gobi Desert, through the Himalayas to India* (London: John Murray, 1898).
Yuan, Rongsu, *Jiao'ao zhi* (Taibei: Wenhai chubanshe, 1969).
Yunnansheng dang'anguan, *Dianyue tielu shiliao huibian* (Kunming: Yunnan chuban jituan, 2013).
Zajontz, Tim, 'Railway imperialism in East Africa: laying the track for exploitation', in *The Oxford handbook of economic imperialism*, eds Zak Cope and Immanuel Ness (New York: Oxford University Press, 2022), 515–34.
Zatsepine, Victor, *Beyond the Amur: frontier encounters between China and Russia, 1850–1930* (Vancouver: UBC Press, 2017).
Zeng, Kunhua, *Zhongguo tielushi* (Taibei: Wenhai chubanshe, 1962).
Zhang, Emma Yu, 'Socialist builders on the rails and road: industrialization, social engineering, and national imagination in Chinese socialist films, 1949–1965', *Twentieth-Century China*, vol 42, 3 (2017): 255–73.
Zhang, Yan and Sun Yanjing, eds, *Jinan can'an* (Beijing: Guojia tushuguan chubanshe, 2014).
Zhang, Zhidong, *Zhang Zhidong quanji* (Wuhan: Wuhan chubanshe, 2008).
Zheng, Guanglu, *Sichuan baolu yundong lishi zhenxiang* (Chengdu: Sichuan minzu chubanshe, 2011).
Zheng, Xiaowei, *The politics of rights and the 1911 revolution in China* (Stanford: Stanford University Press, 2018).
Zhongguo renmin jiefangjun tiedaobing, *Renmin tiejun* (Beijing: Tiedao bingtuan zhengzhibu, 1950).
Zhou, Fu, *Chunqingwang xunyue beiyang haifang riji* (Beijing: Xueyuan chubanshe, 2006).
Zhou, Yuefeng, 'Liening shike', *Qinghua Daxue Xuebao*, vol 5 (2017): 113–28.
Zhu, Congbing, *Li Hongzhang yu zhongguo tielu* (Beijing: Qunyan chubanshe, 2006).
Zhu, Jianjun, *Dezhan Qingdao 1897–1914* (Beijing: Renmin chubanshe, 2010).
Zhu, Miaochun and Xu Feng, *Laogong xueleishi* (Shanghai: Sanlian shudian, 2005).

Index